T5-AFS-427

MICROCOMPUTER SOFTWARE FOR MECHANICAL ENGINEERS

MICROCOMPUTER SOFTWARE FOR MECHANICAL ENGINEERS

Howard Falk

VNR VAN NOSTRAND REINHOLD COMPANY
New York

Library of Congress Catalog Card Number: 86-24551
ISBN 0-442-22658-6

Printed in the United States of America

Van Nostrand Reinhold Company Inc.
115 Fifth Avenue
New York, New York 10003

Van Nostrand Reinhold Company Limited
Molly Millars Lane
Wokingham, Berkshire RG11 2PY, England

Van Nostrand Reinhold
480 La Trobe Street
Melbourne, Victoria 3000, Australia

Macmillan of Canada
Division of Canada Publishing Corporation
164 Commander Boulevard
Agincourt, Ontario M1S 3C7, Canada

16 15 14 13 12 11 10 9 8 7 6 5 4 3 2 1

Library of Congress Cataloging-in-Publication Data

Falk, Howard, 1927-
Microcomputer software for mechanical engineers.

Includes index.
1. Mechanical engineering—Computer programs—
Catalogs. 2. Microcomputers—Programming. I. Title.
TJ153.F26 1987 621.0056 86-24551
ISBN 0-442-22658-6 F1

Contents

Preface

This book describes software packages for use by mechanical engineers. It is written for engineers who need software that can do the job without requiring that they become computer experts or programmers.

The purpose of this book is to present a broad picture of the personal computer packages now available for use by mechanical engineers. Each chapter is devoted to a specific area, such as heat transfer, gear design, drafting, or equation solving, in which a number of software packages are presently offered for use with personal computers. The chapter introductions explain what kinds of design or analysis or other tasks these packages perform, outlining the available choices and comparing the capabilities of the various packages.

Detailed reviews of individual packages follow. The emphasis here is on what the user must know and do to employ the capabilities of the package. Going beyond general description, these reviews also explain what the packages actually will and will not do. Although many packages are covered, there is no attempt here at completeness. In every category covered in the book, many more packages exist than those that have been reviewed. In the fast-moving field of engineering software, many new packages are currently being written and marketed.

A new style of computing

Personal computers have become widely, perhaps universally, accepted as everyday tools for business applications. An essential prerequisite for this acceptance has been the availability of software packages designed for easy use by business people who have little or no background in computing. It is ironic that, although engineers were using computers long before most business people dared approach a keyboard, personal computer software packages for engineering applications have been slow to develop.

When minicomputers and large, mainframe computers are used to solve engineering problems, the assumption generally has been that the user must know and use a computer language such as FORTRAN or BASIC. It also has been assumed that operation of the computer system must be arranged for maximum effective use of computer resources rather than for the convenience of engineering users.

The basic premises of personal computing are quite different. Users have exclusive access to their own computers, therefore the way they employ computing resources is shaped entirely by their own needs and whims. Although personal computer software can be created by the user, it usually is purchased in the form of a package, shaped to perform the tasks the user wants the computer to do.

In this world of personal computing, the extent to which computers can be useful for performing engineering tasks largely depends on what kinds of engineering software packages are available. If there is a package for the mechanism design or hydraulics calculation or drafting task that needs to be done, then the personal computer can be used. If no such package exists, the engineer will probably turn to other means of doing the work.

A critical approach

The approach throughout this book is a critical one. Each application area and each software package is examined from the point of view of engineering users, specifically those users who may use a computer only occasionally, turning to it when particular problems have to be solved.

Toward the use of formal computer languages, the approach in this book is a skeptical one. It is useful for engineers to know how to program, if for no other reason than to gain a feeling of mastery over computers. However, most mechanical engineers are not and probably should not be programmers. Except for those few who decide to take up programming as their vocation, engineers use computers, as they use any other tools, to help them carry out engineering tasks.

Some engineers spend much of their time repeatedly performing certain design or analysis tasks for which personal computers are well suited. For these users, learning the special-purpose languages used by some finite element analysis, statistical, drafting, and other packages can bring advantages both in speed and flexibility of problem solving.

For the most part, what engineers need are software packages that allow them to continue to think in strictly engineering terms while getting the benefit of computing power. Writing programs that offer this kind of capability is not easy, but it is what the field of programming for engineering applications is all about, and it is the task that the vendors of engineering packages for personal computers have taken on, or should have taken on. This book includes many examples of excellent packages that meet these requirements. Where packages fall short, they have been commented on in a critical manner.

HOWARD FALK

MICROCOMPUTER SOFTWARE FOR MECHANICAL ENGINEERS

Chapter 1
Heat Transfer Packages

Engineers who do heat transfer analysis undoubtedly would like to be able to turn to a personal computer software package that can assist them in solving just about any problem that might arise. They would like that package to be so simple to operate that if they have not used it for a month or two, there will be no need to go through a relearning process.

Unfortunately, that ideal package does not yet exist. Several very capable heat transfer analysis packages are available, but they require special user skills such as familiarity with BASIC or FORTRAN programming, or they are written with the expectation that users are willing to wade through arduous and complex procedures to reach their analysis goals.

One exception is the EZTAP package, which is quite easy to learn and operate. However, this package is limited to the solution of steady-state heat transfer problems. It does not deal with the transient solutions so often needed in this kind of design work.

Solving transient problems

Among the packages that can deal with transient solutions, I/TAS provides users with a tool kit of several different computation methods. I/TAS can handle from 85 to 350 thermal network nodes, depending on the particular method the user chooses. The penalty for this freedom, however, is that users are expected to do some sophisticated manipulation of system software and input sequences to get desired results.

The MTAP package consists of a specialized set of routines and defined procedures that operate as a supplement to the BASIC computer language for solving heat transfer problems. They are designed for use by those who are already familiar with BASIC. Through coded lines, users describe the thermal network, materials, initial temperatures, conductances, capacitances, and heat applied to nodes.

TSAP is a capable package that can handle multidimensional heat transfer networks and temperature-dependent materials. However, the developer of this package had to think in terms of FORTRAN files in order to write it, and users are expected to think in the same terms. The processing routines are designed to handle problems with up to 200 nodes and 400 conductances. The manual invites users to revise the routines so they will accommodate still larger networks.

The THETA package provides a wide variety of capabilities and is particularly good in handling material properties with nonlinear temperature dependence. The package gives users considerable control over printed output of numerical results, with a choice of full, or abbreviated, or one-of-*n* outputs for temperatures and heat fluxes.

In addition to numerical results for transient solutions, THETA is able to produce graphical displays and printouts. The plots can each include up to 12 traces. Where the output is to plotting equipment, the software will control four pens and vary the line types used in the plot. The routine does curve-fitting and can also do autoranging to fit the graphs to the available plotting space.

Finite element analysis

All the packages mentioned thus far require users to represent their heat transfer problems by lumped conductive elements (plus capacitive elements for transient analysis) connected together into a thermal network.

An alternative is the more directly physical representation used by finite element analysis packages. These packages are described in greater detail in Chapters 6 and 7, which deal with finite element analysis. A number of these packages, primarily designed to analyze stress in objects and structures, also have the capability to deal with thermal stresses and can be used to perform thermal conductivity analyses.

As an example, consider the FESDEC finite analysis package and its Thermal Conductivity Analysis option, which handles linear heat conduction problems with linear convective boundary conditions. In addition to the structure itself, the user must specify loads on that structure. For steady-state heat transfer analysis these loads can be heat inputs at nodes, distributed surface heat flux, internal heat generation, ambient fluid temperatures, or prescribed nodal temperatures. After the user specifies initial and boundary conditions (which may vary with time), the package will calculate the resulting temperatures and perform thermal stress analyses.

Electronic packaging

Taking a more specialized approach, STAP is a steady-state heat transfer package designed for analysis of heatsinks used in electronic packaging.

Although it includes convenient capabilities for this particular type of design, STAP can actually be used to solve just about any two-dimensional steady-state thermal conduction problem.

User-controlled solutions

Not every heat transfer problem may be satisfactorily solved by any one particular computation technique. To meet the need for a variety of techniques, the I/TAS package gives users a choice of a steady-state solution by the Gauss–Jordan method or an iterative steady-state solution. Users also may choose to run explicit or implicit transient solutions. Gauss–Jordan solutions are limited to 85 nodes. The other methods can solve for as many as 350 nodes.

Other packages deal with the need for flexible computation by having the user adjust the calculation parameters. For example, before making an actual computation run, THETA users are required to make an estimate of the steady-state starting temperature for each case to be run. Users also can choose temperature convergence criteria and place a limit on the number of iterations that will occur during the solution of nonlinear problems. To keep track of the results, there can be screen displays during the computation run, indicating which calculations are in progress and showing partial results.

Parameters that control the method and procedures of the TSAP solutions are entered by the user in a formatted control block. Included here are 14 parameters such as choice of solution algorithm, time steps, and convergence criteria. Users are expected to rely on experience to set these parameters properly.

One of the reasons for such user-controlled adjustments is to avoid computation failures. I/TAS users, for example, are warned that steady-state problems with radiation may not converge. If they do not, users are advised to reduce the damping factor. Under some conditions, THETA calculations can become unstable and fail due to numerical overflow. To avoid this users are advised to employ a calculated damping factor that decreases as the final solution is approached.

Characteristic of the matrix computation methods used by heat transfer packages is that calculation speed and accuracy can be affected by the order in which nodes are entered into the calculation. Some of the packages simply ignore this situation. Others advise users to experiment and find proper data entry sequences that will result in good node-numbering arrangements. The THETA package meets the problem more directly with a routine that automatically renumbers the thermal network nodes for computation. The procedure is known as bandwidth minimization. Use of this module is optional, but the manual recommends it when "larger" problems are being run. A series of passes through the minimization procedure may be needed to achieve significant bandwidth reduction.

Other factors besides node numbering can affect computation accuracy and speed. Thus, when a calculation run is called for, STAP asks the user to specify the largest allowable temperature difference between calculation iterations.

With this package, as with the others, computation accuracy increases as the number of nodes used to represent a heat transfer problem increases. Up to 1,600 nodes can be handled by the IBM/PC version of STAP. However, the time taken by a package to complete its calculations will depend on the number of nodes involved, so some sort of balance will likely be desirable between accuracy and speed of computation. A STAP problem with three devices was run with a maximum resolution of 20 and an allowable temperature difference of 1 between iterations; the calculation ran for three iterations, each one took about 25 seconds to complete.

Entering the data

A good part of the user effort with these packages is in entering descriptions of thermal networks into the computer. This process is handled through a number of different procedures by the various packages, some quite convenient and some not.

One of the simplest procedures is that of the EZTAP package. EZTAP queries the user to enter a pair of node numbers, then the value of the conductance between those two nodes. This is followed by similar prompts for added similar entries. When users feel that every conductance in the network has been entered, they simply press the Return key at the next query. The package then displays totals of the number of conductances and nodes entered and shows the user a table of node connections and conductances for checking and approval. If any have been left out, the user continues the entry process, setting up the necessary additional node pairs and conductance values. To correct entered conductances, the user reenters the appropriate node pair and conductance value.

In contrast, data entry for finite element analysis can be quite complex. To describe a structure using the FESDEC package, a model of it must be built up using the finite elements provided by the package. To this end the user enters coordinates for nodes to define the structure, selects the elements that are to be used, and specifies their material properties. To get around the need to place each element individually, the package includes an automatic mesh generation capability, by which sequences of structure nodes will automatically be covered by designated types of finite elements.

In order of data entry arduousness, most of the packages lie somewhere between EZTAP and FESDEC. Entry into the MTAP package is done in response to sequences of screen-displayed queries. The THETA package presents the user with 15 different detailed input procedures to learn and follow. I/TAS requires the user to enter data in lines of BASIC-formatted code.

TSAP also requires the user to input strictly formatted data, but it gives users the ability to automatically generate input data for a sequence of nodes. These nodes all must have identical initial temperatures, heat capacities, and internal heat sources.

Like a finite-element package, STAP requires the structures being analyzed (heatsinks) to be covered with a network of nodes. However the heatsink shape is assumed to be simple and rectangular. The package prompts the user to enter the physical dimensions of the heatsink in inches, then generates a specified number of nodes to cover. Devices mounted on the heatsink are then located by their coordinate position on this network of nodes.

Stored parameters

Sometimes the information to be entered is standard in the sense that it may consist of catalog or handbook data. As a convenience, some packages come with disk-stored data of this type that users can quickly move into place for computation. Thus the THETA package allows users to draw on prestored material properties to minimize the data entry process.

The STAP package, dealing as it does with heatsinks and electronics, is in a particularly good position to take advantage of stored component data, and some limited advantage is taken of this fact. STAP users can select a heatsink material from a menu list with nine choices, and the package supplies an appropriate thermal conductivity value. A list of four types of electronic device packages is presented during the data input sequence. Users can choose from this list and the package will automatically insert the corresponding junction-to-case thermal resistance and area of heat dissipation into the calculation.

Avoiding input errors

Some of the packages have built-in controls to help users avoid entry of erroneous data. The THETA package is particularly well supplied with such controls. This package checks to see if zero, positive, and negative numbers are being entered where they should be logically excluded. There also are checks to see that nodes and surfaces outside of user-defined bounds are not referenced. Other controls are concerned with such matters as completeness (are all needed data items included in the input?) and logical inconsistancies in the data.

If I/TAS users type in too few or too many variables, the message "REDO FROM START" cues them to reenter the offending line. With the STAP package, entries not in correct format, containing incorrect characters or the wrong number of characters will not be accepted.

Side calculations

Auxiliary calculations, made available to the user with these packages, can help to simplify and extend the user's ability to enter desired data. For example, with the TSAP package input data for temperatures, thermal capacities, and other parameters may be entered in array form. The package contains a routine that can call up and interpolate such array data.

In a similar manner, thermal conductivity, specific heat, and other parameters can be entered into MTAP in the form of tabular data. The package can automatically interpolate the data to calculate the desired parameter. However, that is just the beginning of MTAP's side calculations. The package allows algebraic expressions to be used for conductance values; with these, the user can enter formulas for forced and free convection. There is a routine for determining the variation in solar heating during the course of a day. Another routine determines the equivalent conductance for radient heat exchange. The package allows algebraic expressions to be used for conductance values. With these, the user can enter formulas for forced and free convection. Finally, the main computational "engine" for this package is a matrix routine that solves simultaneous equations. In addition to using this routine in the course of solving heat transfer problems, it is available for separate use as a general-purpose equation solver.

EZTAP

EZTAP solves steady-state heat transfer problems in two or three dimensions where the conductances have constant values. It is easy to learn and simple to operate.

EZTAP runs on IBM/PC, TRS-80, and CP/M computers. It sells for $149 from CYBERTECH, 501 Via Codo, Fullerton, CA 92635. (714) 738-1882.

To set up a heat transfer problem for solution by this package, users must first represent it in terms of lumped conductive elements connected together into a thermal network. Once that step is made, the actual entry of the data for the problem is quite simple.

The main menu gives users a choice of taking input data from a previously stored file or of entering new data. For new data entry, the package queries the user to enter a pair of node numbers, then the value of the conductance between those two nodes.

The package keeps prompting for more entries. When users feel that every conductance in the network has been entered, they simply hit the Return key at the next query. The package then displays totals of the number of conductances and nodes entered, and shows the user a table of node connections and conductances for checking and approval.

If any have been left out, the user continues the entry process, setting up the necessary additional node pairs and conductance values. To correct entered conductances, the user reenters the appropriate node pair and conductance value.

After corrections, pressing the Return key calls up the node table again, and, when this is approved, the package prompts for heat applied to nodes. These data are entered by keying in a node number, then a heat number. At the end of this sequence a table of nodes and heat-applied values is displayed for checking and, if necessary, correction by further entries.

Users then are given a choice of Farenheit, Celcius, Kelvin, or Rankine temperature scales and are prompted to enter boundary temperatures for the boundary nodes. It is up to the user to see that a set of values for conductance, heat rates, and other parameters consistent with those for temperature are used.

If desired, users can display or print out the data entered for the thermal model, or they can save the data in a disk file for further use.

The problem solution run is initiated by a menu choice. The package produces a table of network connections, followed by another showing the heat transfer rate for each node connection. Finally, calculated temperatures for each node are given.

FESDEC: THERMAL CONDUCTIVITY ANALYSIS

This package is a specialized extension of the FESDEC finite element analysis package. It handles linear heat conduction problems with linear convective boundary conditions.

FESDEC runs on HP 9845 and other HP computers. The FESDEC finite element analysis package sells for $8,000; the heat conduction extension for *the package costs an additional $1,000. The vendor is H.G. Engineering, 260 Lesmill Rd., Don Mills, Ontario M3B 2T5, Canada. (416) 447-5535.*

FESDEC with thermal conductivity analysis allows users to specify a structure, establish thermal loads and internal heat generation, and initial and boundary conditions (which may vary with time) and then calculate the resulting temperatures and perform thermal stress analyses.

When the FESDEC user selects the Input option of the package, the choices include "steady-state thermal problems" and "transient thermal problems." Input data are entered into files by responding to screen-displayed queries with data keyboarded in prescribed formats.

To describe a structure, a two-dimensional (or axisymmetric solid) model of it must be built up using triangular and quadrilateral finite elements provided by the package. To this end the user enters coordinates for nodes to define the structure, selects the elements that are to be used, and specifies their material properties. The package includes an automatic mesh generation capability by which sequences of structure nodes will automatically be covered by designated types of finite elements.

Specifying heat loads

In addition to the structure itself, the user must specify loads on that structure. For steady-state heat transfer analysis these loads can be heat inputs at nodes, distributed surface heat flux, internal heat generation, ambient fluid temperatures, or prescribed nodal temperatures. As with the structural data, these load data are keyboarded in prescribed formats as the user responds to displayed prompts.

For transient analysis, case data for surface film coefficients can be entered as well as for the steady-state load types (now boundary condition types) listed above. In this type of analysis, surface heat transfer coefficients are allowed to vary with time.

In addition, users set the initial conditions for the analysis. This can be done by entering a temperature for every node on the structure from the keyboard, by setting a single initial temperature for all the nodes, or by letting the initial condition temperatures be the same as the final temperatures calculated in a previous analysis run.

To specify a calculation run, users set a solution time step, a starting time, and the time for which the output is to be taken.

I/TAS

I/TAS is a general heat transfer analysis package that can handle from 85 to 350 nodes, depending on the type of analysis used. Although interactive data entry is provided, this is very much a do-it-yourself package. Users are expected to manipulate system software, solution methods, input sequences, and node types to get desired results.

I/TAS runs on IBM/PCs and sells for $450 from R. F. Warriner Associates, 3838 Carson, St., #300, Torrance, CA 90503. (213) 540-6299.

To input data interactively, users respond to a sequence of screen-displayed queries with keyboarded data.

There are some input controls. If users type in too few or too many variables, the message "REDO FROM START" cues them to reenter the offending line. If many input errors are made, users can scrap what has been entered and quickly restart the whole input process by pressing the right keys. Or they can complete the data input sequence and then go through an editing procedure to correct the data.

Examples of package use given in the manual include steady-state and transient conduction through a wall, a lumped mass in a spacecraft with heat source and radiation, and radiating plates used for multilayer insulation between a heat source and a component.

Although the manual gives no examples of their applied use, tables can be entered to account for variation in time versus boundary temperature, and in time versus heat input. This entry follows the same query and response procedure, and the procedures are awkward. They could be greatly improved by providing a tabular display on which the user moves a cursor to fill in the needed items.

The package has no built-in printing facilities. Users are instructed to use the print capabilities of the IBM/PC system software to print out whatever files or displayed data they wish. Likewise, the package uses the editing feature (EDLIN) of the system software to edit input data files.

A manual with little regard

The user manual provides a format for inputting data in a batch fashion as if it were being entered using punched cards. This should appeal to data-processing-oriented users.

The package and its manual have been written with little regard for the needs of novices. For example, one of the first interactive input queries calls for entering the total number of nodes, the number of boundary nodes, and the number of arithmetic nodes. Apparently, users are supposed to come to the manual with knowledge of exactly what arithmetic nodes are, since this term is not explained and does not appear in the table of contents. The manual has no index. Back in the midst of Appendix B, readers may discover that an arithmetic node is one with no thermal capacitance.

Choice of analysis

One of the queries to the user asks for the desired type of analysis. Here users have a choice of a steady-state solution by the Gauss–Jordan method or for an iterative steady-state solution. Users also may choose to run explicit or implicit transient solutions. Gauss–Jordan solutions are limited to 85 nodes; the other methods can solve for as many as 350 nodes.

Users are warned that steady-state problems with radiation may not converge. If they do not, users are advised to reduce the damping factor. Users also are warned that the explicit transient solution may require "a large amount of computer time."

Beyond a few such warnings, the package presents itself as a tool kit from which the knowledgeable user chooses appropriate calculation techniques, and with which users, after trial and error, will, it is hoped, obtain meaningful solutions.

Not much in the way of assistance can be expected from the package. For example, as with other procedures involving matrix manipulations, the way that nodes are numbered can affect the speed and accuracy of the calculations. Instead of providing for automatic node renumbering, the package designers expect the users to experiment and find proper data entry sequences that will result in good node numbering arrangements.

MTAP

MTAP is a specialized set of routines and defined procedures that operate as a supplement to the BASIC computer language for solving heat transfer problems. The routines and procedures are designed for use by those who are already familiar with BASIC.

MTAP runs on IBM/PC, TRS-80, and CP/M computers. It sells for $195 from CYBERTECH, 501 Via Codo, Fullerton, CA 92635. (714) 738-1882.

MTAP solves linear and nonlinear heat transfer problems. It can analyze transfer by conduction, convection, radiation, or mass transfer. Both steady-state and transient problems can be handled. The package provides for the use of temperature-dependent material properties and time-dependent boundary conditions.

To set up a problem for analysis by this package, users must first represent it in terms of lumped conductive and capacitive elements connected together into a thermal network.

Facility with BASIC programming is expected from users. The input to the package consists of lines of BASIC computer language code. For example, the first line is a BASIC DATA statement with codes that spell out the number of nodes and tables to be used, among other items.

Through similar coded lines, users describe the network, materials, initial temperatures, conductances, capacitances, and heat applied to nodes. Along the way there are some helpful calculations available through BASIC subroutine calls. For example, the package allows algebraic expressions to be used for conductance values. With these, the user can enter formulas for forced and free convection.

Thermal conductivity, specific heat, and other parameters can be entered into MTAP in the form of tabular data. The package can automatically interpolate the data to calculate the desired parameter. There is also a routine for determining the variation in solar heating during the course of a day. Another routine determines the equivalent conductance for radient heat exchange.

A matrix routine

The main computational "engine" for this package is a matrix routine that solves simultaneous equations. It can be used either in solving heat transfer problems or separately as an equation solver.

When the required data lines have been entered, the user enters a computation statement. Time taken by the package to complete its calculations will depend on the size of the problem and, for transient problems, on the number of steps. The vendor claims that a typical 20-node problem should run in about one minute.

At the end of the computation, there is a display of the calculated data, and the user can choose to print out final temperature results or save them to disk.

Because the package is essentially a specialized collection of BASIC language tools, users have a lot a flexibility in the way its facilities can be used. However, this flexibilty comes at the expense not only of having to do

BASIC computer programming but of having to learn the needed package functions, their limitations, and ins and outs.

STAP

This steady-state heat transfer package was designed for analysis of heatsinks used in electronic packaging but can be used for many other two-dimensional thermal conduction problems as well.

STAP runs on IBM/PC, TRS-80, and CP/M computers. It sells for $69.95 from BV Engineering, 2200 Business Way, #207, Riverside, CA 92501. (714) 781-0252.

Heatsinks are divided up for analysis into matrices of squares (nodes), linked to one another by thermal resistances. Up to 1,600 nodes can be handled by the IBM/PC version of the package. Other versions can handle up to 400 nodes. The larger the number of nodes used in the calculation, the more accurate the results will be.

The package prompts the user to enter the physical dimensions of the heatsink in inches: thickness, *x*, and *y* dimensions. Users than select a heatsink material from a menu list with nine choices, and the package supplies an appropriate thermal conductivity value. If the material used is not listed, users can enter thermal conductivity directly.

Node by node entry

The package then prompts, one device at a time, for descriptions of the devices mounted on the heatsink. These can be heat-generating devices, cooling devices, or simply locations, each of which may or may not have a thermal resistance associated with it. Users specify the number of such devices, give each one a designation, establish the coordinate location of each one, specify the power in Watts dissipated or absorbed by each, and establish thermal resistance values for each.

A list of four types of electronic devices is presented during this data input sequence. Users can choose from this list, and the package will automatically insert the corresponding junction-to-case thermal resistance and area of heat dissipation into the calculation. For devices not on this list, users must supply their own values for these parameters. In any case, the user must input a case-to-heatsink value for each mounted device. The user's manual cautions that this value is highly dependent on the mounting techniques used for devices, yet must be accurate if the calculation results are to be accurate.

Data editing during the entry process is by the use of the Backspace key, allowing the user to erase what has been keyboarded, one character at a time. After data entry, users can still add or delete devices. Changes are made by deleting a device, then adding it back in the desired, changed form. Unfortunately, users have to identify devices to be deleted by reference number (the order in which they were entered), rather than being able to use the device designations they supplied or the location of devices on the heatsink or package types.

If users want to view the entire list of device entries, complete with reference numbers, they can call for a summary display. There does not seem to be any means provided to print out such summaries.

Once entered, sets of such data can be stored and retrieved as needed. This is convenient not only for rerunning the same data (the package provides for calling up such files as input, in lieu of keyboard data input) but for modifying existing data to describe similar device arrays, thus saving input effort.

The package includes some controls over entry of invalid data. Entries not in correct format, containing incorrect characters, or the wrong number of characters will not be accepted. In some cases an error message will be displayed, requesting the user to reenter the data. In other cases the user will simply be returned to the previous menu or prompt with no other error indication.

Setting resolution and iterations

When a calculation run is called for, the package queries the user on the number of nodes and on the longest dimension of the heatsink to be used for the analysis. Users are also asked to specify the largest allowable temperature difference between calculation iterations.

Choices of these parameters determine the accuracy and speed of the calculation. There is some discussion in the user manual on making these choices, but they will obviously involve some experimentation and experience on the part of users. It is hoped that there will always be some choice that can produce sufficiently accurate results in a reasonable time. During this review, a problem with three devices was run with a maximum resolution of 20 and an allowable temperature difference of one between iterations. The calculation then ran for three iterations, each one took about 25 seconds to complete.

Users also are prompted to specify the ambient temperature, and they also may call for sides of the heatsink to be held at specific constant temperatures.

As the calculation runs, residual values for each iteration are displayed. Users probably will need the diversion of watching these values progress toward zero as the long computation process converges. At completion, node temperature values are displayed. There does not seem to be any provision for saving the results.

THETA

THETA does steady-state and transient analyses and can account for radiation, conduction, and convection. All material properties can have nonlinear temperature dependence. Plotted and printed output is available.

THETA runs on Hewlett-Packard 9000 series desktop computers. It sells for $8,000 from H. G. Engineering, 260 Lesmill Rd., Don Mills, Ontario M3B 2T5, Canada. (416) 447-5535.

THETA consists of over a dozen software routines, each of which can be called into action through choices on a main menu display. The routines draw on common data files for their information.

Many menus and formats

EDIT is the routine used for data input and revision. In addition to title description, there are 15 different types of data that can be entered. These include data on body nodes, surface nodes, two types of conduction links, radiation links, coolant circuits, coolant links, and so forth.

The entry and editing procedure for each of these data types is reached through a separate menu, so the package has 15 different data entry menus. In fact, users have to pass through a main menu to a choice of two second-level menus in order merely to choose which of the 15 data-type menus they wish to use.

With all this elaborate presentation, one might expect procedures that would lead users by the hand through the actual data entry and editing process. Nothing of the kind occurs. Instead, the package merely presents users with a blank line at a time for data entry, while the manual informs them that they will have to follow prescribed formats in filling these lines.

There is a different entry format for each of the 15 data types. Some 34 pages in the manual are devoted to spelling out the details of these entry formats. It would be far preferable to relegate the formatting tasks to the package itself, leaving the users with some simple, uniform way to enter their data. For example, the package could display the required format visually on the screen and have users fill the needed data into assigned slots.

There are some data entry convenience features. For example, users can draw on prestored material properties to minimize the entry process. A routine called ARCH controls input and editing of the stored properties.

Like data entry, data editing proceeds on a line-by-line basis. However, for editing, users are presented with a previously entered line on which to key their changes.

A laudable feature of the entry process is the inclusion of many built-in control checks on the keyed-in data. The package checks to see if zero, positive, and negative numbers are being entered where they logically should be excluded. There also are checks to see that nodes and surfaces outside of user-defined bounds are not referenced. Excluded values are not accepted. Instead the user is presented with a display of the offending data, accompanied by a "warning list" and must edit it to correct the errors. Users are on their own in keeping track of the units they are using and making their entries consistant in this regard.

Automatic controls during data entry are supplemented by another set of controls carried out by the package's PRE routine. Here the controls are concerned with such matters as completeness (are all needed data items included in the input?) and logical inconsistancies in the data. The PRE routine also allows users to invoke a Gebhart link calculation to deal with multiple reflection effects.

Bandwidth minimization

A routine called OPTIM automatically renumbers the nodes in the thermal network for the computation process, thereby minimizing the amount of computer memory and time needed to reach a solution. The procedure is known as bandwidth minimization. Use of this routine is optional, but the manual recommends it when larger problems are being run. A series of passes through the minimization procedure may be needed to achieve bandwidth reduction. The package informs the user of the resulting maximum and root-mean-square differences between node numbers at the ends of any link; these are measures of the desired minimization.

Before making an actual computation run, the user sets calculation parameters with the Run module. For example, a user estimate of the steady-state starting temperature is required for each case to be run. Users also can choose temperature convergence criteria and place a limit on the number of iterations that will occur during the solution of nonlinear problems. This module also allows users to specify how much printed output they will get, with a choice of full, abbreviated, or one-of-*n* outputs for temperatures and heat fluxes.

Under some conditions, the calculation can become unstable and fail due to numerical overflow. To avoid this users are advised to employ a damping factor that decreases as the final solution is approached. A damping factor calculation of this type is included as part of the package.

If desired, there can be screen displays during the computation run, indicating which calculations are in progress and showing partial results.

In addition to numerical results for transient solutions, THETA can produce graphical displays and printouts. The plots each can include up to 12 traces. Where the output is to plotting equipment, the software will control four pens and vary the line types used in the plot. The routine does curve fitting and also can do autoranging, or users can choose to scale the plots themselves.

A postprocessor module called Time can be used to calculate the time response for a user-specified group of nodes.

TSAP

TSAP does steady-state and transient heat transfer analysis involving conduction, convection, and radiation. Networks can be multidimensional, and temperature-dependent material properties can be handled. Users are expected to be familiar with FORTRAN programming.

TSAP runs on IBM/PC and CP/M computers. FORTRAN is required. The package sells for $150 from T. Kao, PO Box 314, Roseland, NJ 07068.

Input to this package follows set formats that the user must follow in detail. Users are specifically warned, for example, not to leave any blank spaces in the data unless instructed to do so. Failure to follow the formats exactly can lead to erroneous results.

The input procedure includes the ability to generate input data for a sequence of nodes if they all have identical initial temperatures, heat capacities, and internal heat sources. There is also a special input procedure for nodes whose heat capacity is a function of the node temperature. Similar procedures are available for entering data on conductances.

Parameters that control the method and procedures of the solution are entered by the user in a formatted control block. Included here are 14 parameters such as the choice of solution algorithm, time steps, and convergence criteria. Users are expected to rely on experience to properly set these parameters; there seems to be no guidance in the user manual on this subject.

If desired, input data for temperatures, thermal capacities, and other input parameters may be entered in array form. The package contains a routine that can call up and interpolate such array data. Up to 20 arrays with a total of 200 data items can be accommodated. As with the other input data, these arrays must be entered in a strictly prescribed format.

Another special input procedure handles time-dependent boundary conditions and nonlinear temperature-dependent conductances and thermal capacities.

FORTRAN files

The developer of this package thought in terms of FORTRAN files when writing it, and users are expected to think in the same terms. The processing routines are designed to handle problems with up to 200 nodes and 400 conductances. The manual invites users to revise the routines so they will accommodate still larger networks. In any case, in order to get the processing underway, the user has to link together several of the vendor-supplied files and get them on the right disks.

The output data consists of temperatures for each node in the network. Where transient data are being calculated, there is a set of such temperature data for each calculated time.

Chapter 2
Hydraulics and Pipe Design Packages

The packages discussed in this chapter include those for solving pipeline design problems, analyzing pipe networks, and working with pumps.

Some personal computer pipe design packages, for example, the HP3M and Program Hardy Cross, are written specifically to deal with water distribution. Others, such as PIPE 123, are targeted to piping systems within buildings. PIPE 123 is particularly interesting in that it is written to work with the popular Lotus 1•2•3 package, making use of the Lotus spreadsheet tabular format and operating as a spreadsheet template. Many pipe design packages are more general purpose, allowing the user to design for a variety of liquids and gases. Thus the Automatic Pipe Sizing package comes with data on pipeline conveyance of paper stock and on several gases, including air, steam, natural gas, carbon dioxide, and anhydrous ammonia. Users can work with other fluids if they enter the appropriate densities and viscosities.

Software packages for pipe sizing and pressure and flow analysis using personal computers can offer designers substantial help in getting needed answers. Using a handbook to size pipe involves compromises. For example, with the *Cameron Hydraulic Data Handbook* the user can select the type of pipe, the inside diameter, and the flow, but it then is necessary to interpolate between tabular entries for velocity, velocity head, and head loss. The procedures are time consuming, and the results are often inaccurate.

Working from stored data

A well-designed computer package, such as Automatic Pipe Sizing, allows the user to make accurate calculations for any set of conditions. There is no need to interpolate between tabular entries, and the user has control over variables, such as fluid temperature and absolute roughness, that are normally fixed in manual calculations.

Automatic Pipe Sizing also includes complete dimensional data on stainless steel, steel, and black iron pipe, fittings, and valves. The fact that pipe data deal with a finite number of discrete sizes makes computer storage of such data feasible. Thus, the DPF package from Pixel in Idaho Falls, Idaho, comes with stored files of nominal pipe sizes and schedule numbers as well as files of the resistance coefficients of standard piping components. Using the package involves creating still more files containing fluid names and properties and conversion factors.

Beyond manual calculations

With software packages the designer can do things not feasible with manual calculations. For example, one convenient feature of the FLO series packages from Engineered Software in Olympia, Washington, is the ability to duplicate an already-completed pipeline design. By computer-screen editing, the user can quickly make changes to design pipelines similar to the duplicated one. This feature might be used, for instance, to evaluate system operation with a cooler out of service, to upscale or downscale a design by changing the GPM, or to see how a system behaves when a different fluid is handled. If desired, Engineered Software's PIPE-FLO, a data storage and pipeline design package, will even do an optimum pipe-size calculation.

In the NET-FLO pipe network design package from the same vendor it is a simple matter to close a pipeline in a network to see how the overall flow is affected. Pressure regulating, back pressure, and check valves can be installed easily. Supply and booster pumps can be selected quickly. With this package it is feasible to arrive at near-optimum pipe network designs by varying pipe diameters in successive calculations to achieve reasonable fluid velocity while maintaining proper flow and pressure distribution.

Gas and liquid pipelines

The GASPIPE package calculates steady-state pressures in a gas pipeline and the pump horsepower required for specified flow rates and compression ratios. As with many engineering packages, the main task for the user is to collect the necessary data to feed into the package. Unfortunately, the input method provided by GASPIPE is not very convenient.

Liquid pipelines are the concern of the LIQTHERM package, which calculates steady-state hydraulics for a buried, heated liquid pipeline. Liquid temperature, specific gravity, viscosity, and the pressure along the pipeline as well as the horsepower required at a given flow rate are calculated, considering heat transfer with the surroundings. Heat generated due to friction may be included as an option.

Exemplary data entry

SDNET handles storm drain network analysis for both full and partially full pipes. Minor losses and headwater effects can be included. This package uses an orderly and easy to follow entry procedure that simplifies the difficult job of correctly keyboarding extensive network data.

Data entry is made on full-screen display forms. As the user completes each entry on the form, the cursor moves on to the next entry position. When the entries on a screen are complete the user can review them and make changes before entering the entire screen.

Working from a hand sketch of a pipe network, the SDNET user consecutively numbers each pipe and node, then keys in the total number of pipes. If the Manning number for all pipes is the same, that number is then entered. A data entry screen is then displayed for the first pipe in the system, and the user enters upper and lower node numbers, length, diameter, flow, and Manning number (if different Manning numbers are used) for that pipe. Similar screens are then filled out for each of the other pipes in the system. Node data are entered on a similar set of screens. This system allows for orderly entry of large amounts of data as well as easy checking.

Transient analysis

HYTRAN is exceptional because it is able to simulate transient as well as steady-state fluid flow in a hydraulic network. Liquid and gas flow problems can be solved for a network consisting of pipes, valves, accumulators, pumps, and other head-loss elements. The transient capability allows phenomena such as water hammer, liquid column separation, and pump failure to be handled. These results are obtained by solving the Navier–Stokes equations for conservation of mass, momentum, and energy.

Unfortunately, this package is not designed with user convenience in mind. It assumes the user is familiar with computer techniques as well as engineering principles. Thus, input to this package requires the creation of a specially formatted file. The input data not only must be included but also a set of instructions as to what computational procedures are to be carried out and in what order. In other words, the user has to construct a special-purpose program to tell this package what to do.

Pump selection and performance

The CPPAC centrifugal pump package has a catalog section that uses approximate design calculations to find a design that closely resembles a commercial pump. The idea is to help the user match desired flow and head conditions to the best efficiency points offered by available commercial pumps.

A second section of the package aids the user in revising and redrawing pump curves for changes in speed, impeller diameter, viscosity, number of stages, or specific gravity.

The PUMPCURV package takes a slightly different approach. It predicts the performance of centrifugal pumps with different impeller diameters, speeds, and conditions of destaging and restaging. Calculations are based on the affinity laws for centrifugal pumps. The package also calculates performance of several pumps in series or in parallel configurations.

AUTOMATIC PIPE SIZING

Reviewer: Victor E. Wright, P.E., Louisville, Kentucky.

Automatic Pipe Sizing is a valuable replacement for the hydraulic handbook. It can be used to size piping systems designed to convey fluids and to calculate the characteristics of fluid flow in an already-designed system. Although it costs more than the handbook, it offers better accuracy and more information.

The package runs on the IBM/PC with 128K main memory. It sells for $395 and is available from Computer Technology Specialists, 1855 Avenue of America, #3D, PO Box 1136, Monroe, LA 71210. (318) 323-6060.

Sizing pipe for fluid flow is not a trivial exercise. Even for ideal fluids, the equations are not simple. For real fluids both theoretical and empirical equations are needed. Those of us who size pipe for a living have learned to use manual procedures that give results that are "close enough." This reviewer happens to prefer to use the *Cameron Hydraulic Data Handbook* (Ingersoll-Rand) when forced to size pipe manually. This book contains data on water and other fluids plus a section on hydraulic principles.

Using a handbook to size pipe involves compromises. With the *Cameron* handbook the user can select the type of pipe, the inside diameter, and the flow. Then it is necessary to interpolate between tabular entries for velocity,

velocity head, and head loss. Parameters such as absolute roughness and temperature are fixed. The tabular data are accurate, the procedures are time consuming, and the results are not so accurate.

A well-designed computer procedure such as the Automatic Pipe Sizing package allows the user to make accurate calculations for any set of conditions. There is no need to interpolate between tabular entries, and the user has control over more variables. Thus, this package allows the user to control parameters such as fluid temperature and absolute roughness that are normally fixed in manual calculations.

A universally accepted equation

The process of sizing pipe comes down to selecting a size that will produce a pressure drop or friction loss that meets desired criteria of noise, wear, pump horsepower, and so forth. Although there are several equations that model head and pressure loss values, this package uses the universally accepted Darcy–Weisbach equation, the same one used in the *Cameron* handbook.

The equation itself is handled easily by manual calculation. However, one of the equation terms, the friction factor, is more difficult to deal with. It is commonly available in the form of the Moody chart, or it can be calculated from the Colebrook equation. Unfortunately, for those who use manual solutions, that equation has to be solved iteratively. The Automatic Pipe Sizing package uses the Colebrook equation and solves it automatically by a Newton–Raphson numerical method.

Three operating modes

Automatic Pipe Sizing offers automatic, manual, and pump modes of operation. In the automatic mode users specify allowable pressure drop and allowable velocity in a segment of pipe, along with flow rate and the type of fluid that is flowing. The package selects a standard pipe size, from a pipe schedule that the user chooses, that meets the loss and velocity criteria. The user can choose to disregard either of these criteria by setting the allowable value to a very high number.

In the manual mode, users specify the type of fluid, the flow rate, and the pipe size and schedule. The package then calculates the actual loss and velocity in the pipe segment.

In the pump mode, additional hydraulic data are input, and the package calculates the pump system head. The package facilitates the development of system curves in this mode.

Built-in data

The package comes with data on paper stock and several gases including air, steam, natural gas, carbon dioxide, and anhydrous ammonia. Users can work with other fluids if they enter the appropriate densities and viscosities. For water and the gases listed above, the package calculates the needed viscosity and density. An empirical method is used to determine a pseudo-Reynolds number and friction factor for paper stock.

The package also includes complete dimensional data on stainless steel, steel, and black iron pipe, fittings, and valves. Sizes for stainless steel are included through 30 inches nominal, for steel and iron pipe through 36 inches nominal. The variety of fittings and valves seems adequate for almost any situation.

Package limitations

The package is supposed to be able to run on IBM-compatible computers, but this reviewer was unable to get it to work on a Zenith Z-100 system. The package is not copy protected, so users can freely make backup copies of the disk.

In its current version Automatic Pipe Sizing will not handle flashing condensate, but a future version of the package will. Also for the future is a version that can make use of the 8087 math chip to speed computation. This would be particularly useful for building system curves in a form suitable for plotting.

The package does not allow users to enter a description of an entire piping system. One segment must be sized at a time. The package does not provide for storing output information on a disk. Users can print one sheet of output per pipe segment or can produce printed versions of material displayed on the computer screen.

CPPAC: CENTRIFUGAL PUMP PAC

CPPAC has two sections. The catalog section uses approximate design calculations to find a design that closely resembles a commercial pump. The idea is to help the user match desired flow and head conditions to the best efficiency points offered by available commercial pumps.

The pump curve section of the package aids the user in revising and redrawing pump curves for changes in speed, impeller diameter, viscosity, number of stages, or specific gravity.

CPPAC runs on HP Series 80, HP-41, HP-9816 and IBM/PC computers. It sells for $500 from Gordon S. Buck, 5801 Parkhaven Dr., Baton Rouge, LA 70816. (504) 293-6581.

CPPAC operates through displayed menus, with menu choices made by pressing function keys. For example, pressing function key #2 while the initial menu is displayed will bring up the catalog portion of the package.

Data are then entered in response to a sequence of screen-displayed queries. Following each query, the user types an answer. When this is entered, the next query appears. Key data items include rated and maximum temperatures, specific gravity, viscosity, and vapor pressure.

Catalog calculations

The user is offered the choice of entering the Net Positive Suction Head Available (NPSHA) directly or of inputting vapor and suction pressures from which the package will calculate NPSHA. Pump performance can be specified in terms of discharge PSIG, differential PSIG, or total dynamic head. Users can enter rpm, stages, and number of suction eyes directly or have the package calculate these as well.

To determine the required flange rating, the user is asked to enter an API material code. For those not familiar with these codes, pressing the HELP key will call up a short list. Function key #7 is reserved for HELP. At various points in the procedures, added information for the user will be shown on the screen when this key is pressed.

With all the needed input data entered, the package makes its catalog calculations and displays the results. If desired, the user can call for a printout of this information. Data input by the user can be stored on a disk for later use, thus bypassing the need to go through a detailed query and response sequence. The storage is accomplished by pressing a function key and providing a file name for the data. Changes in the input values of flow, head, NPSHA, rpm, stages, or suction eyes also can be made, and the calculation can be rerun.

Pump curve drawing and revision

User procedures for the second section of the package are similar. However, the user is required to enter a set of four to ten data points so there is some repetitive entry here. Like the other data, these points are entered in a serial query and answer format. If more than a few points are to be entered, this can become a bit unwieldy. It would have been preferable to have this entry onto a tabular screen display.

The entered data include values for flow, head, efficiency, and NPSHA, and the package will display and print curves for differential head, brake horsepower, efficiency, and NPSHA plotted against flow.

These curves can then be revised using a change menu that allows the user to select new values of specific gravity, viscosity, rpm, diameter, and number of stages. When these are complete, revised curves can be drawn.

Calculation limitations

This package is intended for approximate calculations. For example, impeller diameter is determined from an equation that is valid only for speeds between 400 and 3,500. Maximum values for this package include 2,500 for the flange rating, 800°F for the temperature, and 12,000 for suction-specific speed.

Calculated efficiency is assumed not to change with speed or impeller diameter, and NPSHA follows the law of constant suction-specific speed.

When high values of viscosity or flow are entered, the package will give an "OUT OF RANGE" warning. This may indicate that use of a centrifugal pump is not appropriate for the design problem.

The user manual explains these and the other limitations of the package and the calculations used by the package to produce its results.

DPF: PIPE NETWORK HYDRAULIC ANALYSIS

DPF calculates steady-state pressure drop and flow in piping systems. It can find mass flow rates and static pressure drops in looped pipe networks and in those with single and multiple branches. Hydraulic resistance of pipes, fittings, valves, and user-defined components are considered, as are elevation changes, pumps, external flows, and pressures. Darcy's equation is used to determine hydraulic resistances. The linear theory method is used to calculate parallel flow rates.

DPF runs on IBM/PCs and compatible machines. Use of an 8087 math chip is recommended. The package sells for $295 and is available from Pixel, 1066 Claredot Drive, Idaho Falls, ID 83402. (208) 522-0447.

The DPF package comes with some stored information, including files of nominal pipe sizes and schedule numbers as well as files of the resistance coefficients of standard piping components. Using the package involves creating other files of fluid names and properties and of conversion factors.

Package limitations

Users should be aware of the assumptions and limitations built into this package. When a computer with 256K main memory is used, maximum network size will be 100 branches, 999 components, 100 nodes, and 75 tees. All branches must be connected to the network. Stored pipe sizes cover only the range of 0.5 to 36 inches NPS, schedule 5S to XXS or 160. Characteristics of other pipe sizes must be supplied by the user.

Two-phase flow cannot be analyzed correctly. Fluids are assumed to be constant density, steady state, and Newtonian. Gases may be used as long as the pressure drop across any branch is no more than about 10 percent of the total pressure at that branch inlet.

Resistance coefficients for valves, built into package files, are derived from empirical formulas and may not be accurate for some particular units. Use of vendor data is preferred. Hydraulic resistance interaction effects between closely adjacent components are ignored.

For all components except reducers and nozzles, the effect of velocity changes on static pressure are ignored. For all components except pipe, the effect of body forces caused by gravity on static pressure are ignored. If these effects are significant, the designer must calculate them outside the package.

Linear theory method computations sometimes fail to converge. The vendor finds that, with the help of an 8087 math chip, results networks with under 100 branches are usually obtained within a few minutes.

Do-it-yourself input

DPF takes input information in the form of files that the user must prepare outside the package. File preparation can be done with the help of almost any competent editing or word processing software, but it does involve careful adherence to a set of format rules required by DPF. For constant users of this package there will be no problems in this arrangement, but for occasional users it is a decided nuisance since it involves learning and remembering what is essentially a specialized language.

Menu-style and interactive input formats were invented to circumvent this sort of user-unfriendly operation. The designers of this package have chosen to ignore the convenience of such inputs.

Units and output data

Output data are labeled with appropriate units. The user may choose the units to be shown by creating a conversion file as part of the supplied input. If not, the package will resort to its default set of units.

The output data themselves include the hydraulic resistance and pressure drop caused by flow and elevation change for each network component and each branch. Resistance coefficients and pressure drops between branches are given for each tee. Also listed are the number of simultaneous equations used overall and for loops and nodes, as well as the convergence error and number of iterations.

GASPIPE

GASPIPE calculates the steady-state pressures in a gas pipeline and the horse power required for a specified flow rate and compression ratio.

The package runs on IBM/PC/XT/AT computers and most other personal and microcomputers. It sells for $149 from SYSTEK, 963 N. Brandywine, Corona, CA 91720. (714) 734-1351.

For GASPIPE, the user describes a pipeline with data for a set of profile points. At each point the user gives a mile-post position, elevation, and diameter and thickness dimensions.

All input data are entered in response to queries from the package. To describe a given pipeline the user answers a query for its name and the number of profile points to be used. Then the package will prompt for each point and the user replies by keying in the four data items for that point, separated by commas (but no spaces are used). When these data have been entered, the package automatically files the data under the pipeline name.

The package then queries for the remaining needed data: flow rate, specific gravity, viscosity, flow, suction, base temperatures, base and discharge pressures, compressor efficiency, and compression ratio.

Output from the package consists of a figure for horsepower required, along with a calculated pressure for each profile point. This output may be displayed or printed.

HYTRAN: HYDRAULIC TRANSIENT SIMULATION PACKAGE

HYTRAN simulates transient and steady-state nonisothermal fluid flow in a hydraulic network. Liquid and gas flow problems can be solved for a network consisting of pipes, valves, accumulators, pumps, and other head-loss elements. The transient capability allows phenomena such as water hammer, liquid column separation, and pump failure to be simulated. The solution is obtained by solving the Navier–Stokes equations for conservation of mass, momentum, and energy.

This package is not designed with user convenience in mind and assumes familiarity with computer procedures as well as engineering principles.

HYTRAN runs on IBM/PC and compatible computers and on CP/M computers. It sells for $300 from T. Kao, PO Box 314, Roseland, NJ 07068.

Input to this package is done by setting up a formatted input file. Not only must the input data be included but also a set of instructions as to what computational procedures are to be carried out and in what order. In other words, the user has to construct a special-purpose program to tell this package what to do.

In addition to a block of control data (that is, program instructions), users must enter a formatted block of node data. There are special formats for such items as accumulator gas nodes and surge tank nodes, and lots of format rules to remember. Then there is an input block for node-connecting elements with special rules for different types of elements.

The package includes models for pumps and motor drives, so there is a special input data block for pumps. The user also is expected to set up a special data block to establish nonlinear and time-dependent boundary conditions for the package's calculations.

The package is flexible in that it allows users to include their own computation subroutines along with those provided by the vendor.

Once input of data and instructions is completed, the user begins the execution of the program. This involves formatting and moving various files onto two disks, then running first one disk then the other. The output obtained is a tightly packed array of data and annotation terms that requires some study to interpret.

LIQTHERM: HEATED LIQUID PIPELINE HYDRAULICS

LIQTHERM calculates the steady-state hydraulics for a buried, heated liquid pipeline. The liquid temperature, specific gravity, viscosity, pressure along the pipeline, and the horsepower required at a given flow rate are calculated, considering heat transfer with the surroundings. The heat generated due to friction may be included as an option.

LIQTHERM runs on IBM/PC/XT/AT and compatible computers and *sells for $225 from SYSTEK, 963 N. Brandywine, Corona, CA 91720. (714) 734-1351.*

Input data include a pipeline mile-post and elevation profile, pipe diameter, wall thickness, pipe roughness, depth of burial, liquid flow rate, specific gravity and viscosity of liquid, inlet temperature, soil temperature, soil and pipe insulation conductivities, suction and delivery pressures, and the pump efficiency. The viscosities of the liquid at two known temperatures are input to determine the viscosity–temperature correlation. If the pipe is bare (uninsulated), the program automatically accounts for this by using a large number for the insulation conductivity.

The output consists of the temperature, specific gravity, viscosity, and pressure along the pipeline together with the horsepower required. Additionally, the total head (including elevation) at each profile point is also output. This can then be used to plot the hydraulic gradient.

Pressure drop calculations are based on the Darcy–Weisbach equation, using the Colebrook–White transmission factor. Laminar, transition, and turbulent flows are automatically accounted for by the program. LIQTHERM can be used to determine the location of pump stations along a heated liquid pipeline based on a specified allowable pipeline pressure.

NET-FLO: ANALYSIS OF PIPE NETWORKS

NET-FLO determines the balanced flow rate and pressure drop for each pipeline in network pipe systems.

Operation of this package relies on pipe data stored by using a related package called PIPE-FLO. Data on elements such as pumps and filters are also stored, with the help of NET-FLO. Use of this stored information allows a convenient inputting method and easy modification of designs.

NET-FLO runs on IBM/PC and compatible computers and on those that use CP/M system software. It is sold together with the PIPE-FLO package for $1,500 by Engineered Software, 2420 Wedgewood Dr., Olympia, WA 98501. (206) 786-8545.

To describe the the network configuration, the user first assigns a two-character label to each node. Then, on a displayed diagram, the pipes and components between nodes are identified by single numerals and letters, and the user also gives the assumed direction of flow between the two nodes. A separate screen is used to enter the hydraulic grades for nodes that are to have fixed grades.

Entering and editing all this information are simple procedures. At the press of the Return key the screen cursor moves from one data position to another on the display screen. When the screenful of information has been completed it is stored for further use.

Results produced by the package for each node include the demand, hydraulic grade, and pressure. For each pipeline connecting one node to another, the package shows head loss, velocity, and flow rate.

Computer-based design techniques

One of the most productive ways to use this package is to configure a network in which the individual pipelines remain undefined. With the pipeline list produced by NET-FLO in hand, the user can then turn to the PIPE-FLO package to separately define each pipeline. The resulting data are stored in a file that is immediately accessible to NET-FLO for completion of the design process.

In NET-FLO, computer capabilities are used to easily implement design procedures that would be arduous if done by hand. Thus, it is a simple matter to close a pipeline in the network to see how the overall flow is affected. Pressure regulating, back pressure, and check valves can be installed easily.

Booster pumps can be selected by first installing a generic pump (one that develops a set pressure, regardless of the flow rate) into the network. From the total developed head calculated by the package the user can select an actual booster pump for the design. Similarly, network supply pumps can be selected by adjusting the elevation of the supply, then selecting a pump to achieve the desired flow rate at the necessary pressure.

With this package it is feasible to arrive at near-optimum designs by varying pipe diameters in sucessive calculations to achieve reasonable fluid velocity while maintaining proper flow and pressure distribution.

PIPE-FLO: SETTING UP AND USING DATA FOR PIPELINE DESIGN

PIPE-FLO is divided into two parts: a piping design procedure and a procedure to set up and maintain tables of data for piping calculations. It performs head-loss and system-resistance calculations for flow in a full round pipeline conduit. The package also is used to maintain data tables for those calculations and for calculations performed by related design packages from the same vendor.

PIPE-FLO runs on IBM/PC and CP/M computers. It sells for $500, from Engineered Software, 2420 Wedgewood Dr., Olympia, WA 98501. (206) 786-8545.

The basic operating concept of this package is that the user will first set up tables of information about the piping equipment and system that are to be analyzed, then draw information from these tables when the analysis itself is to be performed.

A slow start

The resulting procedure involves a few added steps when compared to packages that go directly to analysis, but the user is rewarded with ready access to the entered data for solving future problems.

A table of data must be provided for each pipeline that is to be included in the analysis. For this, the user fills in data and choices on a screen-displayed pipeline data entry form. These include such items as the pipe material, design flow rate, fluid density, viscosity, entrance and exit elevations and pressures, and pipeline length.

The design calculation draws on stored pipe size schedules. When the user selects a desired nominal pipe size, the package displays the schedules on file for sizes closest to that nominal, and the user chooses which of the schedules are to be used.

To describe valves and fittings that may be part of the pipeline, the user fills data into a valve and fitting display screen form. One of the initial steps in the design procedure is to select the stored table that describes the valves to be used. PIPE-FLO comes with a table listing the characteristics of 28 often-used valves and fittings. If these are initially selected, the names of these standard valves and fittings will be listed on the valve and fitting display screen, and the user can simply enter one or more of these names as items to be currently used. If nonlisted fittings or valves are to be currently used, a name and description parameters for each must be entered on the valve and fitting screen.

Calculation, duplication, and controls

When entries for the pipeline and the valve and fitting data screens have been completed, the package immediately calculates the K values, friction factor, head loss, and exit–entrance pressure difference for that pipeline.

These results are displayed, and the user can then go on to analyze a second pipeline.

One convenient feature is the ability to duplicate an already-completed pipeline design. Using the package's editing capabilities, the user can then easily make changes to get a design for a pipeline similar to the duplicated one.

If desired, the package will do an optimum pipe size calculation. The user indicates the desired maximum fluid velocity, and the package finds a size that will limit the velocity to this value.

The calculated results, along with input information and user comments, can be printed in a report format, but the package does not provide for any graphical display. This would be a useful feature, particularly to provide a system resistance curve.

There are some built-in controls to help the user avoid mistakes. The package will not accept pressures of less than −14.7 PSI to be entered since this corresponds to a perfect vacuum in the gage pressure convention. If the calculated output pressure proves to be under −14.7 PSI, a "???" warning will be displayed to alert the user to the problem.

Setting up stored tables

Using the maintenance feature of the package, users can set up their own stored tables of materials, pipe sizes, and valves and fittings. The package comes with data on steel, so users who want to analyze cast iron or other pipes must compose entries describing these materials. Similarly, entries for valves, fittings, and pipe sizes that do not come with the package can be composed and stored for later use.

Limitations of the solutions

PIPE-FLO makes use of the Darcy equation to calculate head loss. This equation assumes that a noncompressible Newtonian fluid is flowing through the pipe. Users are advised not to try to design systems where the Reynolds number is close to the critical value of 2,100 where flow may be either laminar or turbulent.

To calculate system pressures, the package uses the Bernoulli equation, assuming the fluid is noncompressible with constant density throughout the system and also assuming the velocity of the fluid is the same at all points in the system. In certain applications, such as pump suction and control-valve sizing, the velocity component of the pressure must be taken into account. Users must manually add this velocity component to the final system pressure since the package does not perform this correction.

Care must be taken by users to follow PIPE-FLO conventions about placement of such components as reducers and tees so that correct values of the fitting head-loss coefficient will be calculated under the procedures set up in the package.

PIPE 123: PRESSURE DROPS FOR LOTUS USERS

PIPE 123 sizes piping and calculates pressure drop based on flow. The package is intended for design of piping systems in buildings. It is written for use with the Lotus 1·2·3 multipurpose software, which sells for $495.

This package is low cost for those who already have Lotus 1·2·3. It offers the user the spreadsheet advantage, over manual calculation, of easily producing further answers while changing the input parameters. However, it is not designed for ease of learning. The manual leaves much to be desired.

PIPE 123 runs on IBM/PC and compatible computers with 256K main memory and with the Lotus 1·2·3 package. PIPE 123 sells for $40 and is available from E. Jessup & Assoc., 4977 Canoga Ave., Woodland Hills, CA 91364. (213) 883-9021.

To start up PIPE 123 the user must have both Lotus 1·2·3 and PIPE 123 on disk, ready to be run. Package use begins with keyboard entry of the date and description of the current computation. The first step is to select the desired safety factor for the design and to choose the roughness expected with the type of pipe that is to be employed. There are three choices offered here: rough, fairly rough, and smooth pipe.

Foibles of a filled-in spreadsheet

The user is then presented with a filled-in spreadsheet that lists parameters for a sample pipe calculation. This is the entry form for the package. To enter their own data, users must position the cursor at the desired spots and type over the sample data to replace it with their own.

Use of a spreadsheet is a good idea. It gives the user an overall look at what has been input before a function key is pressed to run the calculations. However, the arrangement used here has the distinct disadvantage of having the input cluttered up with spurious data from the package example. If the user fails to input data at every position of the input spreadsheet, there will be no blanks to signal these omissions. Instead there will be spurious data.

Recognizing that extra care is needed to check the spreadsheet display, the package authors have suggested that users photocopy blank data entry sheets and pencil in the data they intend to enter before they sit down at the keyboard. However, this is exactly the input-shaping function a blank spreadsheet format on the computer display would perform.

Entry of fittings, such as valves, elbows, and tees, is facilitated by a displayed menu that numbers the various choices offered by the package so those numbers can be used as spreadsheet entries.

Output on a spreadsheet

Like the input, the output of this package appears in a spreadsheet format. When the calculation is complete, users have before them the data on flow, length of pipe, fittings, equipment pressure drop, and safety factor that they keyed into the computer. Also shown on the final spreadsheet are pipe sizes, fitting equivalent lengths, total equivalent lengths, friction losses, and friction heads for each section of the system. Overall system friction and pump friction head also are calculated and shown.

A worked-out example is given in the manual, showing a completed output spreadsheet and the piping diagram from which it was derived. However, the package vendor has not gone to the trouble to explain the rules that were used for representing piping system elements as spreadsheet entries, opening the way to misinterpretation, error, and confusion.

If users want to expand the application of this package to piping systems larger than those that will occupy the 30 spreadsheet rows this package provides, they are invited to use their own skills at manipulation of Lotus 1·2·3 to replicate the package formulas and allow more pipe sections to be handled.

PROGRAM HARDY CROSS: WATER DISTRIBUTION NETWORK ANALYSIS

PROGRAM HARDY CROSS can be used to analyze networks with up to 650 pipes, 500 junctions, 225 loops, and 15 supply sources.

The package runs on Wang 2200-VP computers and sells for $1,500 from Rhys A. Sterling, W. 817 Crestview Rd., Rt. 1, Spokane, WA 99204. (509) 448-0536.

PROGRAM HARDY CROSS requires that the user help in the process of defining the loops of the network being analyzed. Users have to identify and enter the number of source loops in the network and assign appropriate signs to the pipe numbers in these loops. Users similarly have to define any network loops that contain cross-over pipes. Care also must be taken to number network nodes in an order that the computer will be able to handle. These are procedures that take some effort to master, and the manual provided with the package is wordy but not too helpful in deciphering what exactly is to be done. The process seems to be a bit like learning to ride a bike: Easy to do once you have learned, but hard to get the things underway.

All at one sitting

An example of a pipe system is given as a sort of rosetta stone, but to follow this example it is necessary to shuttle back and forth between the text and two network drawings, separated from one another by six pages. All in all, a rather disorganized presentation, but patience conquers in the end.

With a large network there can be a lot of pipe numbers and other data to enter. Rather unsympathetically, the manual prescribes that "the entire data file should be defined at one sitting."

From selected points, taken from a pump operating curve, a routine in the package will calculate an operating curve equation for the pump. Another routine allows the user to design a pressurized storage tank.

Sometimes lengthy calculations

The manual indicates that it will take about 20 seconds of computing time to balance a network with ten loops, provided the solution can be obtained in 20 iterations. The package will balk if more than 200 iterations take place for the network loops, or if more than 40 iterations are used to balance the pump-tank combinations, or if over 25 iterations are used to balance the pressure-reducing valves. If you try to get a solution for the maximum 225-loop network and the package goes through 200 network loop iterations, that part of the calculation might run an hour and a quarter, not to speak of the pump-tank and valve computation times.

PUMPCURV: CENTRIFUGAL PUMP PERFORMANCE

PUMPCURV predicts the performance of centrifugal pumps with different impeller diameters, speeds, and conditions of destaging and restaging. Cal-

culations are based on the affinity laws for centrifugal pumps. The package also calculates performance of several pumps in series or in parallel configurations.

PUMPCURV runs on IBM/PC/XT/AT and compatible computers. It sells for $149 from SYSTEK, 963 Brandywine, Corona, CA 91720. (714) 734-1351.

Input data, keyboarded by the user in response to screen-displayed queries, consist of sets of capacity, head, and efficiency data selected from the pump curve. Impeller diameter and specific gravity of the liquid are also entered. Output produced by the package includes total flow, total head, and total horsepower for each of the input data sets.

SDNET: BACKWATER CALCULATIONS IN GRAVITY-FLOW PIPE NETWORKS

This storm drain network analysis package handles both full and partially full pipes. Minor losses and headwater effects can be included. The package uses an orderly and easy-to-follow entry procedure that simplifies the difficult job of correctly keyboarding extensive network data.

SDNET runs on IBM/PC and compatible computers and on CP/M computers. It sells for $375 from Disco-Tech Microcomputer Products, 600 B Street, PO Box 1659, Santa Rosa, CA 95402. (707) 523-1600.

SDNET works from a main menu. Data entry is started by selecting the Create New Job option. A data entry screen is displayed and the cursor moves to the next entry position as the user completes the previous one. When the entries on a screen are complete the user can review what has been entered and make changes before entering the entire screen.

A Minor Loss screen allows the user to select calculations for sudden enlargement or contraction, entrance, bend, angle, junction, manhole or catch basin losses, all for inclusion in the computed results.

Well-organized entry for pipes

Working with a hand sketch of the pipe network, the user consecutively numbers each pipe and node, then keys in the total number of pipes. If the Manning number for all pipes is the same, that number is then entered.

A data entry screen is then displayed for the first pipe in the system, and the user enters upper and lower node numbers, length, diameter, flow, and Manning number (if different Manning numbers are used) for that pipe. Similar screens are then filled in for each of the other pipes in the system.

And for nodes

When entry of pipe data is completed, a General Node Data screen is displayed. The user enters the design HGL at the low end of the system, indicates whether the entrance loss coefficient is to be constant, whether headwater is to be considered in the computations, and whether the package should check for maximum HGL at each node.

An entry screen for the first node is then displayed. The user enters the node type (round, rectangular, angle, or junction), then diameter, width, and angle, as appropriate. The node elevation is entered and, if needed, an entrance loss coefficient, maximum design HGL, and headwater value. If bend and losses are being considered, the user must enter deflection angles between pipes entering that node. A similar screen is then displayed for each of the other nodes of the network. When all the node screens have been completed, the package calculates the slopes for each pipe in the system. These can be viewed at the user's option.

The entered data are stored on disk automatically and can be retrieved as needed by entering a user-assigned job number. Since numbers are easily forgotten, the package displays a user-entered description of the stored data to verify that the desired job is being retrieved. This is a clumsy procedure that could have been improved by using names rather than numbers to identify the job files.

Checking, editing, and running

Once retrieved, the input data can be viewed on the display screen or printed. This provides ample means for rechecking the data for errors, and, if any are found, editing procedures allow the user to go to the data screen in question and make the corrections. Changes made in this way may include adding or deleting pipes from the system

When all the input data are complete and correct to the user's satisfaction, a flow analysis can be called for. During the calculation process, which may be lengthy for larger systems, the package keeps the user informed of its progress with messages. If HGL is being checked, the package will inform the user when and how much the calculated value of HGL exceeds the maximum design value at a given node. The user is then given the option of aborting or completing the calculation process.

A restriction on the systems is that they must be simple tree networks, with a single system outlet.

SYS-FLO: ANALYSIS OF SERIES-PARALLEL PIPE SYSTEMS

SYS-FLO calculates head loss for pipe systems with series and parallel segments. It computes inlet and outlet pressure for each segment and also finds flow rates for the parallel paths.

Operation of this package relies on pipe data stored by using a related package called PIPE-FLO. Data on elements such as pumps and filters is calculated by a curve-fitting procedure. Use of this data allows a convenient inputting method and easy modification of designs.

SYS-FLO is for computers that use CP/M system software. The combined price for SYS-FLO and PIPE-FLO is $1,250 from Engineered Software, 2420 Wedgewood Dr., Olympia, WA 98501. (206) 786-8545.

SYS-FLO has a very simple method for inputting the basic information about pipe systems. The user is presented with a displayed sketch of a system segment on which there is a space for identifying each pipe and component in that segment. The user enters numbers to identify pipe elements and letters for other components. There is space to make ten such entries for each series segment. The sketch for a parallel segment shows a group of six series segments.

Data kept in storage files

The numbers and letters refer to stored information on pipe elements and components. Component characteristics are entered on a System Maintenance display. The user must give three data points for each component on a GPM-versus-feet-of-fluid curve. With these, the package completes the curve (but does not display it) to model the component. The user then can enter new values of GPM or feet-of-fluid and get the model value of the other parameter to see if it checks with the known characteristics of the actual component. New data points then can be tried for a better fit.

A complete set of component characteristics is identified by a file name and remains stored on the SYS-FLO disk. Many such files can be kept on that disk. However, if too many files are stored, the disk capacity may be exceeded and the user will get a disheartening "DISK-FULL" message that signals the loss of all design data entered in the current session.

Pipe elements are described in files stored with the help of the PIPE-FLO package. These are identified by number. As an initial step in running the SYS-FLO package, the user identifies the file from which this information is to be drawn

Iterative calculation

When all the pipe segments in the system have been described and all the pipes and components that make them up have been identified and characterized, calculation of results can be performed. The package will notify the user of any item that is undefined and will not perform its calculations until the needed information has been supplied.

Calculation of results is based on an input (or output) flow rate, pressure, and design velocity specified by the user. The package calculates head loss, inlet pressure, and outlet pressure for each segment of the system. It also calculates these same items, plus a velocity, for the pipes and components within each segment.

Where parallel paths are involved, the user is asked to specify a percent of tolerance for the parallel flow rate determination. This calculation is an iterative one and the calculation time may run on for many minutes. To further limit this iterative process the user can place a ceiling on the number of iterations to be performed.

Easy to change designs

To design with the SYS-FLO package, users will usually set up the system, leaving out the pump, to calculate the head loss. Then the pump is selected, added, and calculations run for the completed system. If it turns out that the system design shows excess pressure, the user can select another pump, throttle some of the valves, or install an orifice to give the desired pressure drop.

A copying feature allows users to save an existing system design while editing a copy of that system. This is a handy device for trying out systems that closely resemble the original one, for example, to evaluate system operation with a cooler out of service, to upscale or downscale a design by changing the GPM, or to see how a system behaves when a different fluid is handled.

Several control features are built into the package. If the calculated flow rate through a component exceeds the range of flow used to specify that component, the user will get an "OUT-OF-RANGE" message. If parallel paths contain pumps with unequal shut-off heads there will be an "UNBALANCED" message. If the calculated head loss for a pump is negative, that, too, will be called to the user's attention.

Chapter 3
Gear Design and Rating Packages

The usual alternatives for engineering design are to take a cut-and-try approach or to use some kind of calculation that leads more directly to a design solution. These alternatives are reflected in packages for personal computers that deal with gear design and rating.

Design approaches

The GEARMASTER Spur Gears package takes the direct approach. It seeks to find the design with the smallest gears for specified loads, those with the largest diametral pitch, fewest teeth, and thinnest face width. The user then uses the package to modify these gears to get desired wear and endurance properties.

For the GEARMASTER user, calculations are interspersed with entry of data items. When the user has entered the desired gear ratio, tooth form, and pinion rpm, the package calculates the minimum number of teeth on each gear and displays the results for approval by the user. If desired, the number of teeth on the pinion gear, the gear ratio, or rpm can be changed and the calculation rerun. Within the calculation procedures there are various user alternatives. For example, when wear and endurance are calculated, results are given for three gear grades. To improve these results the user is offered a choice of altering the diametral pitch, gear thickness, or gear hardness. A fourth possibility is to rerun the calculation using an operating horsepower, assuming that the gears were originally sized using a start-up overload horsepower.

GEARMASTER users are specifically reminded that, following the design procedure, gear rating must be done as a separate process, using another package or separate calculations to verify the acceptability of the design results.

With the SGAP package, the user selects a candidate pinion and gear set, which the package will analyze to determine stresses and check to see if interference occurs. The user can hold the gear diameters constant and quickly vary diametral pitches or hold tooth numbers constant and vary the diametral pitches to evaluate a number of similar alternative designs.

If the gear geometry is not specified by the user, the GEAR1 package will do an initial design and propose a set of geometry data that the user can alter. The user, however, can supply the geometry parameters from the start. Performance data are then calculated. The user can alter the design and make repeated calculation runs, entering only the changed input data items and leaving the other design parameters untouched.

Gear1 uses American Gear Manufacturers Association (AGMA) rating formulas to compute surface durability and bending stress of the gear teeth. This package presents itself as suitable for rating pitting resistance and bending strength of gear teeth. In contrast, the SGAP and the GEARMASTER Spur Gears packages use other, and presumably more rapid but less precise, methods since they primarily aim at uncovering design alternatives. SGAP and GEARMASTER Spur Gears are specifically for spur gear designs,whereas Gear1 is applicable to helical gears as well.

In its design method, the SGAP package uses simpler but more conservative geometry factors than those employed by the AGMA. Users who want an AGMA-based analysis are referred to the HASGAP package available from the same vendor. HASGAP can analyze helical as well as spur gears. It provides limited means for varying the design parameters, but there are some nice built-in features.

HASGAP will analyze standard and nonstandard tooth systems with a range of face widths and diametral pitches. During the calculation, gear diameters may be held constant or allowed to vary for a given velocity ratio. If undercutting is present, the package attempts to correct this condition. If a face contact ratio of less than two is calculated, there will be an automatic attempt to correct this as well.

The SCORING+ and AGMA218 packages are clearly labeled "NOT FOR DESIGN PURPOSES." They are strictly rating packages that can be used for both spur and helical gears. SCORING+ concentrates on frictional factors, including elastohydrodynamic (EHD) film thickness, flash temperature, specific sliding ratios, and Hertzian contact stress. AGMA218 is designed to rate a given gearset following the AGMA 218.01, December 1982 standard, performing both life and power rating.

EHL1 is a more specialized package in the sense that it concerns itself with analysis of lubricating conditions in contact zones, that is, with EHD lubrication and appropriate operating temperatures. The package performs this kind of analysis for bearings as well as for gears. Type of lubricant can be specified by the user or will be selected by the package.

Handling the data

Outside of interpreting the results the computer produces, the main task for the user of gear design or rating packages is to prepare and enter the data on which the computations will be based.

In clarity of presenting on the screen exactly what is expected, the GEARMASTER Spur Gears package excels. Instructions for running this package, as displayed on the computer screen and printed in the user manual, are exceptionally clearly arranged and designed to simplify operation of the package. There is a separate screen display for each parameter that is to be entered, and each such screen includes the kind of explanations that will be appreciated by first-time and occasional users. Fortunately, only a limited number of input parameters are required for this package. Separate screens for eight or nine parameters can be helpful, but if users had to wade through 50 such screens, the process would become very tedious.

That is the situation with packages that do gear rating. These all seem to require users to input 50 or more parameters. The AGMA218 package handles this situation by providing a full-screen entry procedure. Spaces for entering many different parameters are displayed on one screen. The user moves the display cursor to screen locations where data are needed, adds or makes changes as needed. When the screen is entered, all additions and changes take effect.

AGMA218 has a particularly flexible ability to present output information. The user can call for a Reports Menu, which allows summaries of input data, geometry, load data, derating factors, strength, stress, life rating, and power rating or any combination of these summaries to be displayed or printed.

If a Miner's Rule rating is being run, the life rating can be based on an array of up to 50 discrete loads instead of the usual single load. AGMA218 handles this with a special input screen for entering the necessary load array. Load and cycle ratio values are entered for 17 loads. Then a continuation screen is called up to enter 17 more, and a third screen for the last 16 loads.

The SCORING+ package also uses full-screen entry. One screen is used to input 25 gear geometry data items, another for 15 materials and lubrication data items, and the third for 5 load and derating factor items. Data entry for the SGAP package likewise is done on a full-screen entry and editing basis. A list of items to be entered is displayed. The user moves the display cursor to the desired item, using the keyboard arrow keys, then types in the data and presses Enter. The Del and Int keys are used to change already-entered items.

The input method used with the HASGAP package has most of the input parameters displayed in tabular form, with the parameters identified by short letter codes. This is fast and fine for experienced, constant users of the package, but novices and occasional users will find it cumbersome to keep leafing through the manual to identify these codes.

GEAR1 users enter data by responding to a sequence of screen-displayed queries. The GEAR1 manual contains helpful tutorial material on journal bearing design, directed to defining the terms used to describe input and output parameters used by the package. A somewhat expanded version of this material is available in response to specific requests for HELP when the package disk is in use.

Most of these gear packages have little if any ability to present results in graphical form. An exception is the SCORING+ package. When the calculation process is completed, an output options menu is displayed. This gives users the choice of graphical or numerical output. Available graphs include those for EHD and specific film thickness, flash temperature, specific sliding ratio, and Hertzian stress. All are plotted against pinion roll angle.

Helpful checks

The AGMA218 package has some built-in checks for errors in the data that users enter. For example, the package looks over the input data and checks for missing or out-of-range items. If an error is found, the rating run stops, and the user is referred back to the screen where the error was found. The SCORING+ package has similar checks.

The GEARMASTER Spur Gears package has a number of automatic controls aimed at insuring the accuracy and validity of the inputs. There are built-in high and low limits on some items of data that are to be entered. If the value the user keys in exceeds these limits, the entry is ignored and the prompt for data is repeated.

In the case of gear ratio, GEARMASTER assumes that the ratio should be equal to or greater than 1.0. If it is not, a query is displayed that asks the user to approve an entered <1 value or else reenter the ratio.

AGMA218

AGMA218 is for rating the pitting resistance and bending strength of spur and helical involute gear teeth, following the American Gear Manufacturers Association standard AGMA 218.01, December 1982.

AGMA218 runs on IBM/PC/XT computers and sells for $1,995 from Geartech Software, 1017 Pomona Ave., Albany, CA 94706. (415) 524-0668.

AGMA218 rates existing gears or compares the ratings of existing gearsets. It is not a tool for gear design or for assuring the performance of complete gear drive systems.

Full-screen data entry

Doing AGMA calculations with the help of this package beats hand calculation by a mile, but there is still a lot of data to enter. The process is simplified by a full-screen entry procedure. It allows the user to move the cursor to screen locations where data are needed and to add or make changes as needed. When the screen is entered, all additions and changes take effect.

There are four screens of required input data. One has 25 items relating to gear and tool geometry, another has 12 items on material properties and heat treatment. There also is a screen with eight load data items, and one with six items concerning derating factors.

When these data have been entered, the package may prompt for some additional data. For example, added data are called for if the user wishes the package to calculate the load distribution and dynamic derating factors for pitting resistance.

If a Miner's Rule rating is being run, the life rating can be based on an array of up to 50 discrete loads instead of the usual single load. There is a special input screen for entering the necessary load array. Load and cycle ratio values are entered for 17 loads. Then a continuation screen is called up to enter 17 more, and a third screen for the last 16 loads.

The package allows users to store files of input parameters on disk, then call them up as desired. This provides a shortcut for experienced users, since previous, similar data sets often can be quickly edited to provide the input needed for new gear sets.

Checking for input errors

The package has some built-in checks for errors in the data that users enter. When the user has completed data entry and calls for a rating run, the package checks for missing or out-of-range items. If such an error is found, that fact is displayed and the rating run stops, referring the user back to the screen where the error was found. The user then corrects the input error and restarts the rating run. If another error is then found by the package, the same procedure is repeated.

When geometry data are being checked, some error warnings may be advisory only, indicating that the specified gear geometry brings some parameter outside a generally accepted range. Users can act on these warnings or simply decide to go ahead with the computation.

Some incorrect data entries may result in errors later on when the package calculations are carried out. For example, the package will not run ratings if both the internal gear I.D. and the external gear O.D. are non-zero, since this gives the message that the gear is both internal and external. Likewise, if an unacceptable value of reliability is entered on the load data screen, an attempt to run ratings will result in an error message. Error messages resulting from a checking procedure performed by the package immediately after a data screen is entered could provide a more convenient way for users to correct such input mistakes.

With the input information entered, life and power rating calculations can be run. Both are initiated by choices from the main menu. The calculation generally takes a few seconds and a summary of results is displayed. The user can then call for the Reports Menu, which allows the user to display or print summaries of input data, geometry, load data, derating factors, strength, stress, life rating and power rating, or any combination of these summaries.

EHL1

EHL1 analyzes lubrication conditions in the contact zones of gears and bearings.

EHL1 runs on IBM/PC/XT/AT computers and sells for $200 from ESDU International, 1495 Chain Bridge Rd., #200, McLean, VA 22101. (703) 734-7970.

Users are prompted to enter several input parameters by displayed queries and menu lists. If bearing lubrication is to be analyzed, the shaft speed in rpm is entered and the user chooses from a menu of seven types of bearings: radial or thrust ball, spherical, and cylinder, tapered, needle, or thrust roller bearings. For radial bearings the user is queried for outer or pitch diameter and bore; for thrust bearings, the queries ask for pitch circle and rolling element diameter.

Type of lubricant can be specified by the user or will be selected by the package. Outputs include lubricant type, Lambda ratio, film thickness, and operating temperature.

For gear lubrication analysis, users choose whether the gear configuration is to be internal or external. They select from a menu the type of tooth finish to be used. Choices are hobbed, shaved, lapped, ground soft or hard, or polished. Some added input data also are required. These include power transmitted, gear ratio, pinion speed, center distance, face width, pressure angle, and helix angle (for helical gears).

Again, the type of lubricant can be specified by the user or selected by the package. Outputs are the same as those for bearings.

Sets of input data can be stored in user-named files and called up when needed. This can provide a convenient way to run similar designs since the stored parameters can be quickly accepted or changed as desired. Editing of the input data is accomplished by following the same procedure of item-by-item acceptance or reentry.

Tutorial material

The manual contains helpful tutorial material on journal bearing design, directed to defining the terms used to describe input and output parameters used by the package. A somewhat expanded version of this material is available in response to specific requests for HELP when the package disk is in use.

Other aspects of the software presentation are less favorable. Displayed prompts and queries are accompanied by unwanted mysterious symbols that clutter up the screen. The text of the queries is repeated, making the presentation unclear. Users experience long waits between procedure steps. There seems to be no way to conveniently abort the rather lengthy design procedure, except by shutting down the system.

The package analysis is applicable only to conventional rollar bearings and parallel axis gears. Bearing loads are assumed static or only slowly varying. Lubricant flow is assumed to be laminar.

GEAR1

GEAR1 is for design of parallel axis spur and helical gears. The AGMA 218.01 December 1982 standard is used to evaluate surface durability and bending strength.

GEAR1 runs on IBM/PC/XT/AT computers and sells for $400 from ESDU International, 1495 Chain Bridge Rd., #200, McLean, VA 22101. (703) 734-7970.

Users respond to a sequence of screen-displayed queries for input parameters and choices. The basic input parameters are transmitted power, pinion speed, operating temperature, required life, and reliability. Application factors also can be entered for these parameters. These are meant to account for vibrations and variations in performance encountered in practice. Initially the user is advised to use the value of 1.0 for these factors, then to adjust them as experience with the design method accumulates.

The package offers a choice of steel, of malleable, cast, or nodular iron, or of aluminum- or tin-bronze as the material used for the pinion and gear. The user supplies a minimum surface hardness for the selected material. Some typical values are listed in the user manual.

Alternatives for manufacturing method are full fillet, finished or preshaved hobbing, or else shaping. Gear precision is specified on the AGMA Qv scale with 12 as very good and 5 as poor.

Load distribution factors may be calculated by either an analytical or an empirical method. For the analytical method, the user enters the AGMA-defined pitch error. The empirical method requires selecting a type of gear mounting and answering a few questions about the application.

If the gear geometry is not specified, the package will do an initial design and propose a set of geometry data that the user can alter, or the user can supply the geometry parameters from the start. These include pitch, base and outer diameters, the number of teeth, and face width; also helix angle and gear inside diameter, if these are appropriate.

Sets of input data can be stored in user-named files and called up when needed. This can provide a convenient way to run similar designs since the stored parameters can be quickly accepted or changed as desired. Editing of the input data is accomplished by following the same procedure of item-by-item acceptance or reentry.

GEAR1 produces output containing significant performance data parameters and includes warnings where design criteria are exceeded.

Tutorial material

The manual contains helpful tutorial material on gear design, directed to defining the terms used to describe input and output parameters used by the package. A somewhat expanded version of this material is available in response to specific requests for HELP when the package disk is in use.

Other aspects of the software presentation are less helpful. Displayed prompts and queries are accompanied by unwanted mysterious symbols that clutter up the screen. The text of the queries is repeated, making the presentation unclear. Users experience long waits between procedure steps. There seems to be no way to conveniently abort the rather lengthy design procedure, except by shutting down the system.

The rating methods used by this package for surface durability do not apply to scuffling or wear of the gear teeth. Excessive vibration effects are not taken into consideration.

GEARMASTER: SPUR GEARS

GEARMASTER is used to find, for a specified application, the gear set with the largest possible diametral pitch, the smallest pitch diameter, and the fewest number of teeth. It also has procedures to fit a gear set with specific pitch diameters on two fixed shafts and to find the allowable horsepower and torque for a gear.

The GEARMASTER Spur Gear package runs on IBM/PC and HP 9816, 9836, 9835 and 9845 computers. It sells for $200 from Engineering Technologies, PO Box 979, Cary, NC 27511. (919) 467-8960.

Instructions for running this package, as displayed on the computer screen and printed in the user manual, are exceptionally clearly arranged and designed to simplify operation of the package, particularly for the occasional user.

The introductory screen reminds the user that this software is for design work and is set up for average-mass gears under average conditions. AGMA standards should be used as a final rating check.

The main menu offers a choice of three procedures. The first uses tooth strength analysis to find the best pair of spur gears with the smallest possible diameters.

Built-in controls

The package devotes an entire display screen to each item of data the user is required to enter. For example, the entry screen for gear ratio explains that the ratio may be that of spur over pinion gear diameters, or it may be the number of spur over number of pinion gear teeth, or the ratio of pinion to spur gear rpm.

There are a number of automatic controls aimed at insuring the accuracy and validity of the inputs. Thus, the package has built-in high and low limits on some items of data to be entered. If the value the user keys in exceeds these limits, the entry is ignored and the prompt for data is repeated. However, the manual does not make it clear that incorrect values can be changed by positioning the cursor with IBM/PC arrow keys, then using the Del key to erase unwanted characters.

In the case of gear ratio, the package assumes that the ratio should be equal to or greater than 1.0. If it is not, a query is displayed that asks the user to approve an entered <1 value or else reenter the ratio.

Some items of input data are selected from screen-displayed alternatives. Thus the desired tooth form is selected from a list of eight possible forms.

Calculations are interspersed with entry of data items. When the user has entered the desired gear ratio, tooth form, and pinion rpm, the package calculates the minimum number of teeth on each gear and displays the results for approval by the user. If desired, the number of teeth on the pinion gear, the gear ratio, or rpm can be changed and the calculation rerun.

Then the user goes on to enter maximum horsepower transmitted by the gear set. Gear material specifications are entered by first selecting from four available material combinations (cast iron, cast iron; steel, cast iron; steel, steel; steel, phosphor bronze). Then there are selection screens for the grades of steel or cast iron or bronze to be used for each gear. For steel, the package provides a choice of 12 grades including cast, forged-carbon, and alloy grades.

During the review a value of 00000 was inadvertantly entered for the maximum horsepower. The package accepted this value, and this resulted in an error during subsequent calculation.

The package then does its optimization calculations. The Lewis tooth strength equations are used, and the results are displayed, showing the rpm, diametral pitch, number of teeth, pitch diameter, and gear thickness for both spur and pinion gears, along with addendum and dedendum, whole depth, clearance, form factors, pitch line velocity, torque, tangent and radial forces, and actual and allowable stresses.

While the package was being reviewed, an error message indicating overflow during division was displayed during the optimization calculation. This was due to the 00000 hp value previously mentioned. The error was not recoverable. The F10 key, used for ESCAPE with the IBM/PC version of the package, was inoperative in this situation.

Fixed shaft, horsepower calculations

The second main menu choice in this package is a procedure for determining the best pair of spur gears to fit on two shafts at a fixed distance from each other. The user inputs the center distance between shafts, gear ratio, and tooth form. If these specifications will not allow for use of gears with diametral pitches of 1–40, the package suggests slight changes in center distance and ratio.

The pinion rpm, transmitted horsepower, and materials information are entered and a tooth strength analysis-optimization is run, producing the same output parameters as the first procedure.

The third main menu choice is a procedure that allows rating a single gear for strength only or a pair of gears for strength, wear, and endurance. The user specifies a tooth form, pinion pitch diameter, and number of pinion teeth. If the resulting diametral pitch is not on a list of 35 pitches stored in the package, the user is asked to verify the use of the nonstandard pitch or to choose a nearby standard one. The number of spur gear teeth, pinion rpm, gear face width, and materials information are entered. The tooth strength calculation is run, and the results are displayed.

If wear and endurance are to be calculated, the user must enter a surface endurance limit (PSI) value. The package then does a dynamic tooth load analysis, based on the Buckingham equations. The results display calculated dynamic tooth load and allowable loads for wear and endurance. These are shown for three gear grades, and, for each, acceptability or unacceptability for wear and endurance is indicated.

To improve the results, the user is offered a choice of altering the diametral pitch, gear thickness, or gear hardness. A fourth possibility is to rerun the calculation using an operating horsepower, assuming that the gears were originally sized using a start-up overload horsepower.

HASGAP

HASGAP, a helical and spur gear analysis package, calculates bending and surface stresses and bending and surface geometry factors. AGMA equations are used.

HASGAP runs on IBM/PC/XT/AT computers and sells for $250 from Engineering Software, 3 Northpark E, #901, 8800 North Central Expressway, Dallas, TX 75231. (214) 361-2431.

HASGAP analyzes standard and nonstandard tooth systems with a range of face widths and diametral pitches. During the calculation, gear diameters may be held constant or allowed to vary for a given velocity ratio. If undercutting is present, the package attempts to correct this condition. If a face contact ratio of less than two is calculated, there will be an attempt to correct this as well.

Other calculations in the package include those for load-sharing ratio, root fillet stress concentrations, top land thicknesses, allowable power, bending stress, and compressive stress.

Pairs of gears analyzed by this package must have teeth formed by a spur gear hob, and the tooth systems must be involute. Gears must operate on a standard center distance. It is assumed that tooth profiles do not affect the load distribution on the teeth, that notch and stress cycle sensitivities have no effect, and that normal backlash will have negligible effect on the results.

Lots of variables

Input to this package includes up to about fifty parameters. Most of these are displayed in tabular form, with the parameters identified by short letter codes. To decipher their meaning the user refers to the manual. Here a brief description of each input parameter will be found, and these listings run across 34 pages of the user manual. Unfortunately, no complete alphabetical listing of parameters is given, so users have to search as best they can when they want to look up the meaning of a particular displayed letter code.

Parameter input is preceded by a few general screen-displayed queries on the project title, date, whether the gears are helical or spur, whether the load is at the tip or at the highest point of single-tooth contact, whether the tooth system is standard or not (be careful here, since all fine-pitch gears are defined as nonstandard), and whether diameters are to be constant or variable.

If desired, the input data can be saved on a disk-stored file and later reused after editing. Since specific items can be changed as desired during the editing process, whereas the rest are retained as is, this reuse process can be a time saver.

Calculated results are displayed in a compact tabular format, with computed quantities identified by short letter codes. Again, these codes are defined in the manual. Since there are only 26 of them, the fact that there is no alphbetical listing is only half as troublesome as with the input parameter codes.

SCORING+

SCORING+ calculates flash temperatures per AGMA 217.01 and the probability of scoring for spur and helical gears. It also calculates factors such as elastohydrodynamic film thickness, specific sliding ratios, and Hertzian stress that affect pitting, scoring, and wear.

SCORING+ runs on IBM/PC/XT computers and sells for $495 from Geartech Software, 1017 Pomona Ave., Albany, CA 94706. (415) 524-0668.

Data input for this package is done on three full-screen displays, reached from the package main menu. One is for 25 gear geometry data items, another for 15 materials and lubrication data items, and the third for 5 load and derating factor items. This arrangement allows the user to move the cursor to screen locations where data are needed and add or make changes as needed. When an input screen is entered, all additions and changes take effect.

When the calculation itself is started, by another main menu choice, the package checks over the entered data. If any items are missing or out of range, an error message is displayed and the user is directed back to the appropriate data input screen. This checking procedure catches only one input error at a time and will repeat until all have been corrected. Then, the package will move on and display a screen of analysis options. Here the user specifies such items as the number of profile points to be calculated and the driving member of the gearset.

When the calculation process is completed, an output options menu is displayed. This gives users the choice of graphical or numerical output. Available graphs include those for EHD and specific film thickness, flash temperature, specific sliding ratio, and Hertzian stress. All are plotted against pinion roll angle.

A reports menu provides for displaying or printing data summaries on the input data and ratings, as well as tables of data on calculated parameters at five reference points, on rolling and sliding velocities, and on flash temperature and film thickness.

SGAP

SGAP is for analysis and design of spur-gear sets.

SGAP runs on IBM/PC/XT/AT computers and sells for $125 from Engineering Software, 3 Northpark E, #901, 8800 North Central Expressway, Dallas, TX 75231. (214) 361-2431.

Data entry is done on a full-screen entry and editing basis. A list of items to be entered is displayed. The user moves the display cursor to the desired item, using the keyboard arrow keys, then types in the data and presses Enter. The Del and Int keys are used to change already-entered items.

The package allows a choice of four tooth systems: 14.5, 20, or 25 degree full-depth or 20 degree stub. Teeth must be involute. There also is a choice of six loading factors that can be used: a static load or one of five functions of the pitch line speed.

Results can be calculated for a single pitch and face or for a range of pitches and face widths. With the multiple calculations, gear diameters may be held constant and the number of teeth varied or vice versa.

For each specified value of diametral pitch, the package produces a set of results that includes bending stresses for desired face widths and contact stresses. If interference occurs, the package will display that fact.

Chapter 4
Shafts, Bearings, and Power Transmission Packages

Design of mechanisms and machinery necessarily includes consideration of shafts, bearings, belts, chains, and other motion and power transmission elements. A growing number of software packages for personal computers has been written to address these concerns. The following examples should provide a rough picture of what is available.

Packages for design and analysis of shafts deal with several different aspects. For example, the SHAFT3 package is for analysis and design of circular cross-section shafts that are subject to torsion, axial, and bending loads. The package calculates stresses and deflections caused by the loads and by specific support movement or misalignment. It also calculates shoulder-fillet stress concentration factors and evaluates safety margins, based on an equivalent combined stress state.

Other packages are concerned with dynamics. The SHAFT package calculates the unbalanced response and critical speeds of a horizontal shaft. TWIST calculates angle of twist and twisting moment of a shaft and the natural frequencies and mode shapes of torsional vibration.

Three for bearings

Three packages from ESDU International address specific aspects of bearing design. The JNBRG1 package is for design of journal bearings, RLBRG1 is for roller bearing selection and design, and THBRG1 aids in the design of thrust pad bearings.

These packages are interesting in that they include some automatic and flexible design features. With the journal bearing package (JNBRG1), diameter, length, diametral clearance, groove length and width are the required geometry parameters. If the user wishes, the package will run a preliminary design on its own, without any geometry input. The results will then include bearing geometry parameters that the user can either accept or modify.

A convenient technique is used to explore the possibilities for refining journal bearing designs. JNBRG1 users can choose to vary load, speed, feed pressure, diameter, length, or clearance. In each case, they specify the range of interest and the number of steps to be calculated within that range. The package then displays a table showing the variation steps and listing their effects on maximum temperature, minimum film, total flow, and power loss. If the range chosen causes normal design limits to be exceeded, a warning message is displayed along with the tabulated data.

With the thrust bearing package, THBRG1, users can specify the bearing geometry by keying in the number of pads, the pad inner and outer diameters, pad size, drop and percentage land, and runner thickness and diameter. Should the user decline to supply this information, the package will immediately calculate and present preliminary geometry values, which can be accepted or modified by the user.

THBRG1 also offers a parameter-variation procedure. Users specify the range of variation and number of change steps either for load or speed. The package calculates maximum temperature, minimum film, pad pressure, and power loss for each step and displays the results in tabular format. If any of these conditions bring the design parameters beyond normal limits, a warning notice is displayed.

When running the roller bearing package (RLBRG1), the user can choose the bearing type. Available types are single-row deep groove ball bearing, cylindrical roller bearing, angular contact bearing, or a pair of angular contact bearings. However, if left to itself the package makes an initial choice of the single row ball bearing and the design proceeds from that point.

POWER TRANSMISSION I takes in a number of design aspects. This package calculates rotating moments of inertia, drive parameters, shaft properties, shaft stresses, and also does belt and chain drive calculations.

For those interested simply in selecting the proper V-belt for a given application, the VBELT package offers a convenient way to pick out the needed components.

Finally, the CYLINDER AND SHOCK ABSORBER ANALYSIS package was written to aid in the selection of cylinders and shock absorbers for use in situations, such as manufacturing operations, where they move and position objects.

Built-in data

One of the computer capabilities that aids considerably in engineering design is the ability to maintain stored data on the characteristics of materials and components. Several of these packages have made good use of this storage capability.

A file of standard cylinder data comes recorded on the CYLINDER AND SHOCK ABSORBER ANALYSIS package disk. If cylinder selection is called for, these data are used to locate the smallest available cylinder that will meet the user specifications.

The VBELT package comes with built-in data from the Rubber Manufacturers Association on A, B, C, D, E, 3V, 5V, and 7V belt cross sections, and with data from the Gates Rubber Co. on their A, B, C, D, E, 3VX, 5VX, and 8V belt cross sections. A procedure is provided for users to create their own additional files of cross-section data.

Users of the JNBRG1 package are offered a menu of 13 types of lubricants from which to choose, and a similar menu of 13 bearing materials. In both cases, the needed parameters are stored on the package disk and are entered automatically into the appropriate calculations that follow.

For use in the calculation of friction torque, RLBRG1 users are asked to select a lubrication condition from a choice of mist, jet, oil bath, or grease lubrication. A choice of nine different ISO grades of lubricant is offered. The package immediately calculates a viscosity at the specified operating temperature, using stored data. Similarly, with the THBRG1 package, type of lubricant is selected from a menu that accesses stored data on nine ISO grades. Users can add their own choices to the menu by entering information on viscosities at two temperatures, specific gravity, heat capacity, and upper and lower temperature limits.

Helpful diagrams

Some of the packages make good use of diagrams to define desired input data. Thus, to help explain to users exactly what input quantities are needed, the CYLINDER AND SHOCK ABSORBER ANALYSIS package displays a diagram that represents cylinder velocity, distance, and time parameters. A similar diagram, with input parameters labeled on it, is displayed to aid shock absorber inputs. While observing this diagram, the user is prompted for such parameters as propelling force, velocity at impact, weight of moving object, coefficient of friction, and impacts per hour.

With the POWER TRANSMISSION I package, choice of Belt/Chain Drive calculations produces a labeled diagram of a drive to guide the user in providing dimensional data. The package then prompts for a choice of center distance or of length calculations and for the needed input data. For a chain length calculation, these would include the pitch of the chain, number of teeth on the two sprockets, and center distance.

For two-station or angular contact bearing designs, the RLBRG1 package displays a diagram that illustrates the required input parameters: distance between stations, distances to load application points (a maximum of ten such points can be used), axial, radial, and moment loads.

The VBELT package makes use of a unique selection display. Data entry starts with queries from the package for a few items of input information: horsepower, rpm of fast sheave, service factor, and diameters of both sheaves and approximate center distances. The user also names the file of stored belt data that is to be used for selection.

Using data taken from the named belt data file, the package then displays a selection plot. Design horsepower and rpm are the *x*- and *y*-axis quantities, and the plot is filled with numerals representing belt cross-section types. Thus, points where a 3V belt can be used are filled with 3s, those where 5V belts can be used are filled with 5s, and so on. A point defined by the user's input data is left blank, so there is a visual picture of how that operating point is positioned in the overall operating ranges for the types of belts that are displayed.

CYLINDER AND SHOCK ABSORBER ANALYSIS

The CYLINDER AND SHOCK ABSORBER ANALYSIS package is for the selection of cylinders and shock absorbers in situations where they are used to move and position objects.

The package runs on IBM/PCs and sells for $125 from P.E.A.S., 7208 Grand Ave., Neville Island, Pittsburgh, PA 15225. (412) 264-3553.

At start up, the package offers the user a choice of either of cylinder or shock absorber analysis and selection. Results are displayed and can be printed, if desired.

For cylinders, there is a choice of calculations for a given cylinder size or of selection of a standard cylinder based on maximum stroke time or maximum acceleration. The package displays a diagram that represents cylinder velocity, distance, and time. Users are then prompted to select the type of cylinder desired: air, low-pressure hydraulic, or high-pressure hydraulic and to opt for a push or pull stroke. Other parameters called for include maximum stroke time, stopping distance, fluid pressure, weight of object to be moved, coefficient of friction, and maximum fluid velocity. These input data are then displayed in a summary on the screen, and the user can make any needed corrections.

A file of standard cylinder data comes recorded on the package disk. If cylinder selection is called for, these data are used to locate the smallest available cylinder that will meet the user specifications. The package displays the diameter, port size, and rod diameter for this cylinder. Rod buckling is not considered during this automatic selection process.

Calculated results from the cylinder procedures include 15 operating parameters such as velocities, acceleration, stroke time, thrust, and GPM.

Shock absorber procedures

When the shock absorber procedure is run, users start by selecting the desired design criterion and entering an appropriate limiting value. The choices are maximum stopping force, maximum deceleration, or shock-stroke limitation. The package then displays a diagram of a shock absorber with input parameters labeled. While observing this diagram, the user is prompted for such parameters as propelling force, velocity at impact, weight of moving object, coefficient of friction, and impacts per hour.

The input quantities are then displayed together and the user has an opportunity to correct or change any of them. The package then makes its calculations and displays the results, which include stopping force, energy per cycle and per hour, shock stroke and stopping distance, and stroke time.

The user is then prompted to enter a new value for each of the input parameters, one at a time, and will get new calculated results for each new entry.

JNBRG1

JNBRG1 is for design of journal bearings.

JNBRG1 runs on IBM/PC/XT/AT computers and sells for $400 from ESDU International, 1495 Chain Bridge Rd., #200, McLean, VA 22101. (703) 734-7970.

JNBRG1 operates by presenting a sequence of queries to the user. Input queries ask for bearing load and tilt, shaft speed, and oil inlet temperature and pressure.

Users are offered a menu of 13 types of lubricants from which to choose and a similar menu of 13 bearing materials. In both cases, users can add to the listed choices by specifying a set of lubricant or bearing parameters for each new entry on the lists.

There is a choice of four groove designs, and the user must specify whether the groove is to have round or square ends.Diameter, length, diametral clearance, and groove length and width are the geometry parameters for the package. If the user wishes, a preliminary design is run without any geometry input. The results include bearing geometry parameters that the user can either accept or modify.

Sets of input data can be stored in user-named files and called up when needed. This provides a convenient way to run similar designs since the stored parameters can be quickly accepted or changed as desired. Editing of the input data is accomplished by following the same procedure of item-by-item acceptance or reentry.

An output report is displayed. It consists of several screensful of data, and the user moves from one to the other by pressing the Enter key. If desired, the output report can be printed. In addition to the input parameters, this report lists performance characteristics for lubricant temperature and flows, eccentricity, attitude angle, and minimum film parameters, and the total power loss.

A convenient technique is used to explore the possibilities for refining the bearing design. Users can choose to vary load, speed, feed pressure, diameter, length, or clearance. In each case, the user specifies the range of interest and the number of steps to be calculated within that range. The package then displays a table showing the variation steps and listing their effects on maximum temperature, minimum film, total flow, and power loss. If the range chosen causes normal design limits to be exceeded, a warning message will be displayed along with the tabulated data.

Tutorial material

The manual contains helpful tutorial material on journal bearing design, directed to defining the terms used to describe input and output parameters used by the package. A somewhat expanded version of this material is available in response to specific requests for HELP when the package disk is in use.

Other aspects of the software presentation are less favorable. Displayed prompts and queries are accompanied by unwanted mysterious symbols that clutter up the screen. The text of the queries is repeated, making the presentation unclear. Users experience long waits between procedure steps. There seems to be no way to abort conveniently the rather lengthy design procedure except by shutting down the system.

The package assumes that bearing load and temperature are stable, that there is no shaft whirl, and that the lubricant flow is not turbulent.

POWER TRANSMISSION I

POWER TRANSMISSION I calculates rotating moments of inertia, drive parameters, shaft properties, shaft stresses, and does belt and chain drive calculations.

POWER TRANSMISSION I runs on IBM/PC computers and sells for $125 from P.E.A.S., 7208 Grand Ave., Neville Island, Pittsburgh, PA 15225. (412) 264-3553.

The five procedures offered by this package are reached through a displayed main menu. When a procedure is chosen, the user is prompted for input data. This completed, the user can view the entered data and call for needed changes in any item. When changes are completed, results are displayed and can be printed, if desired.

The Rotating Moments of Inertia procedure starts by prompting the user to specify material density (the density for steel is stored on the package disk, densities for other materials must be supplied by the user). It then prompts for type of mass (solid round, hollow round) mass diameter (OD and ID for hollow masses), and length for each mass involved in the rotating configuration. For transferred masses, users enter the weight, mass moment of inertia, and distance from rotating center line to mass neutral axis. The package then calculates and displays totals for weight moment of inertia, radius of gyration, and mass moment of inertia, and will display individual calculations for each mass, if desired.

Drive Calculations include those for parameters of horsepower, torque, angular acceleration, rpm, and time. Each calculation requires appropriate inputs of data. For example, input sets for the angular acceleration calculation can be torque and rpm, or horsepower and rpm, or two rpms and a time interval.

The Shaft Properties procedure takes OD and ID of a steel shaft and calculates section moduli and moments of inertia for torsion and bending, cross-section area, circumference, radius of gyration, and weight per unit length.

The Shaft Stresses procedure takes ID and OD and calculates stresses for torsional shear, bending, and combined bending and shear.

Choice of Belt/Chain Drive calculations produces a labeled diagram of a drive to guide the user in providing dimensional data. The package then prompts for a choice of center distance or of length calculations and for the needed input data. For a chain length calculation, these would include the pitch of the chain, number of teeth on the two sprockets, and center distance. Calculated results include sprocket pitch diameters, theoretical and actual minimum length of chain, number of links, angles of contact with sprockets, and number of links engaged with sprockets.

RLBRG1

RLBRG1 is for rollar bearing selection and design.

RLBRG1 runs on IBM/PC/XT/AT computers and sells for $400 from ESDU International, 1495 Chain Bridge Rd., #200, McLean, VA 22101. (703) 734-7970.

Users have a choice of running a design procedure for an individual bearing station on a shaft, two stations on a shaft, or for an opposed angular contact bearing arrangement. These choices are presented in menu form. When the user makes a choice, a sequence of displayed menus and queries follows to elicit the necessary input data.

The user can fix the bearing type. Available types are single-row deep groove ball bearing, cylindrical rollar bearing,angular contact bearing, or a pair of angular contact bearings. If left to itself, the package will make an initial choice of the single-row ball bearing. Users also are queried to specify bearing bore in millimeters, bearing speed, and operating temperature.

For use in the calculation of friction torque, users are asked to select a lubrication condition from a choice of mist, jet, oil bath, or grease. A choice of nine different ISO grades of lubricant is offered. The package immediately calculates a viscosity at the specified operating temperature. Users are then queried for a required reliability in percent.

For an individual bearing station, users are asked to enter radial and axial loads in Newtons. For the two-station or angular contact bearing case, the package displays a diagram that illustrates the required parameters: distance between stations, distances to load application points (a maximum of ten such points can be used), axial, radial, and moment loads.

The package places upper and lower limits on several of the input parameters. If the entries exceed these limits, they are not accepted and the query for that item is repeated.

Sets of input data can be stored in user-named files and called up when needed. This can provide a convenient way to run similar designs since the stored parameters can be quickly accepted or changed as desired. Editing of the input data is accomplished by following the same procedure of item-by-item acceptance or reentry.

Well-designed output tables

Calculated results are presented in easily read tabular format. Included in the table are the bore, O.D., dynamic and static capacities, film factor P/Cr ratio, and life for four different bearings, selected from data stored on the package disk. If any of these is nonstandard, that fact is noted.

For an individual bearing station, the user selects the desired bearing from the four that are shown. The package then does a performance check on that choice, displaying frictional torque, limiting speed under normal loading, and permissable tilt for the bearing.

For two-station arrangements, a similar table of results data is displayed for each station, assuming a back-to-back arrangement. Then two tables of data are given for a face-to-face arrangement of the bearings.

Tutorial material

The manual contains helpful tutorial material on journal bearing design, directed to defining the terms used to describe input and output parameters used by the package. A somewhat expanded version of this material is available in response to specific requests for HELP when the package disk is in use.

Other aspects of the software presentation are less favorable. Displayed prompts and queries are accompanied by unwanted mysterious symbols that clutter up the screen. The text of the queries is repeated, making the presentation unclear. Users experience long waits between procedure steps. There seems to be no way to abort conveniently the rather lengthy design procedure, except by shutting down the system.

The package does not take into account shaft and housing fits and tolerances, bearing clearances, or axial adjustment for angular contact ball, or tapered roller bearings.

SHAFT

SHAFT calculates the unbalanced response and critical speeds of a horizontal shaft.

SHAFT runs on IBM/PC/XT/AT computers and sells for $350 from The

Structural Members Users Group, PO Box 3958, University of Virginia Station, Charlottesville, VA 22903. (804) 296-4906.

SHAFT presents the user with a sequence of 25 queries that can be answered, one by one, from the keyboard. Alternatively, the user can enter the needed answers without viewing the queries. The package will also accept a batch input from a formatted file put together in advance, and it provides for the user to edit such files.

A lumped parameter model can be used to represent the mass of the shaft. If so, users must specify left and right end conditions. There can be up to 19 changes in cross section and 19 changes in the modulus of elasticity. Users must specify initial positions and magnitudes for all Es, initial positions, and inner and outer radii for all cross sections. If shear deformation is considered, initial positions and magnitudes must be entered for a ll Poisson's values. Up to 19 changes in these values can be handled.

The package will also account for up to 19 changes in mass-per-unit shaft length, 20 axial forces at the left end, 20 concentrated masses other than moment imbalances, and 20 concentrated rotary inertias. Users enter positions and magnitudes for each of these.

Up to 20 moment imbalances can be accounted for; users must specify position, mass, phase angle, and eccentricity for each.

Up to 30 flexible bearing systems can be handled. These may be linear or rotary (tilt) bearings. Spring constants and damping coefficients must then be input by the user, as well as bearing mass or mass moment of inertia. Up to ten other in-span supports also can be used.

Output from this package is in columnar form, with columns of data for location along the shaft, deflection, slope, moment, shear, and angle. For bearing reactions, the output data columns are location, force, force-angle, moment, and moment-angle.

SHAFT3

SHAFT3 is for analysis and design of circular cross-section shafts that are subject to torsion, axial, and bending loads.

SHAFT3 runs on IBM/PC/XT/AT computers and sells for $325 from Engineering Software, 3 Northpark E, #901, 8800 North Central Expressway, Dallas, TX 75231. (214) 361-2431.

The shafts may have any number of supports. The package calculates stresses and deflections caused by the loads and by specific support movement or misalignment. It will also calculate shoulder-fillet stress concentration factors and evaluate safety margins, based on an equivalent combined stress state.

Operation of the package is through screen-displayed prompts and queries to which the user replies with keyboard inputs. After responding to queries about the number of points where there are to be loads, support displacements, and supplied stress concentrations and entering material properties for the shaft, the user is asked to input data on the points of interest.

These inputs consist of strings of numbers, each separated by a comma, that represent asked-for parameters such as shaft diameter, displacement constraint codes, and loadings. The formats are explained in the user manual and will undoubtedly have to be referred to repeatedly by occasional users of the package.

Outputs give a table of *x, y,* and *z* displacements and rotations for each point of interest, then tables of loadings and stresses for each point of interest within each element of the shaft.

THBRG1

THBRG1 is for the design of thrust pad bearings.

THBRG1 runs on IBM/PC/XT/AT computers and sells for $400 from ESDU International, 1495 Chain Bridge Rd., #200, McLean, VA 22101. (703) 734-7970.

Users respond to screen-displayed queries and make choices from menu displays to type in specifications for the desired bearing. Input parameters include starting and steady-running loads, tilt, shaft speed and diameter, oil inlet, and drain temperatures.

Type of lubricant is selected from a menu of nine ISO grades. Users can add their own choices to the menu by entering information on viscosities at two temperatures, specific gravity, heat capacity, and upper and lower temperature limits.

Other choices presented to the user include that of directed jet or flood feed lubricant supply, white metal, copper-lead or aluminum-tin bearing material. Bearing types made available by the package are uni- or bidirectional inclined pads or a tilting pad.

Users can specify the bearing geometry by keyboarding the number of pads, the pad inner and outer diameters, pad size, drop and percentage land, runner thickness, and diameter. If the user declines to supply this information, the package will immediately calculate and present preliminary geometry values, which can be accepted or modified by the user.

Sets of input data can be stored in user-named files and called up when needed. This can provide a convenient way to run similar designs since the stored parameters can be quickly accepted or changed as desired. Editing of the input data is accomplished by following the same procedure of item-by-item acceptance or reentry.

Printed and displayed reports

After the design calculations are performed, an output report is presented that repeats the input parameters and gives results including power loss and running torque, film thickness, maximum pad temperatures, starting and working pad pressures, and required lubricant supply rate.

A parameter-variation procedure is available. Users specify the range of variation and number of change steps for either load or speed. The package calculates maximum temperature, minimum film, pad pressure, and power loss for each step and displays the results in tabular format. If any of these conditions bring the design parameters beyond normal limits, a warning notice will be displayed.

Tutorial material

The manual contains helpful tutorial material on thrust pad bearing design, directed to defining the terms used to describe input and output parameters used by the package. A somewhat expanded version of this material is available in response to specific requests for HELP when the package disk is in use.

Other aspects of the software presentation are less favorable. Displayed prompts and queries are accompanied by unwanted mysterious symbols that clutter up the screen. The text of the queries is repeated, making the presentation unclear. Users experience long waits between procedure steps. There seems to be no way to conveniently abort the rather lengthy design procedure, except by shutting down the system.

The package assumes that the bearings are dimensionally stable under both temperature and load conditions. It takes no account of turbulent lubricant flow. The pivot position for tilting-pad bearings is assumed to result in stable pad inclination.

TWIST

TWIST calculates angle of twist and twisting moment of a shaft and the natural frequencies and mode shapes of torsional vibration. For extension systems, consisting of masses, springs, or uniform segments, the package will compute steady-state axial displacement and force, natural frequencies, and mode shapes of longitudinal vibration.

TWIST runs on IBM/PC/XT/AT computers. It sells for $350 from The Structural Members Users Group, PO Box 3958, University Station, Charlottesville, VA 22903. (804) 296-4906.

Input to the package consists of responses to 24 screen-displayed queries. Some of the queries are in the form of short menus. For example users are asked to specify boundary conditions by choosing from fixed-fixed, fixed-free, free-free, or free-fixed boundaries. (Ends with applied loads, concentrated masses, or springs are treated as free boundaries.) In addition there may be a lot of detailed, formatted data to enter.

Thus, users may enter their own transfer matrices in a format prescribed in the user manual. Up to 50 lumped segments can be handled. Users have to indicate initial and end locations, spring constants, or shaft stiffnesses for each segment. Changes in shear modulus of elasticity or polar moment of inertia also have to be entered. If dynamic analysis is to be done, added items to be entered include initial locations and magnitudes of E, areas, polar moments of inertia, mass moments of inertia, mass density, as well as initial locations and elastic moduli of foundation. Up to 49 changes in the elastic modulus of foundation can be accomodated.

Added data to be entered for distributed loading includes specifying initial and end locations and magnitudes of uniform torques and initial and end locations and gradients for ramped torques. If there is thermal loading, users must supply initial and end locations and temperatures, as well as magnitudes of the coefficient of thermal expansion.

Up to 50 discs or concentrated masses, 50 concentrated applied torques, and 50 branch springs can be handled, and detailed entries of data must be made if these are used. Up to ten gear chains also can be included by entering detailed data for these as well.

Input formats are generally simple, consisting of numerical entries separated by commas. The package makes no provision for calculation of the input values, and there is little help in correcting erroneous inputs. The package will, on request, provide a printed summary of the input for the user to read and check.

Alternatively, the user can enter the needed input items without viewing the queries. The package also accepts a batch input from a formatted file put together in advance, and it provides for the user to edit such files.

Output is in tabular format, listing columns of locations, angles and twisting moments for the static response and for each natural frequency mode. The frequency of each mode is printed.

For extension problems, the output columns list location, extension, and axial force for the static response and for each natural frequency mode. The frequency for each mode is printed here as well.

VBELT

VBELT aids in selection of V-belt components.

VBELT runs on IBM/PC/XT/AT computers and sells for $195 from Engineering Software, 3 Northpark E, #901, 8800 North Central Expressway, Dallas, TX 75231. (214) 361-2431.

Data entry starts with queries from the package for a few items of input information: horsepower, rpm of fast sheave, service factor, diameters of both sheaves, and approximate center distances. The user also names the file of stored belt data that is to be used for selection.

A unique selection display is then shown on the screen, using data taken from the belt data file named by the user. Design horsepower and rpm are the x- and y-axis quantities, and the plot is filled with numerals representing belt cross-section types. Thus, points where a 3V belt can be used are filled with 3s, those where 5V belts can be used are filled with 5s, and so on. A point defined by the user's input data is left blank, so there is a visual picture of how that operating point is positioned in the overall operating ranges for the types of belts that are displayed.

If the selection display results are not satisfactory, the user can select a different belt type from a menu display.

The package then displays the pitch line velocity for that belt and a calculated belt length for the specified center distance. The user enters a desired standard belt length, and the package responds with a calculated center distance and number of belts needed. A summary of the results follows.

It can be displayed or printed, or the user can return to the input procedures to modify parameters and choices.

The package comes with built-in data from the Rubber Manufacturers Association on A, B, C, D, E, 3V, 5V, and 7V belt cross sections, and with data from the Gates Rubber Co. on their A, B, C, D, E, 3VX, 5VX, and 8V belt cross sections. A procedure is provided for creating additional files of cross-section data.

Chapter 5
Cam, Spring, and Mechanism Design Packages

Personal computers can be effective tools for designing a variety of different types of mechanisms, and software packages like MICRO-MECH and PLANET-PC are proving the point to many designers.

The MICRO-MECH package is for analysis of planar mechanisms with revolute and slider joints. It performs kinematic and dynamic force analyses, outputting position, velocity, acceleration and force data, and can provide graphical animation of the mechanisms.

Up to ten points on the mechanism can be selected for display. The package shows the paths of these points by connecting straight lines between the positions calculated during the analysis. Successive positions of the mechanism are displayed rapidly to give an animation effect, and the speed of the animation display can be controlled by the user.

The MICRO-MECH design process is speeded by a modification module that allows users to run an analysis, then modify the mechanism or analysis parameters and quickly do a rerun. MICRO-MECH gives the user some added assistance by catching some of the problems that can arise during the analysis and signaling their presence by screen-displayed messages. If, for example, the position of the mechanism changes too greatly from one iteration to another, the package will assume that something is wrong, stop the analysis and display a "Mechanism Does Not Assemble" message. There also are messages for dead points, where the mechanism stalls, and for situations where the dynamic force analysis does not produce a valid result.

PLANET-PC performs essentially the same overall functions, handling systems of linkages that can include such elements as masses, springs, and dampers that are connected by pin or slider joints. It provides a simulation of the operation of the mechanism, calculating displacement, velocity, and force for each simulated time step.

To provide a ready summary, PLANET-PC can display or output to a plotter graphs of functions such as *x, y* components of displacement, velocity or force, angular displacement or velocity, and torque.

Designing and selecting springs

Packages specifically written for spring design and selection all seem to have capabilities for handling helical springs. For example, SPRING1 is for the design of helical compression and extension springs under static or dynamic loads.

Good use of computer-stored data is made in the SPRING1 package. Materials options offered by the package include music wire, oil-tempered wire, hard-drawn wire, chrome-vanadium, and chrome-silicon wire. Materials data for these choices comes recorded on the package disk; properties are approximated by curve-fit formulas. The user also can enter material properties for other types of wire.

In addition to helical spring calculations, the Spring Design package performs design and analysis procedures for Belleville springs. This package operates from a combination of menus, prompts, and other displays and includes helpful diagrams. For example, on selecting Helical Spring from the main menu, the user is offered a choice of design or of three analysis procedures. If design is chosen, prompts for shear modulus and end type follow. Then a diagram of a coil spring is displayed to illustrate needed dimension data, and the user is prompted for the mean diameter, free length, working deflection, and deflection force. Extension springs can be handled by entering extended length instead of free length.

Spring Design starts its calculations by considering a one-coil spring solution. The display of results includes a theoretical wire diameter, shear stresses at working deflection, and solid length. If these stresses are not satisfactory, the user presses a key to get results for a two-coil spring and can press once again for three-coil results.

The SpringStar package allows the user to specify flat springs, determine their ability to sustain static loads, alter the specifications, and recalculate as needed. It performs similar functions for helical springs.

The flat spring procedure allows calculations for cantilever or simple beam springs. The user enters four of the following five parameters: load, deflection, thickness, length, and width. The procedure then calculates the fifth parameter and the stress level.

Plate cams

As an example of a design package in another area of mechanism design, consider the CAM package. CAM is for designing plate cams, and it will calculate results for five follower types. The output of this package is a table of cam profile and cam cutter-center coordinates, listed in polar and rectangular coordinates. These are displayed and stored in a file for possible NC machining use.

When using the CAM package for a design that involves a translating flat-face follower, the user enters nine input parameters, including rise, dwell, fall and reference angles, cutter radius, follower rise, cam base circle diameter, and cam rotation angle delta. These are explained on a diagram in the user manual. Inputs for the other types of followers are similar, with a few added items of input data for each in addition to the parameters just listed.

Describing the mechanism

When a mechanism is made up of a number of interconnected elements, it is necessary to let the computer know just how they are interconnected. The approach taken by MICRO-MECH is to describe one element, then go on to another element connected to the first one, and so forth. When the user defines an element, the package displays prompts calling for the name of the link and a previously defined reference link to which it is connected. Names are short, one to four characters only. The first link to be defined is generally a fixed link. As each link is added in this way, the package prompts for the type of joint between the new and the reference link.

The PLANET-PC package takes an overview approach. First the user tells the computer how many elements of each of the allowable types are in the mechanism. Then for each element, taken in just about any order the user wishes, other elements connected to it have to be identified.

Both approaches seem to achieve roughly the same results with about the same amount of effort. In both cases the user is advised to start by making a manual sketch of the mechanism. For the PLANET-PC package the sketch is mandatory since numbers of element types must be accurately counted.

Data entry for mechanism design packages can be a lengthy process. For example, in the PLANET-PC package, for most of the element types there are three or four parameters to enter. Rigid bodies require 6, damping elements up to 7, bearings 7–12. For a mechanism with 20 elements, users will find themselves preparing and entering perhaps a hundred parameters.

Much of the input effort with these packages will probably be in obtaining the necessary parameters. The entry process itself is well organized. For each element, there is a display of needed input items, and the user moves from one item to the next.

The PLANET-PC user manual gives some guidance on how to calculate input parameters. The MICRO-MECH package even displays little calculation formulas. However, these calculations are not performed by the packages. Nor are frequently used constants, such as material damping factors, stored by the packages.

Full-sceen entry

The SPRING1 package has a convenient full-screen data entry procedure. It presents the user with an input display screen on which entries have to be filled in for a list of clearly identified parameters. Moving to the various items, by using up and down arrow keys, along with the Ins and Del keys, the users enter and edit the data as needed. Items for spring end-type and material are chosen from short displayed lists of possible choices.

The SpringStar package gives the user a typical set of input data items to work with. For example, for compression springs the screen will show data for spring diameter, wire diameter, and modulus in torsion, along with values of two loads and two corresponding spring lengths and a choice of spring-end type. Starting with the first item the user enters values (or accepts displayed values) for each item on the input list. This process can be repeated, if needed, to correct the entries.

CAM

CAM is for designing plate cams. Results can be calculated for five follower types.

CAM runs on IBM/PC/XT/AT computers and sells for $125 from Engineering Software, 3 Northpark E, #901, 8800 North Central Expressway, Dallas, TX 75231. (214) 361-2431.

On start up, the user is presented with a menu offering a choice of five follower types: translating flat face, translating roller, oscillating flat face, oscillating rollar with angle specified, and oscillating rollar with radial rise specified. The display cursor is positioned at the desired choice with the arrow keys, the user presses Enter and the corresponding data input screen is displayed.

For the translating flat face follower, the user enters nine input parameters, including rise, dwell, fall and reference angles, cutter radius, follower rise, cam base circle diameter, and cam rotation angle delta. These are explained on a diagram in the user manual. Inputs for the other types of followers are similar, with a few added items of input data for each in addition to the parameters just listed.

The output of the package is a table of cam profile and cam cutter-center coordinates, listed in polar and rectangular coordinates. These are displayed, stored in a file for possible NC machining use, and can be printed using the IBM/PC PrtSc key.

MICRO-MECH

MICRO-MECH is for analysis of planar mechanisms with revolute and slider joints. It performs kinematic and dynamic force analyses and can provide graphical animation of the mechanisms.

MICRO-MECH runs on IBM/PC/XT/AT computers and sells for $395 from Ham Lake Software, 631 Harriet Ave., Shoreview, MN 55112. (612) 483-0649.

MICRO-MECH is operated by making selections from displayed menus and responding to displayed prompts. The first choice on the main menu is to define a new mechanism.

Link by link

The manual advises users to start by sketching the mechanism they wish to define. On this sketch links, joints, points, and other mechanism elements are named and local coordinates are set up for each link.

After being prompted for a short name for the mechanism, the user is presented with three choices: define a new link; close a loop between previously defined links; or continue with the definition process.

The link-defining choice calls up prompts to name the link and a previously defined reference link to which it is connected. Names are short, one to four characters only. The first link to be defined generally is a fixed link. As each link is added in this way, the package prompts for the type of joint between the new and the reference link. The user then assigns a name to the joint; provides coordinates for joint center locations, rotation angles for revolute joints; slider origin, angle, and initial displacements for slider joints. Names of the links, and information on the joint where a loop is formed are entered in a similar manner.

Continuing the data entry process, the user defines a gravity vector, then mass data including center of gravity, weight, and moment of inertia. Significant points on the links are defined by naming them and entering coordinates.

Entry of data on known forces is aided by a menu that provides procedures for entering information on several different types. Forces can be entered with constant or time-varying magnitudes. Effects of hydraulic cylinders, springs, viscous dampers, and joint friction are some of the types of forces that can be accommodated here.

Data on independent joint elements are entered with the help of another menu. Types of joint motion that can be defined include constant increment, velocity, and acceleration as well as periodic and constrained-linear motion. The user chooses the desired type and then is prompted for such items as velocity, acceleration, increment, delay time, offset, and phase angle, as appropriate.

Mechanism description data can be saved in a disk-stored file with the file name assigned by the user.

Making modifications

During the original entry procedure, the package allows the user to back up from one entry step to any previous one, then work forward again. This provides a crude method for correcting entry errors that affect mechanism topology: do it over again from the point where an error was made.

From the main menu of the package, users may call for a modification module. At this choice, a summary listing of links and loops is given. This module allows parameters of joints and links to be changed, but it does not give the user any means to make changes in the basic topology of the mechanism.

The modification module allows users to run an analysis, then modify the mechanism or analysis parameters and quickly do a rerun.

Analysis procedures

One of the last steps in the data entry process is to specify parameters for the analysis process. Users choose the desired number of mechanism positions that are to be analyzed. They select the desired time interval between the position "snapshots." They also select the desired level of analysis: position-only, velocity, acceleration, force.

Mechanisms with up to six loops can be analyzed. There also is a limit on the number of elements. Multiply the number of variable joint elements times one plus the number of independent joint elements; this product cannot be greater than 120, a respectable number, considering that this product comes to just eight for a four-bar mechanism.

Some problems that arise during the analysis are signaled by screen-displayed messages. If, for example, the position of the mechanism changes too greatly from one iteration to another, the package will assume that something is wrong, stop the analysis and display a "Mechanism Does Not Assemble" message. There also are messages for dead points, where the mechanism stalls, and for situations where the dynamic force analysis does not produce a valid result.

Animation display

Up to ten points on the mechanism can be selected for display. The paths of these points are displayed by connecting straight lines between the positions calculated during the analysis. Successive positions of the mechanism are displayed rapidly to give an animation effect. The speed of display can be controlled by the user.

Displayed material also can be printed. A report generator module controls this process, producing plots of various parameters, which can be annotated with titles and labels. A tabular listing of results also can be displayed or printed.

PLANET-PC

PLANET-PC handles systems of linkages that can include elements such as masses, springs, and dampers and which are connected by pin or slider joints.

PLANET-PC runs on IBM/PC/XT and Columbia computers. The price is available on request from Engineering Software, 2 Bloor St., W #100-202, Toronto, Ontario M4W 3E2, Canada. (416) 226-6080.

The analysis performed by this package is based on the Planet programs originally developed at the University of Waterloo.

The manual suggests that the user start by sketching the system to be modeled. Only simulation elements allowed by the package can be used. These include: masses, rigid-arms, position-velocity drivers, bearing joints, dampers, springs, force-torque drivers, and rigid rods.

Users must carefully number the elements in the sketch. For each different element-type several parameters will be needed. For example, for masses these include the magnitudes of the masses, displacements, velocities, rotational inertia, and specification of the constraint paths.

Reference connections

When the package is started up, the user is asked for some general job-description information, then the package main menu is displayed. Selections from this menu are made by pressing IBM/PC function keys.

Choosing Input/Add Data File calls up the data entry menu. The Network Specifications choice queries the user on the quantities of each of the the various types of simulation elements that are to be used. The remaining menu choices provide for inputting information on each type of element: element number, parameters, and references to identify other elements to which this one is connected. There are some detailed rules to follow in connection with this last item.

For most of the element types, there are three or four parameters to enter. Rigid bodies require 6, damping elements up to 7, bearings 7–12. For a mechanism with 20 elements, users will find themselves preparing and entering perhaps a hundred parameters, so this input procedure can be time consuming.

The user manual gives some guidance on how to calculate input parameters, but these calculations are not included in the package. Nor are frequently used constants, such as material damping factors, stored by the package. Constraint types for mass elements and rigid arms, driver element types, and bearing and spring types must be entered as numbers taken from tables in the user's manual.

Most of the input effort probably will be in obtaining the necessary parameters. The entry process itself is well organized. For each element, there is a display of needed input items, and the user moves from one item to the next.

The entered data can be viewed or printed with a Print/Display Element option on the main menu. There is also a Modify/Delete choice that allows users to add, remove, correct, or change these element descriptions after they are entered.

A squeezer mechanism is given in the user manual as an example to illustrate the element numbering and data entry procedures.

Graphical and printed output

In the final step in the input procedure, the user selects the integration method. Either a Runge–Kutta or Crane–Klopfenstein method can be chosen. The latter is said to be potentially more accurate but longer running. Users must specify initial and final times and time-step size for integration. The Crane–Klopfenstein method requires added specifications on truncation error limits and time-step limitations. A formula and table for calculating the time step are given, but this calculation is not included in the package. Again, users must resort to their own calculators.

The output data can be displayed or printed. Displacements, velocities, and forces are produced for each element, at each time step in the simulation. Depending on the number of time steps involved, that can be a lot of data.

To provide a ready summary, the package can display or output to a plotter (model unspecified), graphs of functions such as x, y components of displacement, velocity or force, angular displacement or velocity, or torque.

Variable driving

If an applied force-torque or position-velocity driver is to be variable, the user must write a FORTRAN subroutine. The manual states that a list of variables for this purpose is given in an appendix, but that appendix could not be located.

Limitations on systems that can be modeled include motion constrained to paths no more complex than can be described by a second-order conic equation. This includes linear, circular, ellitical, and parabolic paths. The package does not model motion deviation due to impact.

Kinematic chains in which masses are directly connected to one another can not be modeled, but bearings or zero-length spring-damper combinations can be used to model the joints in such a chain. The package does not model linkages composed only of massless springs and dampers.

All rotating bodies in a modeled mechanism must have parallel axes of rotation.

SPRING1

SPRING1 is for the design of helical compression and extension springs under static or dynamic loads.

SPRING1 runs on IBM/PC/XT/AT computers and sells for $125 from Engineering Software, 3 Northpark E, #901, 8800 North Central Expressway, Dallas, TX 75231. (214) 361-2431.

SPRING1 presents the user with an input display screen on which entries have to be filled in for a list of clearly identified parameters. Moving to the various items, by using up and down arrow keys, along with the Ins and Del keys, users enter and edit the data as needed. Items for spring end type and material are chosen from short displayed lists of possible choices.

Materials shown include music wire, oil-tempered wire, hard-drawn wire, chrome-vanadium wire, and chrome-silicon wire. Materials data for these choices comes recorded on the package disk; properties are approximated by curve-fit formulas. The user also can enter material properties for other types of wire.

Input data sets can be stored for future use. This allows calculations to be rerun quickly after editing changes are made to the input data.

Results are displayed as a list of clearly identified items that include input parameters along with solid height, total number of coils and number of active coils, spring rate, and safety factor for yielding.

Spring Design

Spring Design performs design and analysis procedures for helical and Belleville springs.

Spring Design runs on an IBM/PC and sells for $125 from P.E.A.S., 7208 Grand Ave., Neville Island, Pittsburgh, PA 15225. (412) 264-3553.

This package operates from a combination of menus, prompts, and other displays. For example, on the selection of Helical Spring from the main menu, the user is offered a choice of design or of three analysis procedures. If design is chosen, prompts for shear modulus and end type follow. Then a diagram of a coil spring is displayed to illustrate needed dimension data, and the user is prompted for the mean diameter, free length, working deflection, and deflection force. Extension springs can be handled by entering extended length instead of free length.

When these entries have been made, the package displays them on the screen, and the user can accept the entries or make corrections on this screen.

The package then goes on to calculate results for a one-coil spring. The display of calculations includes a theoretical wire diameter and shear stresses at working deflection and solid length. If these stresses are not satisfactory, the user presses a key to get results for a two-coil spring and can press once again for three-coil results. Displayed results can be printed, if desired.

Analysis procedures

A Stress Analysis procedure determines shear stress, given the mean diameter and force on a helical spring. This allows the user to calculate exact stresses for commercially available wire diameters. A Buckling procedure calculates the critical axis load for hinged and built-in ends. The user inputs spring rate, free length, and mean diameter. The rate of a spring and its natural frequency can be calculated with the help of another procedure. Users input wire diameter and mean spring diameter and number of active coils.

Calculations for Belleville springs include axial load and stress, given outer and inner diameters, thickness, H and deflection. Alternatively, deflection and stress, or spring thickness and stress can be calculated from appropriate input parameters.

SPRINGSTAR

SpringStar allows the user to specify helical and flat springs, determine their ability to sustain static loads, alter the specifications, and recalculate as needed.

SpringStar runs on IBM/PC/XT/AT computers and sells for $91 from TexStar Software, PO Box 15937, Austin, TX 78761.

As the main menu of this package shows, there are separate procedures for helical compression, extension and torsion springs, as well as a procedure for flat springs.

The user selects the desired spring type then enters specifications on a data input screen. For compression springs, the screen will show data for spring diameter, wire diameter, and modulus in torsion, along with values of two loads and two corresponding spring lengths and a choice of spring-end type. Starting with the first item, the user enters values (or accepts displayed values) for each item on the input list. This process can be repeated, if needed, to correct the entries.

When satisfied with the input data, the user calls for calculation and results are displayed. Included for compression springs are the spring rate, free length, solid height, inside, mean and outside diameters, and total and active coils. There is also a three-column display of length, load, and corrected stress levels.

The procedure for extension springs is for springs made of round wire. It includes a provision for design with initial tension. The torsion spring procedure is also for springs made of round wire with small initial tension.

The flat spring procedure allows calculations for cantilever or for simple beam springs. The user enters four of the following five parameters: load, deflection, thickness, length, and width. The procedure then calculates the fifth parameter and the stress level.

The user's manual for this package is minimally adequate, but it gives no vendor phone number. What is a user to do when problems arise?

Chapter 6
Finite Element Analysis Packages I

As long as finite element analysis was done only on large computers, the applications it found in engineering necessarily were limited. Today many finite element packages run on personal computers. By bringing this capability to the desk of virtually every engineer, these packages open the way to much more widespread employment of finite element techniques. More important, they challenge the inventiveness of engineers to find new uses for these techniques in tackling design problems.

This chapter is not the only one in which finite element packages appear. Chapter 7 makes a detailed comparison of three state-of-the art personal computer finite element packages. Packages covered in these two chapters are those that offer a variety of elements intended to model many different types of objects. Because of their interesting features, the special-purpose CAEPIPE and CAEFRAME packages are also discussed in this chapter.

There may be some misconceptions about using personal computers for finite element analysis. Because these computers are physically small does not mean that they cannot handle large problems. Today's minicomputers and mainframe computers are faster than personal computers, and their capacity to store data is generally greater. However, the differences between the small machines and their larger cousins usually amount only to about an order of magnitude.

Later in this chapter we will take a look at the actual capacities and speeds offered by personal computer finite element packages, but let us begin by considering the question of computing speed.

How important is speed?

A personal computer like the IBM/PC generally does its computational number crunching more slowly than a mainframe computer or a minicomputer like the DEC VAX or a Prime machine. That means that a small problem that runs for 10 or 20 seconds on the PC might be completed in just one second on a larger computer. A large problem that could run for an hour on the larger machine might grind away for 10 or 20 hours on the PC.

In terms of inconvenience and disruption of work activity, the differences in these situations may not be very significant. Seconds of waiting time will not make a significant difference. Where long minutes or hours of computing are involved, the user will probably turn to other tasks while the problem runs.

The difficulty with computer runs that last for hours is that an analysis might have to be repeated several times, with changes incorporated each time, before the needed design information is obtained. Then runs that take ten hours each can easily add up to a week of design time, and the larger machine may be needed.

One of the main advantages of a personal computer is that the user has complete control over what the computer is asked to do and when it is used. There is no waiting for programmers or data processing managers and no need to compete with other users for shared resources. In many situations these advantages will override the drawback of slower computational speed for the small machine.

A side benefit of the personal computer is that it allows vendors of software packages to make their packages easy to learn and use. Personal computer packages are expected to be self-explanatory, allowing users to sit down at the computer keyboard and interact with screen-displayed questions and instructions to accomplish whatever they desire.

As we shall see, the extent to which the packages described in this chapter manage to reach the goals of easy learning and use varies. Overall, though, their achievement is impressive, reducing the user's chores dramatically compared to what most mainframe and minicomputer finite element software has demanded.

Finite element equations

Finite element packages usually have to solve second-order differential equations in order to find the static and dynamic properties of structures and systems. The general form of those equations includes an unknown vector with acceleration, velocity, and displacement components. Known quantities include mass, damping and stiffness matrices, and a vector of applied forces and moments.

To obtain static solutions, only equations involving the stiffness term have to be solved. Dynamic analyses involve the solution of the other terms as well.

Structures and systems that are to be analyzed by the finite element method are represented by node points to which elements are connected. An element is defined by its deflection capacity, which is expressed in terms of degrees of freedom. In a three-dimensional structure, these can be thought of as translations along the three axes and rotations about these axes. Therefore, there can be six degrees of freedom at a given node. If an element is restrained from movement, one or more of these degrees of freedom will be removed (zeroed).

The number of equations that have to be solved to analyze a given structure corresponds to the active (non-zero) degrees of freedom. The number of rows and columns in a matrix that represents those equations will also correspond to the active degrees of freedom. Therefore, degrees of freedom are a measure of computation time since larger matrices will take longer to solve.

Generally, the non-zero elements of the stiffness matrix will tend to cluster near the main diagonal. The distance of the furthest non-zero element from that diagonal is termed the *semibandwidth* of the matrix. A matrix with a small semibandwidth can be solved relatively fast, and with the use of a relatively small memory space. Thus the limits on problem size will depend on the semibandwidth of each problem as well as on the number of nodes or degrees of freedom involved.

Various techniques are used to simplify the needed matrix calculations. As much advantage as possible is taken of symmetry; this allows half the stiffness matrix to be immediately discarded. Zeros in the main diagonal of the matrix also can be discarded. Other programming techniques strive to optimize the speed of disk or memory access. Some of these techniques are fairly standard, others may be the individual inventions of software package designers. Depending on the ingenuity with which such techniques are applied, different packages may run similar problems more or less rapidly.

In some packages the ability to handle large problems can be limited by the precision of the arithmetic the computer employs. When so-called single-precision arithmetic is used, roundoff error can produce answers that are increasingly inaccurate for larger problems.

Main memory and disk storage

One way to solve the system of equations that will yield the key answers is to place all the necessary information in the main memory area of the computer. This memory is directly, rapidly accessed by the central processing circuits of the computer, and the solutions therefore can be completed at the maximum speed the computer can muster. The drawback to this use of main memory is one of size. Even the 640 thousand bytes of main memory (a byte is approximately the amount of memory required to store one alphabetic character) allowed as a maximum by the IBM Personal Computer is not sufficient to accommodate and process solutions for larger finite element problems.

A second approach makes use of the computer's disk storage. This storage can be substantially larger than the main memory. For instance, a so-called hard disk unit may provide 10 million or more bytes of storage. In this approach, the solutions are computed in blocks, taking one block of equations at a time from disk memory into main memory for solution. Then the size of problems that can be handled is limited only by the size of available disk memory and the time the user is willing to wait for a solution to run. Although finite element software is designed to minimize use of disk storage space and to speed access to and from the disk, disk-based solutions are much slower than those based on main memory.

The essential tradeoff is a fast but limited-size main memory solution versus a much slower disk-based solution that can handle very large problems. In an attempt to get the best of both these approaches, at least one vendor (COADE) has structured its packages so they will run smaller problems in main memory but will automatically shift to a disk-based solution when larger problems are presented.

Finite element packages that run on personal computers often place limits on the number of nodes, elements, and degrees of freedom that can be handled. Most of these limitations refer to main-memory based solutions. They are simply the limits on the equations that can be handled within the computer's available main memory space. Problems that fit within these limits may run in a few seconds or in a matter of minutes.

Easing data input

Many finite element packages for personal computers use interactive dialog between the computer display and the user to shape the way information is entered and package capabilities are employed. With this kind of interaction, the user is prompted to provide all the data the package requires and to make all the alternative choices the package provides. Interactive input of data and commands is useful not only to the beginning user but also to those who do finite element analysis only occasionally and are likely to forget commands and procedures between sittings at the computer. However, the step-by-step format imposed by interactive use can become confining for the constant user who soon becomes very familiar with the necessary procedures, is looking for the most efficient way to do problem solving, and may prefer to issue direct commands. It is possible to offer users the best of both worlds by providing interactive menus with the option of using direct commands. The SUPERSAP package approaches this kind of operation by offering menus as an optional feature.

Some of the packages continue to use the batch entry procedures familiar to users of finite element programs that run on larger computers. For these, the user must prepare the necessary input items in correct format, then enter them into the computer as a group. At least one of the personal computer packages allows for the use of word processing to set up this kind of batch input.

One notion promoted by some finite element package vendors is that the user can come to the computer keyboard with just a rough sketch of the system that is to be analyzed. However, dealing with truss and beam elements requires the user to come prepared with data on cross-section properties, and various material properties also are required as input to many analyses. These generally include such properties as Young's modulus, Poisson's ratio, weight and mass densities, and inertias. Where anisotropic behavior is to be represented, elastic constants are needed. For heat transfer calculations, inputs may include convection and radiation properties. Thermal stress analysis requires thermal expansion coefficients and stress-free temperatures.

Built-in computer files can ease the input burden on the user by collecting and storing this kind of information. The storage capability is often referred to as a library or database. In some packages, data the user puts in for one element can be drawn from the database for use with other, similar elements. When packages are directed to very specific types of design, a lot of very useful information can be kept in the database. For example, the CAEPIPE package from SST carries complete descriptions of standard parts in its database. From this database, pipes and piping components, complete with all dimensions and properties, are supplied at the user's request.

The units of input and result quantities are an important concern. For correct results, units of input parameters not only must be correct but also consistent with one another. Most of the personal computer finite element analysis packages rely on the user to monitor correctness and consistency of units. However, the packages from SST (CAEPIPE and CAEFRAME) prompt the user for the right units for input data and display the correct units for the output data. Users can elect to input either in English units (based on inches or feet) or in metric units (based on centimeters or meters). Regardless of what input units are used, the packages will display results in whatever units the user selects, making all necessary conversions internally.

Another way that computers can ease the laborious process of entering data is by providing internal controls or checks on the accuracy of the input. Both logical and typographical errors can be checked in this way. Packages may, for example, refuse to accept negative values for area or for Young's modulus. In connection with the entry of nodes and elements, the package can check to see that there are no isolated nodes. Visual checks of a graphical display of the input model are especially valuable in uncovering errors such as missing or misplaced elements.

Generating nodes, elements, and meshes

To simplify the process of inputting node and element data, some of the packages allow the user to input data for nodes and elements, then replicate that data to reproduce those nodes and elements as needed to form the desired structure. For example, if a beam is to be defined by six identical nodes with interconnecting elements between them, the user only has to input data for the first element and the first and last node locations. The package will then replicate the data for all the desired nodes and elements.

Objects to be modeled by these packages can be defined in a very general way by placing nodes in space to define the object's surface and then connecting them to appropriate elements. The process of setting up this network of nodes and elements is called mesh generation. In personal computer packages, as in similar software for larger computers, automatic mesh generators for specific types of objects are frequently contained in "preprocessors" that come with the packages.

Of the packages discussed in this chapter, only one provides a variety of preprocessors. The SUPERSAP package has ten, and these preprocessors are selected so the shapes they produce can be used individually or combined with one another to serve as models for all the mechanical and structural problems the package designers could envision.

Looking at graphics

The packages described in this chapter come with various graphic capabilities. Some packages provide no graphics at all or have the user rely on an external computer-aided design (CAD) package to provide graphics.

Nevertheless, graphics are an important feature for a finite element analysis package. In the first place, they provide a valuable check on the correctness of the model the user inputs into the computer. Sometimes the user may leave out a node or misplace an element. Even if many nodes and elements are generated automatically, the user will still want to see that the desired shape has actually been generated. It is therefore almost essential to view the shape that results from the input data and to note how that shape is covered by nodes and elements. Errors in the input data, even gross errors, are not easily recognized in numerical form, but they can be dramatically obvious in a graphical view.

To examine the graphical display of the model from every aspect, a number of packages, particularly those that deal with three-dimensional models, allow the user to translate and rotate the image. In addition, it is helpful to be able to view images larger than the screen area, to zoom in on details of the model and to pan across to reveal off-screen portions.

To connect the actual data to what the user sees on the graphical shape, some means of correlating the two is needed. This is usually provided by displaying node and element numbers on the graphics. Some packages also provide the reverse capability. While viewing the complete model, users of packages from SST can ask to see material properties on a certain element. The desired data will be displayed, and the indicated section of the model graphics will be highlighted.

It is convenient to have both graphics and data shown on the display screen at the same time. One way of doing this is to provide a split or windowed screen with separate areas for graphics and data; this is provided in the Images 3D package from Celestial Software. Packages from SST use two separate display monitors: a monochrome monitor for data and a color monitor for the graphics.

Display of the deformed model, after loads have been applied, is another valuable feature. This allows the user to see where the largest deflections take place and to observe the overall way the model responds to the loads. Although the same information is included in the output data listings, one picture can be worth a thousand lines of numbers. Some packages provide magnified viewing of the deformed portions of the model so these changes can be viewed more easily.

Animated graphics display is a desirable feature. It can lead to a clearer understanding, for the designer, of such characteristics as the vibrational response of structures.

Color display also can be used to advantage. For example, in the CAE-PIPE package, result values that meet code limitations are shown in yellow, whereas those that exceed the code are shown in red. In the Images packages, colors are used to distinguish between element and message types. The SUPERSAP C-Plot module draws stress, deflection, and temperature contour plots in multiple colors.

Data at the output

Because all the packages described in this chapter use the finite element method, they produce similar basic output data. For static analyses, the basic outputs are deflections, reaction forces, strains, and stresses. For dynamic analyses, the outputs are more varied: resonant frequencies and mode shapes, transient responses, and frequency responses.

The usefulness of these packages can be enhanced by giving users control over combinations and selections of the basic output variables. For example, maxima and minima can be sorted out and listed for displacements and stresses. Results can be compared to relevant standards, deviations from allowed values can be listed, and outputs for different load cases can be compared.

Graphics can be used to present these output data as well as to show views of the models. The Finite/GP package from COADE, for example, provides contour plots of stresses, displacements, and variables of heat transfer analysis.

Transfers to and from larger computers

In several cases, the packages described in this chapter are versions of software originally designed to run on larger computers. Thus, both the SAP-86 and SUPERSAP packages are derived from the SAP IV software that has been in use on mainframe and minicomputers for several years.

As long as the files of input data built up in the personal computer versions are identical in format to those used with the larger computers, there is little difficulty in transferring models, input on a smaller machine, to be run on a larger one. This can be particularly valuable where large models are to be run repeatedly.

Even when there is no version of the package that runs on larger computers, it is possible to transfer models from personal computers by translating the model data files into formats that can be accommodated by the larger computers. Thus, for the Images 3D packagc from Celestial Software, there are translation routines that allow the user to move models built on a personal computer over to larger machines that run such finite element programs as Strudl, Nastran, Ansys, and Stardyne.

Becoming a user

Most of the packages described in this chapter operate on an interactive basis. That means that when data are needed from the user, a request for them is displayed. If a command is needed, the user is notified by the software and is usually presented with a choice of available alternatives. This interactive dialog between user and computer eliminates the need for the user to memorize procedures and command names, and speeds the process of learning to use the package. Some packages also provide screen-displayed HELP messages to aid the learning process.

For those users who are generally familiar with finite element analysis, vendors who have interactive packages estimate it will take from 20 minutes to a few days to begin to run practical problems. In contrast, estimates of time to learn from vendors of those packages that do not provide interactive dialog range from several days to two weeks.

Most of the packages provide sample problems to help the new user understand how they work. These are helpful, particularly if problems similar to those sampled are what the user wants to run, but they are no substitute for a well-written manual.

Beyond learning the machanics of package operation, users should be aware that the finite element method is not as straightforward as we are sometimes led to believe. It is an approximation method that allows users to easily generate bad models as well as good ones. The object is to create models that reasonably represent the mechanics of the objects that are being analyzed.

There can, for example, be differences in the way mass is distributed, depending on the number of nodes and elements used in a model. Thus it may be possible to model a given structure using 20 nodes and get answers that correspond closely to what would be measured in an actual, physical structure. However, if the same structure is modeled with ten nodes, the answers may be inaccurate.

Many universities offer short courses in finite element analysis that provide excellent perspective on the pitfalls that may be encountered in using these models.

Hand holding and updating

All of the package vendors will field questions from users over the phone. In some cases the vendors encourage users to send in printouts of their input and results so problems that present difficulties can be rerun and further analyzed. Experience with the responsiveness of a vendor is the only way to gauge the extent to which they will actually prove helpful so, if this is a matter of concern, prospective users would do well to discuss it with colleagues who are already using the package under consideration.

Finite element packages, like any other software, are subject to continued change and improvement. Several of the package vendors indicated that they issued updated disks every six months or once a year. The cost to users for these updates ran from the cost of disk and postage to about 10 percent of the purchase price.

CAEPIPE, CAEFRAME: RICH IN FEATURES

CAEPIPE and CAEFRAME solve static problems for 2D and 3D structures and systems. CAEPIPE analyzes networks made up of standard piping ele-

ments and fittings. CAEFRAME analyzes truss and frame structures. User-oriented features include a graphics display that utilizes color to highlight results, stored libraries of standard elements, and the ability to automatically convert units and monitor their consistent use. The specialized nature of these packages permits fast operation, with larger problems running only a minute or two to completion.

The one-time license fee for CAEPIPE is $6,900; for CAEFRAME the fee is $4,900. The packages run on IBM/PC and compatible machines. Required equipment includes 640K main memory, 8087 math chip, monochrome adapter and display, Techmar Graphics Master board, color graphics monitor, and a graphics printer. The packages are available from SST Systems, 335 W. Olive Ave., Sunnyvale, CA 94086. (408) 773-1171.

With the CAEPIPE package, up to 450 elements can be used. In piping systems with many branches, only about 200 nodes can be used. The limitation is in the available computer main memory size, a maximum of 640K in the case of the IBM/PC. With CAEFRAME, 2,700 degrees of freedom can be accommodated for a beam problem, or half that number for a plate or shell problem, depending on the bandwidth of the stiffness matrix.

Elements for CAEPIPE include the piping itself, elbows, standard fittings, elastic supports and hangers. For CAEFRAME the elements include beams (six degrees of freedom per node), trusses (three degrees of freedom per node), plate elements (six degrees of freedom per node) and spring elements (six degrees of freedom).

Standard pipe components

The packages have built-in libraries of standard components. CAEPIPE has standard piping materials and the ITT–Grinnell pipe hanger catalog. The program automatically selects the appropriately sized hanger once its location on the pipe network is specified. It also checks for conformity with ANSI standards.

CAEFRAME has stored libraries of standard AISC structural sections, pipe sections, standard pipe, and structural material properties. The package checks for conformity of the structures to AISC codes.

Types of loads for CAEPIPE include sustained dead weight and pressure, seismic G loads, thermal expansion, and anchor movement. These can be applied in any combination permitted by ANSI standards.

The CAEFRAME package handles applied displacements, seismic G loads, nonuniform distributed loads (triangular, trapezoidal, and so forth), thermal loads, and concentrated loads at the nodes or at any location along the elements. Up to 99 loads and 99 load cases can be used with either package.

According to the vendor, speed of operation for CAEPIPE running a complex piping problem with many branches and 200 nodes was under two minutes per load case. More typical problems with about 50 elements run about 25 seconds per load case. CAEFRAME runs a problem with 30 elements in about 25 seconds for the first load case. Problems with 100 elements run in just under two minutes per load case.

Packages take care of units

The user can elect to have input and output either in English or metric units. The input and output selections are independent. Thus, the user can elect to have the package prompt for input entries in English units (based on inches or feet) or in metric units (based on centimeters or meters). Regardless of what input units are used, the packages will display output results in whatever units the user selects, making all necessary conversions internally.

CAEPIPE can automatically generate series of runs and elbows. CAEFRAME includes automatic node and element generation. Both packages include numerous checks to detect inconsistent input data. If, for example, the user attempts to input a pipe element and connect it to an incorrect fitting, the CAEPIPE package will immediately bring that discrepancy to the user's attention.

Color: dual screen display

Two display screens are used, one for interactive text and data display, the other for color graphics and data. As the user enters the input data on the text monitor, a graphical representation of the resulting model takes shape on the color screen. In this way the user can see immediately when obvious input errors are made and can verify that the data represent the desired model.

The graphic display can be rotated. Users also can pan across images larger than the display area and zoom in on display details. The color display also highlights the node or element the user is currently inputting or examining.

Information on the model is coordinated with the display, and is available for user inquiry. For example, with the CAEPIPE package, the user can ask to see material properties on a certain six-inch section of pipe. The desired data will be displayed and all the sections of the model that use that six-inch pipe will be highlighted on the graphics screen. This provides an added method for checking on the correctness of the model.

Deflected shapes of models also can be displayed on the color screen, along with result data selected by the user. Thus, the user can elect to view all pipe elements that are subjected to stress above, say, 7,000 psi. The model will then be shown with these elements highlighted and with the calculated pressure numbers displayed next to these elements.

The user can control a magnification factor on deflections so that, if too small to be clearly seen, they can be emphasized as needed. On the display of the deflected model, maximum values are highlighted. The highest stress value is shown in a box.

A BROWSE command allows the user to call for selected results to be displayed on the text screen while corresponding elements of the model are highlighted on the color screen. For instance, the user can ask to see the forces on element number 33. These will be listed on the text screen while element 33 is highlighted on the color graphics screen.

Output data include stresses, displacements, forces, moments, spring forces, support loads, and stress ratios (calculated over allowed values).

The vendor finds that users familiar with piping or frame design can learn to use these packages in about four hours. To supplement the manual and the interactive procedures, HELP information that guides the user with screen-displayed messages is provided.

Support is available from the vendor by telephone. It is provided with the packages, along with updates, for 90 days. After that time, the monthly charge for further updates and support is about 1.5 percent of the license fee.

EASI4: FOR STRUCTURES AND TEMPERATURES

EASI4 performs static analysis and will shift automatically from an analysis based on main memory to one based on disk access. The package also has thermal capabilities. For example, those who want to analyze welding stresses can enter the differential temperatures and thermal coefficients, and the package will do that analysis.

EASI4 sells for $500; an improved version that uses the MegaBASIC computer language sells for $600. Both packages are available to run on computers that use the CP/M-80, CP/M-86 and MS-DOS (that includes the IBM/PC) operating systems.

Floppy disks used with these packages must have capacities of at least 150K; 128K main memory is required, 256K if the MegaBASIC version is used. The vendor recommends that users have their IBM/PCs equipped with the full 620K of main memory to handle as many problems as possible within that memory. The package is available from EASI Software, 2891 Lavonia Center Rd., Lima, NY 14485. (716) 346-2022.

EASI4 uses main memory based finite element solutions and can solve typical plane frame structures having about 160 nodes on an IBM/PC. On so-called 8-bit personal computers, such as the Apple II, structures having about 80 nodes can be handled. EASI versions written in the MegaBASIC language can solve structures with 2,000 or more degrees of freedom.

Any problem that is too large to solve within main memory is automatically shifted to a disk-based solution. Then the equations are brought into main memory a few at a time for solution. This is considerably slower, but the complexity of the structures that can be handled is limited only by the available disk storage space. With a hard disk, very large structures whose solutions may run for 20 or 30 hours can be handled.

EASI4 has rod elements, three-node and four-node membrane elements, and three-node and four-node plate elements.

Interactive data entry

To prepare data for input to the package, the user has to break down the structure into coordinates, then enter the materials properties. Data entry is interactive. The user enters data in response to questions from the program. After entry is completed, the user can go back and edit what has been entered. There is no built-in checking of the input data, except for a graphical check on the shape of the input structure.

The package can produce simplified views of the input and deformed structure, with elements represented by asterisks. The displays can be rotated and reproduced on a printer.

There is a stored table of materials. For each material type, the user enters cross-section and material properties. Then when the structure is defined, the user references the entries on this table. When applying loads, the user has to calculate the moments for point loads. There can be uniform loads on the rod elements.

Some of the larger aircraft companies have put their engineers on these packages to get them familiar with finite element analysis. The main reason for this choice has been the ease of use. According to the vendor, it usually takes engineers about 20 minutes to begin running problems on this package.

EASI Software takes phone calls from users who have problems. About once a year they issue an upgrade of the packages. The current upgrade includes a new manual and disk, at $50 for current users.

FINITE/GP: FOR A VARIETY OF PROBLEMS

Finite/GP handles static analysis for heat transfer and fluid flow as well as for structural problems. The package includes mesh generation and substructuring capabilities. It also provides carpet plots of output data for the heat transfer and fluid flow analyses.

There is a $3,500 one-time charge for the package and a lease arrangement for those who wish to try out the package. Sections of the package are offered separately. For example, a section is offered on axisymmetric analysis and another on 3D frame analysis.

Finite/GP is for use with the IBM/PC and compatible computers. Additional equipment required is an 8087 coprocessor chip, 512K main memory, color graphics adapter, and a dot-matrix printer. The package sells for $3,500 from COADE, 8552 Katy Freeway, #320, Houston, TX 77024. (713) 973-9060.

Finite/GP solves for plane stress, plane strain, axisymmetric stress, plane heat transfer (convection or conduction), axisymmetric heat transfer (convection or conduction), flat plate bending, potential fluid flow (Poisson equation solving), and will also find 3D beam and truss solutions.

Structural elements available include the beam, three-noded and four-noded quad elements, also a six-noded triangle and an eight-noded quad.

The package uses a disk-based solution technique and will handle up to about 2,000 degrees of freedom. Problems using 2,000 constant strain triangles have been sucessfully run by the vendor. The same problems can be run using six-noded triangles, and fewer elements would be required.

The vendor finds that the average problem will run under ten minutes on this package. Problems that are so complex as to be near the limits of the package are said to run about 20–30 minutes.

All input data are stored in the same allocated memory area. Node, element, and load data all compete for the same space, so the number of load cases that can be handled depends on the size of the problem. Like nodes and elements, the number of loads and load cases is limited only by available memory.

For heat transfer calculations, the required inputs are convection, conduction and thermal expansion coefficients, and stress-free temperature.

Data input is interactive, so it can be performed by answering screen-displayed questions. An alternative batch-mode input, using a word processor to format the input data, is also provided.

Graphical and data output

The mesh generation technique used is a conformal mapping. The user can start with a simple sketch of the object that is to be represented and use the mesh generation facility to get a usable model. Substructuring is provided so different shapes can be attached to one another for analysis.

Both input and deformed structures can be graphically displayed. In heat transfer analysis, graphical output includes contour curves and surface (carpet) plots of temperatures and gradients. Numerical data output for heat transfer includes temperatures, potentials, and gradients.

Fluid flow analysis yields velocities, potentials, and pressures. Structural analysis produces the usual stress and displacement information. Carpet plots of these results also can be displayed. From a menu the user chooses which plots are to be shown.

The package provides no built-in transfer of files to larger computers. However, a file is reserved by the package for handling the formats, and users able to do so can specify the necessary reformatting.

User questions are taken over the phone. With the package purchase the user gets one year of updates. That includes two major updates plus several minor ones. After the first year, users must make a separate support agreement with the vendor.

FRAME 2D AND SAFE: SIMPLE AND EASY

Frame 2D is for static analysis of two-dimensional structures. Similar analysis of three-dimensional structures is done with the SAFE package. Both are quite simple to learn and use. Only one type of element, a beam, is available for Frame 2D. SAFE offers a larger variety of element types and can handle models with up to 600 nodes and 600 elements. The capacity of Frame 2D is only about half as great. The packages allow fairly complicated loading and run reasonably fast.

Frame 2D costs $495. SAFE costs $795. The packages run on IBM/PCs and compatible computers, require 128K main memory, two disk drives, and a graphics adapter. Use of an 8087 math chip is recommended. The packages are available from Engineering Software Co., 1405 Porto Bello, Arlington, TX 76012. (817) 261-2263.

Frame 2D uses straight-line beam elements. These have been used to model structures such as airplane wings, frame structures, even C-clamps, since curved surfaces can be approximated by straight-line segments. It will handle from 75 to 150 nodes and about 300 elements.

A larger variety of elements is offered to the SAFE user. These include both plane and space elements for trusses and frames, a pipe/shaft element, plane stress triangle and rectangle, a rectangular plate for bending applications, and elastic foundation elements for beams and slabs.

Models with up to 600 nodes and 600 elements can be handled by SAFE. Up to 25 sets of different element characteristics can be used in any model.

Types of loads in Frame 2D include those at node points (*x*, *y*, or moments) and distributed pressure loads perpendicular to the surface of the elements. The package will accommodate as many load cases as can be handled in disk storage.

Frame 2D uses main memory based solutions. A structure with 36 nodes and 55 members ran for about three minutes to a solution.

Frame 2D is interactive, cueing the user on the data items to enter. If the geometric data are entered in incorrect format, the package will not accept them. As with most finite element packages, users must come to it prepared to input a number of characteristics of the structures beyond the geometry. For example, to represent the cross-section properties of beams in Frame 2D, the user enters a flexural inertia quantity (I) and a section modules (I/C). The flexural inertia is used to calculate the element stiffness for the matrix computation. The section modulus is used to calculate the maximum bending stress following the matrix solution. Frame 2D will handle 30 different cross sections within one structure.

Using the graphics

Once users key in the data they can view a graphics representation to see if everything has been correctly entered. The package will run without a graphics card, but to view the models a graphics card must be used. In fact, if there is no graphics adapter installed in the computer, the package will sense this fact and the interactive dialog will not even offer the user the option of viewing graphics displays.

The package also provides displays of the deformed structures after loads have been applied, allowing the user to see the deformation and get an idea of where the largest deflections take place.

SAFE has some added graphics capabilities. It allows the user to specify a desired viewing angle, then display the structure as seen from that angle. It also includes panning of images and zooming in increments of 20 percent increase or decrease in size.

The Frame 2D package includes a calculation of the factor of safety for each member. The user can input an allowable stress. Then the stress in each member is compared to that allowable stress and a factor of safety is calculated. All that is tabulated. Then at the end of the process, the worst-stressed elements are pinpointed and identified so the designer can take remedial action, if needed.

Frame 2D data results include displacements and rotations at the nodes with maximum and minimum node deflections noted. For the elements, axial and shear forces, bending moments, maximum stress, and factor of safety are listed. Maximum and minimum stress, moment, shear, and force in the structure are noted.

Easy to learn and use

The effort needed to learn to use the packages is minimal. The vendor has had few requests for assistance from users who seem to be able to run the package after about a half day of experience. The package disk includes sample problems, and many users start by going directly to these sample problems to learn how to use the program.

The vendor's policy is to try to answer all users' questions the same day they call. To this point, there have been no upgrades on either package, but when and if there are, they are to be made available to all users at nominal cost.

IMAGES 2D, 3D: DYNAMIC ANALYSIS WITH A SPLIT SCREEN

Images 2D and Images 3D provide static and dynamic analysis of 2D and 3D structures and systems. A variety of different elements are available. There is no mesh generation, but facilities for replicating nodes and elements are quite flexible. Many internal consistency checks of data input are provided to aid the user. Complex loading cases can be handled easily with these packages. The 3D package has a convenient split-screen graphics capability.

The 2D package costs $195 for the static analysis version and $395 for the static plus dynamic versions. The 3D is $595 for the static version, $995 for the combined version. A companion program that performs AISC/ASME code checks of structural members sells for $200. The packages will run on the IBM/PCXTAT and compatibles. A color graphics adapter is required. For the 3D version, the 8087 math chip is required; it *is recommended for the 2D package. The vendor also recommends a hard disk for the 3D package. Images 2D requires 256K main memory. Images 3D requires 512K main memory. Images 2D and 3D are available from Celestial Software, 125 University Ave., Berkeley, CA 94710. (415) 841-7175.*

Images 2D will handle structures with 100 nodes, 150 beam elements, or 50 triangular plate elements. The total degrees of freedom it can accommodate is 300, with three degrees of freedom per node. Images 3D will handle 300 nodes; elements handled can include 300 beams plus 300 plates plus 300 springs with a total of up to 1,800 degrees of freedom for both static and dynamic analyses.

Six types of elements are available for the 3D package: truss, beam, membrane plate, membrane-plus-bending plate, a single-node spring, and a two-node spring. The 2D package has all the above elements, except for the membrane-plus-bending plate and the two-node spring. Up to 50 different cross sections are available for each package for use with the beam and truss elements.

There are five different types of loading: concentrated loads or moments, gravity loads that can be automatically generated on request, enforced or specified displacements, temperatures for calculation of thermal expansion, and distributed loads. Both the 2D and 3D packages have these. In addition, the 3D package has pressure loads and tapered distribution loads.

Five separate load cases in 2D and up to 99 in 3D can be built up and used for each design. Those can be summed and factored as desired. A built-in routine sums and applies factors to the load cases.

Generating nodes and elements

In Images 2D there is node and element generation. The user specifies the endpoints, and the package will interpolate to fill in elements on a straight line. Users also can specify an element and then replicate it by specifying the nodes to which it will be attached.

The 3D package generates nodes and elements using pattern, repeat, and fill capabilities. There are many other preprocessing features built in. A complex model such as a piping elbow can be generated from a single node without need for the user to calculate any coordinates. There is the ability to rotate and offset and do scaling factors in any direction.

The user can input nodes in any convenient fashion. Then a minimization routine will renumber the nodes for calculation purposes to shrink the size of the calculation that will be needed. If the structure should prove unstable, or for any other reason changes are desired, the package provides a cross-reference list that allows the user to find the node location of any particular degree of freedom.

The package also checks to see that the model the user inputs can run. It will detect user input errors such as forgetting to specify a needed material property or creating a node and not connecting it to anything.

Split-screen graphics

The original geometry in complete form, or with nodes only or elements only, can be displayed. A zoom capability can focus the display on specific areas of complex models.

The 3D package has a split-screen capability. As the data for the model are entered in response to prompts on one side of the screen, the user sees a graphic representation of the model growing on the other side.

In static analyses, plots can be displayed of the deflected shape. For dynamic analyses, mode shapes can be plotted and animated.

The data output repeats the input geometry, stiffness information, and load cases. Output static analysis data include deflections, global loads, local loads, beam stresses, and restraint reactions. Dynamic results include natural frequencies and mode shapes, along with participation factors for seismic analysis.

For the 3D package there are translation files that allow the user to build a model on a personal computer, then run it on larger machines that run Nastran, Ansys, Strudel, and Stardyne.

Both packages are menu driven, so the user has virtually nothing to memorize in the way of commands. Pressing the H key will display HELP instructions when users are puzzled.

According to the vendor, someone who understands the basics of finite element analysis modeling techniques can be running problems on these packages within an hour.

Program updates are provided free of charge for six months. After that, users can get an annual maintenance agreement. Support is by telephone. Often problems that come in over the phone are run by the vendor to check so that proper advice can be given to the user.

MSC/PAL: STRESS AND VIBRATION ANALYSIS

MSC/pal performs both static and dynamic analyses on structures that can be modeled by beams, plates, and associated elements. Among its features is the ability to handle tapered beams. Several preprocessors are included to ease the input of some common structural problems. The package comes with a comprehensive manual and a dozen assorted sample problems. An updated version of this package, called MSC PAL 2, is reviewed in the next chapter.

MSC/pal sells for $995. The package will run on the IBM/PC with two disks and a color graphics monitor or on a Compaq computer. The analysis (but not necessarily the graphics) will run on virtually any IBM/PC compatible machine. With an 8087 math chip the package is said to run about five times as fast, so it is highly recommended. The package is available from The MacNeal-Schwendler Corp., 815 Colorado Blvd., Los Angeles, CA 90041. (213) 258-9111.

Up to 300 nodes can be used to define the geometry of an MSC/pal problem. In static analysis, depending on the problem, there can be up to 1,800 degrees of freedom overall, but for most structures, the allowable overall degrees of freedom will be substantially lower. In dynamic analysis, 250 flexible degrees of freedom are allowed. Those have to be reduced by the package to 125 dynamic degrees of freedom for normal modes and transients. For frequency response, the package is limited to 225 flexible degrees of freedom that are then reduced to 100 dynamic degrees of freedom.

Static and dynamic elements

Beam elements can be pin-ended rods whose neutral axes do not have to be coincident with the grid-point locations of the structure. Constant cross-section beams can be defined by their cross-sectional areas. Tapered rectangular beams also can be defined, with a different cross section at one end than at the other. And pipe-type beam elements can be defined by their inside and outside diameters. If the user defines spring constants at one end and the connectivity of the element, the package will generate an equivalent beam. A triangular plate element can act either as a discrete (Kirchoff) triangle or as a membrane element.

For dynamic analysis there is a damping element that can represent three-dimensional viscous damping. The same element also can be used for single-axis damping. A link element functions as a rigid link with six scalar springs at each end. A single-axis spring element for lumped parameter vibration studies can be extended to act as a 3D stiffness element.

All loads are applied to nodes. There is no provision for distributed or gravity loads. As many load cases as desired can be set up.

Users have to know the material properties of the structure. Isotropic material is handled. If the user provides E and G or E and Nu, the program will calculate the third value. The user must also provide a mass density value for dynamic analysis problems.

Preprocessors and graphics

Specialized preprocessors are provided to help the user input data for transverse beam bending problems, torsional shaft systems, four types of roof trusses, planar frame problems, and three different types of plate analyses.

The package has a built-in capability that allows similar nodes and elements to be replicated in a single dimension, but there is no mesh generation and no substructuring capability.

MSC/pal is run basically from commands that the user must know and use. However, there is some interactive input connected with the specialized preprocessors.

Input geometric data can be given either in cylindrical coordinates or in rotated rectangular coordinates as well as in the basic rectangular coordinates. The package does not provide internal checking on the correctness of the input data.

The package displays the input geometry, allowing the user to rotate and examine it. It also displays the deformed geometry, and most appropriate for a dynamic analysis package, it also provides animated views.

An MSC/pal input file can be translated into an MSC Nastran file to be run on a larger computer.

The MSC/pal manual not only covers installation and operation of the software, it also explains the method of analysis used and the options available to the user. An applications section of the manual goes through 12 sample problems step by step and compares classical analysis results to those of finite element analysis.

SAP86: MAINFRAME-STYLE ANALYSIS

SAP86 does both static and dynamic analyses of 2D and 3D structures and systems. It is a personal computer version of the SAP IV software used on mainframe computers. Files can be uploaded and downloaded from mainframes that use SAP IV.

SAP86 data input is not interactive (an interactive version is planned). Users prepare the input in batches, mainframe computer style. It has no built-in graphics, requiring the use of an external CAD package for graphics display.

The package sells for $995. It runs on the IBM/PC and compatibles. Required equipment is 384K main memory, the 8087 math chip, two floppy drives or one floppy with a hard disk. SAP86 is available from Number Cruncher Microsystems Inc., 1455 Hayes St., San Francisco, CA 94117. (415) 922-9635.

Elements for use with SAP86 include 3D beam and truss, thin shell or thin plate, 3D solid, 3D pipe, and 3D thick-shell, 2D plane stress and plane strain, 2D membrane, and 2D axisymmetric. There is also a boundary element.

Groups of elements can be combined in only limited ways, due to floppy disk storage limitations. One grouping includes 3D beam and truss, 2D elements, the thin shell and plate elements plus the boundary element. Another group that can be combined is the 3D solid, thick shell, and the boundary element. A third grouping is the 3D beam and truss, pipe elements, and the boundary element. Types of loading provided by the package include pressure, gravity, and thermal loads in addition to concentrated loads.

The limit on nodes is 600–800, depending on the type of problem being run. Elements, degrees of freedom, and load cases are limited only by the available storage.

Dynamic analysis capabilities include computation of modes, frequencies, and a response spectrum. There is also computation of time histories by modal superposition or by direct integration.

Static analysis of a thin shell structure with 55 nodes and 52 elements was run by the vendor in about 3 3/4 minutes. Mode and frequency analysis of a plane frame composed of 3D beams with 110 nodes and 189 elements ran in just over 23 minutes.

Batch input files

The package uses batch input. The user prepares a data file, which is then read by the package. Along with geometry information, the package requires input of fairly standard properties. For some elements, the stress matrix has to be input. Generally, though, what are required are Poisson's and Young's moduli and the shear modulus. For beams, the inertias and cross-sectional areas are required.

SAP86 provides node and element generation. The user specifies desired beginning and end points, and the package fills in between. There is no substructuring capability.

There is a check mode that verifies that needed node-point data have been entered and that the limits of the program have not been exceeded, but no internal checking to see that entered values are within reasonable ranges.

External packages for graphics

SAP86 has no graphical display capabilities. These are provided by an interface to the MicroCAD and AutoCAD graphics packages. Input and deformed structures can be displayed. There is no animation capability.

Outputs include stresses, stress histories, displacements, nodal forces, moments, mode shapes, and frequencies. Time history response can be obtained using modal superposition or with direct integration of the equations of motion. The package will also do response spectrum analysis.

The file input is identical to that used in the public domain SAP IV code. There is no communications facility in the SAP86 package, but users can upload and download files from larger machines that use SAP IV. Files can be downloaded and run on the personal computer, or larger problems can be uploaded to run more rapidly on the larger machine.

If the user has a general knowledge of finite element analysis, it will probably take about two weeks to begin to run simple problems. The package has no interactive capability and therefore no built-in teaching facility.

Telephone consulting is provided to get users started and to help solve any problems with the software that may arise but not for continuing consultation on finite element analysis. Minor updates are provided at the cost of postage and floppy disks.

SUPERSAP: PLENTY OF ELEMENTS AND PREPROCESSORS

The SUPERSAP package solves both static and dynamic finite element problems using a disk-based solution method. It makes a wide variety of elements available to the user.

There are ten preprocessors, each with its own mesh generation facility, and structures prepared with these can be combined to form many different model shapes. Input data can be stored for convenient use. Output data can be flexibly combined.

The complete SUPERSAP package sells for $1,500. Users can purchase the processor unit only, without preprocessing, graphics or postprocessing, for $850. It runs on the IBM/PC with 640K main memory, 8087 coprocessor, and a 10-million-byte disk. It also runs on the Pixel computer. SU-

PERSAP is available from Algor Interactive Systems, Essex House, Essex Square, Pittsburgh, PA 15206. (412) 661-2100.

Using a 5-million-byte disk memory, SUPERSAP can handle problems with 2,000 and more degrees of freedom. A beam problem with 800 degrees of freedom, containing about 350 elements, and a 3D problem with about 400 nodes were run by the vendor; each took about two hours to reach a solution. A small truss problem with only 12 degrees of freedom ran in about two minutes.

A large variety of element types are available, including 3D beam, pipe, and truss (with two nodes), membrane, plate on shell (with 3–4 nodes), brick (with 8–21 nodes), boundary, rigid link, and direct stiffness elements, 2D axisymmetric, and plane strain and stress elements (with 3–8 nodes). Also available are 2D solid and 3D brick heat transfer elements.

Beam area properties are defined by an area that resists axial force, two areas that resist shear deflection, a constant to represent the torsional moment of inertia, and two moments of inertia to represent the two directions of flexure. Beam orientation can be arbitrarily assigned by the user.

With most of the elements, orthotropic material behavior can be represented; the truss, beam, pipe, and brick elements are isotropic.

Loads can be point, pressure, constant acceleration, critical buckling, centrifugal, weight, center of gravity, or thermal. Number of load cases is limited by available disk memory.

Material property inputs for the elements include Young's modulus, Poisson's ratio, weight density, and mass density. Truss and beam elements require cross-section area properties.

Comprehensive preprocessors

Ten different modeling preprocessors are provided. For example, the RADGEN preprocessor produces shapes such as pressure vessels and containers. Models can be constructed in parts by these preprocessors, then manipulated to match one another and be joined together to form more complex models. The concept is to provide enough of these preprocessors so that they can serve for models of all the structural and mechanical problems the package designers could envision.

An Edit feature in the package is used to construct and edit the models. Data entry is on an interactive basis. Once entered,the data for an element or a configuration can be recalled from storage, modified, and reused as needed. For example, stored data relating to elements include material properties, area properties, acceleration constants, and base load constants, as well as the element data.

There are a number of built-in data checks. The package will, for example, refuse to accept negative values for area or for Young's modulus. Both logical and typographical errors are checked.

Graphics and postprocessing

Graphic display includes both input and deformed models. The user enters and sets up the model and then can view it to verify the setup process. The user can then make spot checks at various points to further verify its correctness. To correlate the graphic display with the data, node and element numbers can be shown on the graphics. No animation is provided; that capability is planned for future versions.

Postprocessing provisions allow the user to create various stress, deflection, and load case combinations. These results can then be displayed graphically.

Problems can be transferred to and from larger computers, such as DEC VAX and Prime machines, that run versions of the SUPERSAP package.

Users experienced with other finite element packages will likely be able to use this one with just a few days of practice. With interactive data entry and selection, the user does not have to memorize commands, and this simplifies the learning process.

User questions are handled by telephone. An update disk is supplied every six months at added cost.

Chapter 7
Finite Element Analysis Packages II

ANSYS-PC/LINEAR, MSC PAL 2, and LIBRA are three recently released packages that represent the current state of the art in finite element software for personal computers. They can tell us a good deal about both the achievements and the immaturity of this type of software.

ANSYS-PC/LINEAR is essentially a personal computer version of a traditional command-operated finite element language. It is a large package that does static and modal analyses for finite element models whose size is limited by computer memory capacity and the willingness of users to wait through lengthy solutions. Equations that represent the models are solved by a wave-front procedure that processes elements one after another. Very large models can be handled, but they must be constructed so they are not too wide to pass through the waveform processor.

ANSYS users must make at least 10 million bytes of hard disk storage available for the package on an IBM/PC/XT, and twice that amount on an IBM/PC/AT. The package will not run unless a math chip (8087 or 80287) is installed and the computer is equipped with at least 512K bytes of main (RAM) memory.

MSC PAL 2 is a finite element language with some limited user-oriented interaction features. It performs static, modal, transient, and frequency response analyses. Models can have up to 1,000 nodes and no more than 2,000 degrees of freedom. The limit is based on the fact that the matrix coefficients used in the computation must fit into the computer main memory.

For PAL 2, users must make at least 5 million bytes of hard disk storage available on their personal computers. Like the ANSYS package, this one will not run without a math chip, and 512K is the minimum main memory size.

The LIBRA Structural Analysis package is thoughtfully designed for user convenience. The package can handle finite element models with up to 4,000 degrees of freedom, performing static analyses only. A hard disk is needed to handle larger problems and, like the other two packages, this one requires at least 512K main memory. Use of a math chip is optional.

User-oriented operation

The authors of LIBRA have obviously asked themselves how they could draw upon the capabilities of a personal computer to provide a simple-to-use finite element analysis procedure. What they came up with was a package that presents a coherent set of displayed options through which users can develop a finite element model and then perform an analysis of that model, a procedure aimed at ease of learning and use. In keeping with this user-oriented approach, the LIBRA manual gives a clear step-by-step explanation of how to use the package.

All this is in marked contrast with the other two packages. ANSYS and PAL 2 are powerful, able to handle dynamic analysis and large models, but they do not take sufficient advantage of the user interface capabilities of the personal computer. Instead, users are constantly drawn into matters that are rightfully the sole concern of the package developers. In these packages users are expected to be familiar with the peculiarities and limitations of each command. Arbitrary file names may have to be memorized. Insufficient use has been made of techniques such as menu selection, query-response sequences, full-screen entry and editing that are now standard for most personal computer packages.

By providing a great number of commands, the package authors apparently feel they have been freed from the responsibility of settling on what they believe to be the best procedures for making use of the packages, then putting together organized, easily followable ways of using them.

ANSYS input procedures are carried out with the help of about 130 different commands. There seems to be a different command for just about everything. Users must turn to another set of commands to select appropriate results, sort the output data, produce printouts and plots, and to perform other related operations. In all, there are about 80 such output-related commands.

MSC PAL 2 gets along on a smaller number of commands, providing the user with some 48 commands for model creation. Its authors have made a start in preparing a more user-oriented interface. For a few specific simple structures, interactive model preparation routines are available. Models are loaded and set up for solution by the use of some 20 additional commands. Transient response analysis is controlled by 12 more commands. The package also provides interactive keyboard entry procedures that can be used to set up both static and dynamic solutions.

Defining model geometry

Keyboard input for ANSYS conforms to card-punching, data processing approaches. Thus, if users choose free-format input, characters are keyed into column positions 1–80, with commas separating the data fields. Fields each contain exactly 16 characters.

Local coordinate systems, used for generating and locating nodes, can be Cartesian, cylindrical, or spherical (these can be circular or elliptical), or there can be toroidal coordinates. A separate coordinate system associated with each element is used to input directions for orthotropic materials, applied pressures, and stress output.

Users can locate the nodes for a model by giving the right command, then keying in individual node locations and numbers. However, for a model of any significant size, this will be an excessively lengthy process. ANSYS-PC/LINEAR therefore provides commands for automatically filling in between any two nodes. Nodes along quadratic curves can be generated through three given points. Lines and curves made up of nodes can, in turn, be used to generate surfaces filled with nodes. Symmetry reflection also can be used to speed line and surface generation processes. Sets of nodes can be transferred from one coordinate system to another. There also is an automatic feature for defining nodes at surface intersections. Each of these procedures is initiated and specified by coded command entries from the keyboard. Once nodes are in place, there are another set of command-operated procedures for specifying elements and putting them in place.

MSC PAL 2 model generation proceeds through commands that allow the user to set up nodes in space, arranged so as to take the physical shape of the object that is to be analyzed. The positions of nodes can be defined one at a time. They can also be generated wholesale, along lines or curves or across figures, defined by other nodes. Six different commands control this definition process, based on rectangular, polar, cylindrical, or spherical coordinate systems.

Users then specify, with appropriate commands, the types of finite element they wish to use to give substance to the cloud of nodal dots they have created and the material properties these elements are to have. To position elements on the model, attach them to the nodes and define boundary conditions, additional commands are given.

Alternatively, some specific simple types of PAL 2 models can be set up by interactive means. Users are presented with a sequence of screen-displayed prompts that ask for dimensions, material properties, and other characteristics of the structure. There are routines of this type for simple beams, for stepped shafts, trusses, and frames, for rectangular, circular and beam-stiffened plates, and for singly-capped cylinders.

A similar, but more general, interactive approach is used in LIBRA. Users can set up their models one element at a time by first selecting the desired element from a displayed list and specifying the material to be used from another list. The nodes to which the element is to be attached are then assigned numbers and coordinate locations by responding to screen-displayed queries. When many elements of the same type are to be used, there is a mesh generation procedure that also operates from screen-displayed queries through which the user describes the area to be covered, desired node spacing, and other needed parameters.

No concurrent graphics

It might be nice if node and element placement could be done while the user viewed a graphical representation of the object actually being created, but none of the three packages provides this kind of viewing. Instead, users must go through geometry-description processes guided by a hand sketch of what is desired, hoping that what they are doing corresponds to that sketch.

PAL 2 can produce graphical displays that show model geometry, and node and element numbers. These can be examined before solutions are run to verify that the geometry is as intended. Once displayed, the models can be translated, rotated, scaled up or down, and deliberately distorted for better viewing. Model outlines also can be viewed; the package draws all lines connected by an odd number of elements, all other lines are omitted.

Shapes deformed by static or dynamic loading can be displayed. If desired, they can be superimposed on the undeformed versions. Animated displays of deformed shapes from static and normal mode analyses also can be shown. Users can set the animation speed.

To portray static analysis results, PAL 2 can display plots of element stress or nodal displacement. For transient response analysis, there can be plots of node displacement and node reaction force versus frequency or of element stress. For quadrilateral and triangular plates, stress, rotation, and displacement, contours can be plotted. These are available for static and normal mode response analyses. These plotting routines are controlled by screen-displayed menus from which choices are made by pressing Function keys.

Structures and objects created by ANSYS-PC/LINEAR input commands also can be displayed in graphical form. Plotting commands allow users to locate and magnify displays, to create perspective displays, hidden-line plots, deliberately distorted plots, and plots with areas filled in by color. If desired, element outlines can be separated slightly from one another for better visual definition.

ANSYS also can construct contour plots, based on output data parameters such as displacement, stresses, or safety factors, and input temperatures.

At the time of this review the graphics display features of LIBRA were not yet fully available. However, LIBRA does allow both initial and loaded-deformed models to be viewed. Display options are exercised from a separate graphics menu. The choices for initial model geometry include showing node numbers, element numbers, displaying all lines or only edge lines of the model, and using dashed or solid lines for the display. Displayed models can be enlarged or reduced along coordinate axes. Users also can choose the viewing point for the display, and they can set zoom and rotation factors. Planned, but not yet available, is a stress contour plotting capability.

A choice of elements

Despite the differences between the three packages, all offer about the same number of elements for model construction.

ANSYS gives the user a choice of 13 different element types. Spar elements are used to model trusses, links, springs, cables, and so forth. They are available in both 2D and 3D versions. Shell elements include an axisymmetric conical shell and a quadrilateral shell. Beam elements include 2D and 3D elastic beams and 2D and 3D tapered unsymmetrical beams. A 2D isoparametric solid element is used to model solid structures in two dimensions. There also is a 3D version of this element. There is a generalized mass element, and an element whose elastic kinematic response can be defined by stiffness or mass coefficients. There also is a spring element with longitudinal and torsional capability.

MSC PAL 2 provides about 14 elements in all. Included are constant and variable cross-section beams, circular tube beams, and curved beams. A rigid link connects through springs. Users have a choice of quadrilateral or triangular plate elements. For lumped parameter vibration studies, there is a spring element, and a viscous damper element. Support stiffness can be represented by a 3D element.

LIBRA elements include a 2D and a 3D spring, a 3D beam, two triangular shell elements, a quadrilateral shell element, two solid 3D elements, and an axisymmetric shell element. In addition, two triangular and two quadrilateral elements can be used for plane stress and strain and for axisymmetric stress analysis.

Variety in loading

When the elements for the model have been satisfactorily specified, it is time to define the loads that will act on it. PAL 2 seems to offer the greatest variety of loading arrangements.

PAL 2 loads can be applied in the form of enforced displacements of a single node, specified individual nodes, series of nodes, or arrays of nodes. Forces and moments can be applied to nodes in similar ways. Loading can be applied on a per-unit-length basis to a line of nodes or on a per-area basis to a sequence of nodes. Pressure loads can be applied to a specified group of nodes, to multiple sets of nodes, or to a bounded field of nodes. There also can be centripetal acceleration loading around a specified point of rotation, and there is a command that defines forces due to a steady-state acceleration such as gravity. For dynamic analysis, models can be loaded by enforced accelerations of a single node, specified individual nodes, series of nodes, or arrays of nodes.

ANSYS-PC/LINEAR loads may be displacements, forces, pressures, accelerations, or temperatures. These loads all can be defined as acting at specified nodes. In the case of temperatures and pressures, they can be defined alternatively as acting on elements. All elements with mass can have body force loading such as gravity or spinning.

Loading of LIBRA models is accomplished through a tabular format display that has six degree-of-freedom columns. Each row represents one node, and a separate load can be entered for each degree of freedom at each node. Global "body" forces can be applied to the entire model, with *x, y, z* directional and rotational components. Users can call for calculation of reaction loads.

Processing speed

With any one of these packages it will take a substantial amount of time, tens of minutes, to run sizable models. For many users, such long running times will not be a critical factor since the user can start the computation then walk away and attend to other matters. Unlike a large computer, which can hold up many users when it is tied up with one task, a personal computer is usually at the disposal of a single user, and that user generally is free to make operational arrangements that meet personal needs.

The PAL 2 vendor supplied some timing figures for 3D models of a plate with edges clamped all around. A 10-by-18 (180) node model used 97 percent of memory. This model took 10 minutes of computation to form a global stiffness matrix. The first load subcase ran in 4 minutes, 20 seconds. Subsequent load subcases ran 35 seconds each. A second model with 10-

by-10 nodes used just 47 percent of available memory. The running times were 5 minutes, 30 seconds for the global stiffness matrix, 1 minute, 55 seconds for the first load subcase, and 20 seconds each for subsequent subcases.

To give a picture of ANSYS-PA/LINEAR processing speed, the vendor supplied running times for a connecting rod, described by 242 three-dimensional elements. This model had a maximum computation wave-front size of 159 degrees of freedom and a static analysis ran for about 41 minutes. Another model, made up of 288 thin-shell elements, had a maximum computation wave-front size of 118 degrees of freedom. The static solution run for this model took 58 minutes to complete.

Example problems given in the user manual include analyses of a simple 2D building frame, a concrete dam, and a flat circular plate with a hole in the center. There is a seismic analysis of a simply supported beam and one of a *u*-shaped bar that employs 64 3D solid elements.

LIBRA running times supplied by the vendor included those for a support bracket with 810 nodes and 1,620 degrees of freedom. On an IBM/AT with a math coprocessor, the static analysis for the bracket ran in 25.55 minutes. A pipe elbow with 343 nodes and 2,058 degrees of freedom took 40 minutes. A somewhat larger pipe elbow problem with 553 nodes and 3,318 degrees of freedom ran in 65 minutes. Running times for the same problems using an IBM/XT with a math coprocessor were close to three times as long.

Larger models and computers

For static models with more than 10,000 degrees of freedom, the ANSYS package uses the Guyan reduction procedure to reduce the number of unknowns before running a modal analysis. This procedure involves some approximations that affect the accuracy of the results. Users are given the choice of having the package make the necessary choices about which parts of the model to retain or of making these decisions partly or completely themselves.

Users who wish to run models, created with MSC PAL 2, on larger computers can turn to a routine that converts MSC PAL 2 files into similar files that can be used directly by the MSC/NASTRAN package, which runs on mainframes and minicomputers.

Optimizing the calculations

MSC PAL 2 is able to rearrange node numbering to allow solutions to run faster and use less memory. The capability is called bandwidth minimization. This minimization procedure is built in and automatic. LIBRA also has a bandwidth minimization routine, but it is optional and users have to specifically call for its use.

ANSYS uses a somewhat different computation approach than the other two packages. Its solution speed and memory use are affected by the order in which elements are arranged. ANSYS does not provide any built-in computation to optimize that order. Instead, it gives users a set of commands they can use to do it themselves. The manual discusses some of the considerations users should take into account in this process, leaving the rest up to user experience and experimentation.

Dynamic analysis

One of the strong points of the MSC PAL 2 package is the variety of dynamic analysis procedures it can perform. An interactive procedure to set up a dynamic analysis presents the user with a sequence of 66 messages and prompts designed to select the items that will appear in the output results.

PAL 2 frequency response analysis calculates the steady-state responses of the model to sinusoidal forces. Here the user sets the range of frequencies to be used and the phase angles for the forces. Proportional damping also can be specified. For transient response analysis, the package first does a normal mode analysis, then uses the results to find the response of the model to time-varying forces such as earthquakes or wind gusts.

ANSYS-PC/LINEAR can perform a simple modal analysis that produces natural frequencies and mode shapes. Alternatively, modal analysis may be performed with loading by a seismic or force spectrum or by a power spectral density.

The LIBRA package is not currently equipped to perform dynamic analysis.

ANSYS-PC/LINEAR

ANSYS-PC/LINEAR is a computer language for finite element analysis. It uses about 130 different commands and is designed to find linear static and modal solutions. Models can be very large, provided they are of the right shape.

ANSYS-PC/LINEAR runs on IBM/XT/AT computers. It leases for $300/month from Swanson Analysis Systems, Johnson Rd., PO Box 65, Houston, PA 15342. (412) 746-3304.

The general procedure with this package is first to define a model of a structure or object by locating nodes. The resulting geometry is then covered by suitable two- or three-dimensional elements. Loading is applied to the model. The desired solutions are run, and the results can then be modified to produce added information.

The package uses a computation method that can accommodate very large models. The number of nodes are limited only by the available size of computer main memory. Equations that represent the models are solved by a wave-front procedure that processes elements one after another. Models must be constructed so that the degrees of freedom handled by this procedure at any moment never exceed 288 for static analysis or 200 degrees of freedom for modal analysis.

In other words, very large models can be handled by ANSYS-PC/LINEAR. However, they must be constructed so they will be able to pass through the waveform processor. Larger models will take longer, perhaps much longer, to solve than smaller models.

The vendor cited an example of a connecting rod described by 242 three-dimensional elements. This model had a maximum computation wave-front size of 159 degrees of freedom, and a static analysis ran for about 41 minutes. Another model, made up of 288 thin-shell elements, had a maximum computation wave-front size of 118 degrees of freedom. The static solution run for this model took 58 minutes to complete.

Example problems given in the user manual include analyses of a simple 2D building frame, a concrete dam, and a flat circular plate with a hole in the center. There is a seismic analysis of a simply supported beam and one of a *u*-shaped bar that employs 64 3D solid elements.

ANSYS-PC/LINEAR software takes up about 3 million bytes of disk memory. Files created in the course of running the package will take up additional storage space, so users must have at least 10 million bytes of disk storage available for an IBM/PC/XT and at least 20 million for an IBM/PC/AT. The rest of the system must also be well equipped. The package will not run unless a math chip (8087 or 80287) is installed in the computer and the processor has at least 512K bytes of main (RAM) memory. For graphics, a medium-resolution color monitor also is required.

A command for everything

Input procedures are carried out with the help of about 130 different commands. There seems to be a different command for just about everything. For example, the title is assigned by giving the /TITLE command, the desired analysis type by giving the KAN command. Element types are defined by the ET command. If parameters for stiffness types are to be included, the KEYOPT command is used. Currently defined element types can be listed by the ETLIST command, and types can be deleted from the list with the ETDEL command.

Keyboard input for this package follows the venerable card-punching, data processing spirit. Users can choose "free-format" input, where characters are keyed into column positions 1–80, with commas separating the data fields. Fields each contain exactly 16 characters. The package also allows "nonrestrictive" data input. This means that the rigid requirements of formal FORTRAN computer language data input are relaxed so that users do not have to distinguish between entry of real and integer numbers, can use scientific exponent notation quite freely, and so forth.

Material properties

ANSYS PC/LINEAR users can enter material properties as constants, as polynomial functions of temperature, or in tabular form, where property data points and corresponding temperature points are specified. Properties used in connection with the various elements include elastic modulus, coefficient of thermal expansion, Poisson's ratio, mass density, and shear modulus.

Four coordinate systems

One of the data entry tasks faced by the user is to describe the geometry of the finite element model. A first step is to set up coordinate systems. Although all node points are ultimately defined on an overall Global Cartesian coordinate system, ANSYS-PC/LINEAR makes use of three more specialized sets of coordinate systems as well.

There are local coordinate systems used for generating and locating nodes, and here the coordinates can be Cartesian, cylindrical, or spherical (these can be circular or elliptical), or there can be toroidal coordinates. These coordinate systems can be rotated on command.

For input and output of diaplacement and force components, nodal coupling, and constraint equation definitions, nodal coordinate systems are used.

An element coordinate system associated with each element is used to input directions for orthotropic materials, applied pressures, and stress output.

Input aids

Recognizing that a considerable amount of repetitive material may often be input, the package provides for automatic repetition of commands. This allows an input command sequence to be generated from a single command. Users also can set parameters that allow items, such as frequently used constants, to be entered when the user provides the corresponding parameter names.

When material properties data are entered in table format, there is a command that allows the temperature table to be filled by sucessive temperature data items, each incremented by a specified value.

Data checking

An optional Check routine causes the package to examine entered data for inconsistencies such as zero lengths or areas for elements. It also checks for consistent data input and correctness among items such as degrees of freedom sets and boundary conditions. The check run produces warning and error messages that then can be used to edit and correct the input data.

One check made by the package is an examination of the main diagonal terms of the element stiffness matrices. The largest of these terms for each element are compared and, if the ratio of the maximum to minimum of these is too large, a warning message is given.

Unfortunately, the manual does not provide a list of exactly which data characteristics are checked and which are not.

Swarms of dots

Finite element models are made up of various types of elements whose positions in the overall model space are determined by (x, y, z) node locations. Each node is also assigned a unique identification number.

Users can locate the nodes for a model by keying in individual node locations and numbers. However, for a model of any significant size, this will be an excessively lengthy process. ANSYS-PC/LINEAR therefore provides for various types of automatic generation of node coordinates and numbers.

Nodes can be filled in automatically between any two already-specified nodes. Nodes along quadratic curves can be generated through three given points. Lines and curves made up of nodes can, in turn, be used to generate surfaces filled with nodes. Symmetry reflection also can be used to speed line and surface generation processes. Sets of nodes can be transferred from one coordinate system to another. An automatic feature defines nodes at surface intersections.

Element choices

Once the object or structure to be analyzed has been defined by nodes, it must be fleshed out by elements. This package offers the user a choice of 13 different element types.

Spar elements are used to model trusses, links, springs, cables, and so forth. They are available in both 2D and 3D versions.

Shell elements include an assymetric conical shell and a quadrilateral shell. Beam elements include 2D and 3D elastic beams and 2D and 3D tapered unsymmetrical beams.

A 2D isoparametric solid element is used to model solid structures in two dimensions. There also is a 3D version of this element.

A generalized mass element is defined by a single nodal point. It has both concentrated mass components in coordinate directions and rotary inertias about coordinate axes. There also is an element, described by a matrix of constants, whose elastic kinematic response can be defined by stiffness or mass coefficients. This matrix element is assumed to relate to two node points. In addition, there is a spring element with longitudinal and torsional capability.

Where many individual elements must be used to cover a particular object or structure, it can be convenient to automatically replicate the same element many times. ANSYS-PC/LINEAR allows users to create symmetrical patterns of elements that reflect any already-established pattern.

Some of the elements allow higher-order displacement shapes. Thus, linear elements may allow a parabolic deformation along an element edge.

Displaying the model

Structures and objects created by ANSYS-PC/LINEAR input commands can be displayed in graphical form, a step that is essential to guarantee the correctness of most models. Plotting commands allow users to locate and magnify displays, create perspective displays, hidden-line plots, deliberately distorted plots, and plots with areas filled in by color. If desired, element outlines can be separated slightly from one another for better visual definition. Scaling of plots can be done by a combination of automatic and manual procedures.

Loading the model

When the elements for the model have been satisfactorily specified, it is time to define the loads that will act on it. ANSYS-PC/LINEAR can solve only one load case at a time, but the loads may be displacements, forces, pressures, accelerations or temperatures. These loads can all be defined as acting at specified nodes. In the case of temperatures and pressures, they can be defined alternatively as acting on elements. All elements with mass can have body force loading such as gravity or spinning.

Solutions and contours

The output of an ANSYS-PC/LINEAR analysis computation run includes data on rotational and translational displacements and also may include stress and strain data. Shear deflection effects also can be calculated.

Once obtained, computed data can be manipulated by algebraic operations. The manual gives an example where maximum stress safety margins are evaluated.

Contour plots can be constructed by the package, based on output data parameters such as displacement, stresses, safety factors, and input temperatures. The contour plot lines are determined by linear interpolation, within each element, from nodal values.

As with input operations, users must turn to a set of commands to invoke output procedures such as selecting appropriate results, sorting the output data, producing printouts, plots, and performing added calculations. In all, there are about 80 such output-related commands.

Ordering the elements

Solution speed and memory use are affected by the order in which elements are arranged. ANSYS-PC/LINEAR does not provide any built-in computation to optimize that order. Instead, it gives users a set of commands they can use to do it themselves. The manual discusses some of the considerations users should take into account in this process, leaving the rest up to user experience and experimentation.

Large system solutions

For static models with more than 10,000 degrees of freedom, the package uses the Guyan reduction procedure to reduce the number of unknowns before running a modal analysis. This procedure involves some approximations that affect the accuracy of the results. Users are given the choice of having the package make the necessary choices about which parts of the model to retain or of making these decisions partly or completely themselves. Some guidelines are given for choosing the master degrees of freedom that are to be retained, but, here again, the process is evidently one that takes some experience and experimentation.

Modal analysis options

ANSYS-PC/LINEAR can perform a simple modal analysis that produces only natural frequencies and mode shapes. Damping is not taken into account and no stress output is calculated.

Alternatively, modal analysis may be performed with loading by a seismic or force spectrum or by a power spectral density. In this type of analysis, the mode shapes are converted into actual displacement shapes through the spectrum, and stress output is produced.

The output for seismic loading includes eigenvalues (natural frequencies) and eigenvectors (mode shapes), reduced mass distribution, a table of frequencies versus participation factors, as well as mode coefficient ratios and equivalent masses for each seismic direction. Participation factors also may be modified to achieve a rocking spectrum effect.

Limited user orientation

The authors of the package obviously are conscious of the need to provide interaction and instruction to users. Four different HELP commands are used. The DOCU command lists all the commands curently available to the user. DOCU,ALL provides the same list, with command parameters added. DOCU, NAME gives a more detailed description of a specific command named by the user. Finally, STATUS provides a summary of the current specification settings for the part of the package curently in use.

All this does not, however, change the package from one oriented to expert users to one that can readily accommodate more casual or occasional users. This would require an overall package organization that was more specifically aimed at ease of learning and use. Instead, too many elements of the large computer orientation of the parent package, ANSYS, have been retained.

Users are constantly drawn into matters that are rightfully the concern of the package developers. Arbitrary names are used for files and processing routines. A main data input routine is named PREP7, which creates FILE3, whereas solution runs create FILE12.

It is a command-driven package. Users are expected to know which commands to give to achieve the processing they desire and to be familiar with the peculiarities and limitations of each command. No significant use has been made of the many user-oriented techniques such as menu selection, query-response sequences, and full-screen entry and editing that are now standard for most personal computer packages.

By providing a great number of commands, the package authors have been freed from the responsibility of settling on what they feel are the best procedures for making use of the package, then putting together organized, easily followable ways of using it.

The package makes no attempt to manage the use of units for length, force temperature, and other parameters, leaving it up to the user to see that a consistent set of units is used.

MSC PAL 2

MSC PAL 2 is an enhanced version of the MSC/pal package reviewed in Chapter 6.

MSC PAL 2 runs on IBM/PC/XT/AT computers and sells for $1,995 from The MacNeal-Schwendler Corp., 815 Colorado Blvd., Los Angeles, CA 90041. (213) 258-9111.

MSC PAL 2 is a big package. The software itself, including example problems, takes up about two million bytes of disk memory. Files created in the course of running the package will take up another one to three million bytes, so users must have at least five million bytes of disk storage available on their personal computers. The package will not run unless a math chip (8087 or 80287) is installed in the computer and the processor is equipped with at least 512K bytes of memory.

If users want to display MSC PAL 2 models, they must have an IBM color monitor and color graphics plug-in card, or the equivalent.

Bigger models, better facilities

The first job of this package is to help users create computer-stored files that define models of the objects or structures that are to be analyzed. Graphical displays generated from these files can be used to verify the geometry. The files are then put through processing runs to perform desired calculations, first for static, then for dynamic analyses. Such runs are followed by examination of results in numerical and graphical form.

MSC PAL 2 can handle models with up to 1,000 nodes (increased from the 300 nodes MSC/pal can handle) and having no more than 2,000 degrees of freedom. The limit is based on the fact that the matrix coefficients used in the computation must fit into the computer main (RAM) memory. An appendix in the reference manual explains the various parameters that limit the size of the problems that can be handled and gives some examples of large problems and the percent of memory they use.

The vendor supplied some timing figures for 3D models of a plate with edges clamped all around. A 10-by-18 node model used 97 percent of memory. It took 10 minutes of computation to form a global stiffness matrix for this model. The first load subcase ran in 4 minutes, 20 seconds. Subsequent load subcases ran 35 seconds each. A second model with 10-by-10 nodes used just 47 percent of available memory. The running times were 5 minutes, 30 seconds for the global stiffness matrix, 1 minute, 55 seconds for the first load subcase, and 20 seconds each for subsequent subcases.

As with the previous version, MSC PAL 2 is able to rearrange node numbering to allow solutions to run faster and use less memory. The capability is called bandwidth minimization. However, in this present version, the minimization procedure is built in and automatic. Previously, users had to order it performed as a separate calculation.

Compared to its predecessor, MSC PAL 2 puts more element types, load types, and coordinate systems at the users disposal. Replication commands for model creation have been added. Users have more control over output data (they can now control graphics with function keys), and new plotting capabilities are available. The package still has no ability to keep track of engineering units, so users are responsible for keeping the units consistent for the data they work with.

Limited interaction

The MSC PAL 2 package provides the user with a computer language designed specifically for finite element analysis. Users create models by entering some 48 language commands. For simple structures, a set of interactive model preparation routines is available.

The models are loaded through the use of some 20 additional commands and can be set up for solution runs in the same manner. Transient response analysis is controlled by 12 commands. However, the package also provides interactive keyboard entry procedures that can be used to set up solutions.

Models by command

For those who are not familiar with the command-based methods used by MSC PAL 2 and its predecessor, they will be briefly described.

Essentially, the user must set up nodes in space, arranged to take the physical shape of the object that is to be analyzed. The positions of nodes can be defined one at a time. They also can be generated wholesale, along lines or curves, or across figures, defined by other nodes. Six different NODAL POINT LOCATION commands control this definition process, based on rectangular, polar, cylindrical, or spherical coordinate systems.

Users then specify, with an appropriate command, the first kind of finite element they wish to use to give substance to the cloud of nodal dots they have created. For most elements, users also must specify, through a MATERIAL PROPERTIES command, how elements should respond to loading.

To position elements on their object, users employ commands such as ANGULAR ORIENTATION or VECTOR ORIENTATION. Commands such as CONNECT and ATTACH specify the points to which the element will attach. Each element, by its nature, attaches to a specific number of points. These can be nodes or points that are not nodes but that are fixed in the object defined by the nodes. Boundary conditions are defined, where constraints are required, by the ZERO and DISPLACEMENT APPLIED commands.

It might be nice if node and element placement could be done while the user viewed a graphical representation of the object actually being created, but this is not what happens. Instead, users must go through this process guided by a hand sketch of what is desired, hoping that what they are doing corresponds to that sketch.

Available elements

MSC PAL 2 provides about 14 different elements. Included are constant and variable cross-section beams, circular tube beams, and curved beams. A rigid link connects through springs.Users have a choice of quadrilateral or triangular plate elements.

For lumped parameter vibration studies, there is a spring element and a viscous damper element for which users can set the damping rate. A three-dimensional stiffness element can be used to represent the stiffness properties of structures that support the objects under analysis.

If users wish, they can specify the actual stiffness matrix that is to connect two nodes of a model by using the K MATRIX command to enter 36 constants. Such matrices normally are generated by the package when the user selects a particular type of element. A similar M MATRIX can be used to specify the mass at a node. Structural mass also can be specified for rigid bodies or components of such bodies using the MASS command.

Element replication

With the help of the GENERATE CONNECTS and DO CONNECT commands, multiple copies of an element can be attached to cover a defined area or sequence of nodes.

Model-description files can be created with the help of the user's own editing software or with an editing capability called Input that is part of the MSC PAL 2 package.

Interactive model creation

Recognizing that some users at some times may not want to use the MSC PAL 2 language commands to construct finite element models, the authors of this package have provided interactive model-creation routines for a number of simple structures.

There are routines of this type for simple beams, for stepped shafts, trusses, and frames, for rectangular, circular and beam-stiffened plates, and for singly capped cylinders.

Users are presented with a sequence of screen-displayed prompts that ask for dimensions, material properties, and other characteristics of the structure. In most cases, users also are asked about the node configurations they want to use.

The routines produce MSC PAL 2 files that describe appropriate models, and these can be used as input to the load and solution definition portions of the package.

Commands for loading

Loading of the model is defined by setting up a file of load commands. Loads can be applied in the form of enforced displacements of a single node, specified individual nodes, series of nodes, or arrays of nodes using DISPLACEMENTS APPLIED commands. Forces and moments can be applied to nodes, in similar ways, using FORCES AND MOMENTS commands. For dynamic analysis, models can be loaded by enforced accelerations of a single node, specified individual nodes, series of nodes, or arrays of nodes, using the various ACCELERATIONS APPLIED commands.

Loading also can be applied on a per-unit-length basis to a line of nodes or on a per-area basis to a sequence of nodes. Pressure loads can be applied to a specified group of nodes, multiple sets of nodes, or a bounded field of nodes.

There is a command for centripetal acceleration loading around a specified point of rotation. A GRAVITY APPLIED command defines forces due to a steady-state acceleration.

Static, mode, and transient analyses

During the process of creating the model, the ANALYZE command is used to indicate those elements of the model for which internal force and stress analysis is to be performed.

To initiate an analysis run, the user first specifies the file that contains the desired loading information. For static analysis, users have several choices to define the amount of output information that will be displayed, stored, or printed. Runs are initiated by the SOLVE command.

Dynamic analyses require additional set-up commands. Normal mode analysis can be carried out by the QR method, or users can opt for the Jacobi method of eigenvalue extraction. An ORTHOGONAL EIGEN-

VECTOR CHECK sets a threshold for outputting eigenvector cross products.

For transient response analysis, the package first does a normal mode analysis, then uses the results to find the response of the model to time-varying forces such as earthquakes or wind gusts. Commands used here set the time range, define the excitation at each point in time, select the modes that will be used in the transient calculations, and define damping percentages for the selected modes.

Frequency response analysis calculates the steady-state responses of the model to sinusoidal forces. Here the user sets the range of frequencies to be used and the phase angles for the forces. Proportional damping also can be specified.

An interactive procedure to set up a static analysis presents the user with a sequence of 18 messages and prompts designed to select the items that will appear in the output results. A similar but more lengthy interactive procedure, with 66 steps, can be used to specify a dynamic analysis.

Users can obtain more efficient analysis runs by using transformation commands, such as ACTIVATE and ELIMINATE, during the process of defining their models. These transformations use Guyan reduction to approximate mass calculations while keeping stiffness calculations exact. The commands allow users to specify exactly which degrees of freedom are to be eliminated. Strategies to be employed are briefly discussed in the reference manual.

Function keys for plotting

With a routine called View2, graphical displays can be produced that show model geometry, node, and element numbers. Models can be examined before solutions are run to verify that the geometry is as intended. Once displayed, the models can be translated, rotated, scaled up or down, and deliberately distorted for better viewing.

Model outlines, with interior lines removed, also can be viewed. However, this is not the same thing as hidden line removal. The package draws all lines connected by an odd number of elements; all other lines are omitted. This means that a figure such as a cube would be invisible since all edges connect to an even number of elements.

Shapes deformed by static or dynamic loading also can be displayed. If desired, deformed shapes can be displayed superimposed on the undeformed versions. Animated displays of deformed shapes from static and normal mode analyses also can be shown. Users can set the animation speed.

For quadrilateral and triangular plates, stress, rotation, and displacement contours can be plotted. These are available for static and normal mode response analyses.

Using a second graphical routine, Xyplot2, users can call for additional graphical displays. Static analysis and normal mode analysis results can be shown with plots of element stress or nodal displacement. For transient response analysis, the package will produce plots of node displacement and node reaction force versus frequency or of element stress. Curves may be scaled either automatically or manually.

Both plotting routines are controlled by similar screen-displayed menus from which choices are made by pressing function (F) keys, and in both, selection of data to be displayed is made from a main menu by naming the file in which they will be found. The user then goes to additional menus to produce the specific type of desired plot. Printouts of plots, on dot-matrix printers, are available.

For larger computers

Users who wish to run models created with MSC PAL 2 on larger computers can turn to the Adcap2 routine. This converts MSC PAL 2 files into similar files in a format that can be directly used by the MSC/NASTRAN package, which runs on mainframes and minicomputers.

The art of modeling

Recognizing that experience is necessary to model accurately using finite element techniques, the reference manual for MSC PAL 2 includes a section that presents guidelines for element selection, static, and dynamic analyses. Included here are many useful rules of thumb for avoiding computation inaccuracies.

Much of the static analysis discussion has to do with ways of getting around inherent inaccuracies in triangular plate elements. In discussing normal modes analysis, the manual gives guidelines for making correct choices of computation options.

In addition to a user and a reference manual, the package comes with an applications manual that spells out the use of MSC PAL 2 for 21 applications problems. Included here are analyses of beams, plates, cylinders, a mass-damper-spring system, a two-story building, suspended mass, automobile ride and suspension analyses, earthquake analysis, and rocket transient response.

LIBRA STRUCTURAL ANALYSIS

LIBRA can handle finite element models with up to 2,000 degrees of freedom. Users choose from a list of 14 element types and can perform static analyses only. The package is thoughtfully designed for user convenience, but graphical capabilities are incomplete.

LIBRA Structural Analysis runs on IBM/PC/XT/AT computers and sells for $1,495. A graphics postprocessing package sells for $250. The vendor is Intercept Software, 3425 S. Bascom Ave., Campbell, CA 95008. (408) 377-4998.

The authors of LIBRA asked themselves how they could draw on the capabilities of a personal computer to provide a simple-to-use finite element analysis procedure. What they came up with was a package that presents displayed options through which users can develop a finite element model and then perform an analysis of that model.

In keeping with this user-oriented approach, the LIBRA manual gives a clear step-by-step explanation of how to use the package, and it is provided with a suitable index so users can look up needed information quickly.

Sample problems provided with the package include an analysis of a rectangular plate with a hole in the center that is modeled with 199 nodes and 56 quad elements.

The package is essentially operated from screen-displayed menus. The first menu offers users the opportunity to create a new model file. Once such a file is stored in the computer, the analysis features of the package can be set up and run and output produced.

One element at a time

Model creation starts with geometry input. To build a model, piece by piece, the user begins by entering a 1 in response to the prompt ELEMENT NUMBER? The package then displays the 14 possible choices of element types, and the user enters a number corresponding to one of these.

The next prompt asks for a material name, and the user enters a description such as AlloyB56. The package will assume that this choice of material remains the same for subsequent elements unless told otherwise by the user.

Node numbers, for the nodes to which this element will be attached, are then asked for. Different element types each require a particular number of nodes. The package will accept a given node number only once for each model. If the user mistakenly repeats a node number, the package will reject it and ask for another.

Spatial coordinates for the entered nodes are then requested. For each node, the user enters three numbers corresponding to its *x, y, z,* coordinates. The package checks to see that these node locations are appropriate for the type of element being used. For example, nodes for a flat element must lie in the same plane. If not, they are rejected, and the user must enter new coordinates. Depending on the element type being used, there may be a few more geometry questions to answer.

The user then can proceed to choose and enter the next element and the nodes to which it is attached.

Many elements at once

If many elements of the same type are to be used, a mesh generation procedure can be employed. As before, the package asks for the element type and material to be used. Then a prompt asks how many elements are to be lined up for the model. For a two-node element, a beam, or a spring, only one line-up number has to be entered. For elements such as shells that are to be laid out over a planar surface, line-up numbers along two sides have to be given. For solid brick elements, three line-up numbers are required. From these numbers, the package calculates how many elements will be needed.

The user is then prompted to give a number for the first element and a value for the numbering increment from one element to the next. The same information is requested for node numbering.

To place the elements, the package will accept the coordinates of corner nodes that define the outside boundary of the area or volume the generated elements occupy. As an alternative, the user can enter node coordinates to fix the position of the first element.

In addition, the user must specify the desired spacing of the rest of the nodes to which generated elements will be attached. This spacing can be equal from one node to the next or may increase or decrease. As with the one-element-at-a-time procedure, there will be additional geometry questions to answer, depending on the specific type of element involved.

Incomplete graphics

In a package so obviously oriented to user convenience, it is unfortunate that graphical displays of models cannot be viewed as an integral part of the model creation process. Instead LIBRA graphics are handled as a separate optional feature.

At the time of this review, the graphics display features of LIBRA were not yet fully available. However, LIBRA does allow both initial and loaded-deformed models to be viewed. To get the displays, a Hercules plug-in graphics board is added to the computer.

Display options are exercised from a separate graphics menu. The choices for initial model geometry include showing node numbers, element numbers, displaying all lines or only edge lines of the model, and using dashed or solid lines for the display. Displayed models can be enlarged or reduced along coordinate axes through user choice of "aspect ratio factors." Managing the distortions introduced by changing these factors requires some experimentation by the user. Users also can choose the viewing point for the display, and they can set zoom and rotation factors. Planned, but not yet available, is a stress contour plotting capability.

Fourteen element types

Available elements include a two- and three-dimensional spring, a 3D beam, two triangular shell elements, a quadrilateral shell element, two solid 3D elements, and an axisymmetric shell element. In addition, there are two triangular and two quadrilateral elements that can be used for plane stress and strain and for axisymmetric stress analysis.

Using the material library

The material name, such as AlloyB56, used in element placement procedures refers to a description stored in the LIBRA material library. To locate this description, the package prompts the user to enter the name of the material library file where the specified material is to be found.

If no such description is stored, the user must enter one before analysis of the model can proceed. Through a Material Definition menu, users can store new material descriptions, edit existing descriptions, and search the stored libraries. Menu choices include displaying a list of the names of all materials stored in the libraries and searches of libraries to match a specific material name.

Setting options, conditions, and loading

Before actually running the analysis, users must supply some added information. For example, LIBRA has a bandwidth optimization routine that can reduce calculation time for the analysis, and users can choose to employ this routine. Other available options include a data check on the model file, removal of singularities from the model, and deletion of selected data from the output the package will produce.

Boundaries and constraints for the model are specified on still another menu. Displacements can be prescribed and degrees of freedom fixed for a single boundary node, a set of such nodes, or a number range of such nodes. Degree-of-freedom constraints can be imposed for specified pairs of nodes.

Loading of the model is accomplished through a Solution Data menu. Here loads can be applied to specified nodes. The package displays a tabular format with six degree-of-freedom columns. Each row represents one node and a separate load can be entered for each degree of freedom at each node.

Global body forces also can be applied to the entire model with *x, y, z* directional and rotational components. Users can call for calculation of reaction loads. Multiple, numbered load cases can be set up to be used in solution calculations.

The entered options, conditions, and load cases are stored in files that can be recalled by name and revised with an editing routine. Another editing routine allows stored model geometry files to be edited, adding, renumbering, and deleting elements, moving and renumbering nodes, and changing other parameters.

Tabular output

Analysis is done by a separate program that must be called directly from the computer's operating system. This program prompts the user to name the data file and other files such as those for bandwidth optimization that are to be used. The package then runs the analysis and deposits the results in output files named by the user.

Tabulated output results can include node displacements, element stresses, and reaction loads.

Chapter 8
Simulation Packages

Most personal computer simulation packages seem to expect that users will be willing to train themselves to become simulation programmers. None of the package authors has, as yet, fully used the interactive capabilities of personal computers to provide simulation facilities that can be handled by the occasional user, equipped only with an engineering knowledge of the work at hand.

GPSS and SLAM are simulation languages that have been used on large computers for some time and are now available in personal computer versions. These languages are general purpose in the sense that they can be used to model a wide variety of systems. To model with one of these languages, the user has to juggle perhaps a hundred different functions and routines, understanding many special rules for employing and combining them.

The large computer versions of these languages were written under the assumption that the user would put together a set of input instructions, then enter these into the computer for a run, receiving output material at the end, in an essentially batch-processing environment. For the most part, the personal computer versions of these languages, GPSS/PC, GPSSR/PC, and SLAM II/PC seem to carry on this same assumption. These packages do have some PC-oriented features, but these are add-ons to what continues to be a basic batch-processing approach.

Programmed objects and routes

PCModel is also a simulation language but of a more specialized sort. In the simulations carried out by this package objects are moved through processing stations. Each stop takes a defined length of time, and the whole process follows a defined route. That limits the applications to manufacturing, transportation, and other situations where what is to be simulated can be described by the particular capabilities this package offers. PCModel displays are spectacular, giving the user a moving picture of the objects of interest as they pass from one position to another.

To get PCModel into operation, the user has to write a routing file, with commands drawn from a set of over 40 directives and instructions. As with any computer language, there are formats and syntax rules to learn. One example simulation shown in the manual involves the movement of objects through two testing stations and one adjustment station. Setting up this simulation required the writing of over one hundred lines of statements in the PCModel language. To use this package, one has to become a PCModel programmer and put in the study and practice time needed to attain this status. Although many engineers would find it helpful to run and use one of these models, few are likely to be willing to go through the apprenticeship required to be able to build their own models.

Systems that can be described by formal state and output equations can be simulated with the help of the Psim package.Actally, Psim might be more accurately described as consisting of pieces of a package that the user has to assemble in order to do a simulation. First, two procedures have to be set up to express the particular equations being used. Second, the Psim main program, including these procedures, has to be compiled and linked to other Psim calculation and graphical display routines. Then the user can begin to run the simulation.

Tutsim is a package that makes analog computing available on personal computers. What users of this package have to know is mainly how to put together analog-block models.

PC-oriented features

The GPSS/PC package makes good use of the IBM/PC keyboard. With this package, selected keys take on some new or added functions. For example, a most useful feature is the ability to set up function keys easily so they will produce whatever package operations the user desires. There are also audible signals, various beeps, that tell the user when the package is ready to take input, is finished with its current computation, and when there is a syntax error in the input. The package displays single-character prompts, which cue users on when to enter such items as labels, verbs, and different types of operand fields. The package checks each statement for correct format when it is entered. There is also a HELP facility; by pressing the ? key while typing any part of a statement, the user can get a display of the valid options available for that specific part.

The GPSSR package includes interactive debugging facilities, designed to help the user locate problems while a model is running. The package also has an interactive report mode through which the user can decide what output information is to be displayed.

SLAM II/PC has trace reports, shown during execution, to reveal errors that are connected with individual events in the model.

Large computer compatibility

One of the advantages of having a personal computer package derived from software originally designed for larger computers is that the two may be compatible, so that input from the personal computer can be run on larger machines and vice versa. For example, GPSSR/PC is designed to be a substantial subset of the versions of the GPSS simulation language used on larger computers and is said to be compatible with the descriptions of the GPSS language found in most textbooks.

When it comes to compatibility more can be less. The authors of the GPSS/PC package have gone to great lengths to add features that are not available in large computer versions of GPSS, giving experienced users added flexibility. However, this also means that models set up with other versions of the language cannot simply be read and run by this package.

Graphics displays

Most of the simulation packages have limited, if any, graphics capabilities. Thus, GPSS/PC has a PLOT command that allows users to view selected variables as the simulation proceeds, and the GPSSR/PC package allows users to see a histogram display of any selected variable.

An exception is the PCModel package with its excellent visual depiction of the processing systems it simulates. The objects in these systems are represented as little reverse-video rectangles, each containing a single symbol designating the type of object. These objects move on the screen along user-defined paths and between user-defined graphical symbols. The speed of movement can be adjusted in fine steps by the user to view what is happening or to let the process run at full speed to accumulate statistics on the system. Meanwhile, at the bottom of the screen, a digital process clock runs, and there also are ongoing displays of process counts and statistics.

The art of simulation

To use personal computer-based simulation packages such as GPSS/PC, GPSSR/PC, and SLAM II/PC, those who are not already familiar with the large computer software from which they are derived will need to learn a completely new language. Competent use of the package also requires some background in statistics, needed to validate models and interpret results. However, those will just be first steps for neophytes in the art and practice of computer simulation.

The real challenge is not simply to set up and run models. It is to use these packages so that the models bear a useful resemblence to the real situations that are to be simulated. Above all, users need considerable experience in working with the models before they can handle the kind of verification, validation, and modification needed to come up with meaningful results.

GPSS/PC

GPSS/PC is an adaptation of the GPSS General Purpose Simulation System software that has been in use for several years on larger computers. It processes models at the rate of about 300 simulation block entries per second.

GPSS/PC runs on IBM/PC and compatible computers. It sells for $900 from Minuteman Software, PO Box 171, Stow, MA 01775. (617) 897-5662.

Although GPSS/PC has interactive capabilities not found in the large computer version, it is still essentially a special-purpose computer language. There are 11 general commands to initiate overall package operations, 16 control statements, and 44 statements used to create specific types of simulation blocks.

Users must learn how to program in this language by mastering its commands, statements, and procedures. This includes some special use of the keyboard. Selected keys take on some new or added functions beyond those familiar to IBM/PC users. A feature that can be most useful is the ability to set up function keys easily so they will produce whatever package operations the user desires.

Some of the features of GPSS/PC are additions to what was available in large computer versions of GPSS giving experienced users added flexibility. It also means, however, that models set up with other versions of the language cannot simply be read and run by this package.

Building a model

To build a simulation model, the user writes sequences of GPSS/PC statements in much the same way as statements in BASIC or other computer languages are written. These models, and the processing statements that accompany them, can be used immediately or kept in disk storage for subsequent use.

Statements are made up of parts, called fields, that must be correctly sequenced and formatted. There are statement number, label, verb, operand, and comment fields. There are expression fields in which mathematical and logical expressions can be handled. Some 26 operations and functions can be used in these expressions. There also are over 45 system numerical attributes (SNAs). These "state variables" can be used to control and keep track of the course of a simulation.

Editing a line

The EDIT command allows users to alter a statement called up by its statement number. Keys and key combinations take on special editing functions when this command is given. Like other operations, these functions can be set up as single Function-key actions. There is a separate DELETE command to remove unwanted statements.

Built-in controls and help

GPSS/PC is supplied with a number of mechanisms with which the package spots potential problems and reports them to the user. There are audible signals, various beeps, that tell the user when the package is ready to take input, is finished with its current computation, and when there is a syntax error in the input.

To guide the user in correctly entering statements, the package will display single-character prompts, which cue users on when to enter labels, verbs, different types of operand fields and so forth. Of course, users must learn the meanings of these prompts as part of learning to program in GPSS/PC language. The package checks each statement for correct format when it is entered.

There is a HELP facility. By pressing the ? key while typing any part of a statement, the user can get a display of the valid options available for that specific part.

When the package runs out of memory during a simulation it will notify the user, who then has to know how to take appropriate action and avoid the loss of important data.

Error messages are mainly understandable phrases and those that may be puzzling can be translated by referring to the user manual section where they are explained.

Plotting and reporting

The package has a PLOT command that allows users to view selected variables as the simulation proceeds.

A standard output report can be produced by the REPORT command at the end of a simulation. This displays but does not print the information. To get printed results, users must run a separate GPSSREPT program that comes with the package.

GPSSR/PC

GPSSR/PC is designed to be a substantial subset of the versions of the GPSS simulation language used on larger computers and is said to be compatible with the descriptions of the GPSS language found in most textbooks.

GPSSR/PC runs on IBM/PC and compatible computers. The package can make use of, but does not require, an 8087 math chip. GPSS/PC sells for $750 ($250 to educational customers) from Simulation Software, 760 Headley Dr., London, Ontario, N6H 3V8 Canada. (519) 679-3575.

The users' manual has a nice introductory section that discusses a model of a car wash as an example of how the GPSSR/PC language works and tells how to interpret the formatted printouts the package produces. In the course of this discussion, the reader gets a sense of the difficulty and intricacy involved in setting up a simulation so that it bears some resemblence to the real situation that is supposed to be simulated.

Users of this package, who are not already familiar with the large computer GPSS software from which it is derived, will need to learn a completely new language, but that will just be the first step for those who are neophytes in the practice of computer simulation. Competent use of the package requires some background in statistics, needed to validate models and interpret results. Users also will need considerable experience in working with the models before they can handle the kind of troubleshooting needed to come up with meaningful results.

Fortunately, a number of textbooks are available from which beginners can learn to use the GPSS language and thus be in a position to start to use this package. There is no listing of such textbooks in the package, however.

Setting up a simulation

Users first write a set of GPSSR/PC language statements to describe the model they wish to run. This is done with the user's own text editor or word processing package. The resulting file is then run by the GPSSR/PC package, which includes interactive debugging capabilities. These allow the user to stop a simulation and step through any designated number of blocks, checking the simulation results there. This facility does not seem very easy to use, nor do the results seem simple to interpret. The package itself provides no means to edit the language statements; this must be done by going out to the external text or word editing software.

GPSSR/PC makes use of about 39 block-creation statements, 6 control statements, and 17 definition statements. There also are some 57 different standard numeric attributes (SNAs) that can be used with the package. When variables are defined, about 12 logical and mathematical operations can be used to set up the expressions.

Two versions of the software are included in the package. One uses 16-bit addressing and is designed for use with computers that have 128K bytes of main memory. The other version, for handling larger models, is designed to access up to 1024K bytes of memory. The 128K version is said to be more memory efficient and to run faster.

The package allows users to see a histogram display of any selected variable. GPSSR/PC also allows users to select which statistics will be displayed and what information on the final state of the model will be included in the report the package shows.

PCMODEL

In the simulations carried out by PCModel objects are moved through processing stations. Each stop takes a defined length of time and the whole process follows a defined route. The display screen can show the movement of the objects in simulated time. The simulation typically runs 100 times faster than real time.

PCModel runs on IBM/PC/XT/AT computers. It sells for $450 ($200 to educational customers) from Simulation Systems Software, 2470 Lone Oak Dr., San Jose, CA 95121. (408) 270-2300.

This package is designed to model the movement of manufactured items through an assembly process. It also can be used to model traffic movements, distribution, and similar situations where objects move along routes. Users build a graphical model of the process that will show the motion of the objects in their surroundings. They also write a sequence of flow statements in the PCModel simulation language that will bring the model into action.

In action, the objects are represented as little reverse-video rectangles, each containing a single symbol designating the type of object. The objects move on the screen along user-defined paths and between user-defined graphical symbols. At the bottom of the screen a digital process clock runs, and there also are ongoing displays of process counts and statistics.

The route can be quite complex, involving bypasses, crossovers, and loopbacks, and there can be conditional branching. The package can anticipate collisions of the objects and will automatically avoid these by delaying object movements. A limit to model size is that there cannot be more than 32,000 grid points in the graphics model of the process. Up to 100 jobs with 65,000 objects per job can be defined. Failures can be simulated by inserting blockages in the route.

Statistical data are collected by the package on the processes that are modeled. These include figures on the number of active objects being simulated, the number completed, and utilization statistics at up to 21 points along the route.

Users can control the simulation process by putting the simulation clock into a second-by-second mode or into one where the clock advances immediately to the next time when there is an object movement. The speed of simulation can be varied from maximum to real-time speed in 35 steps, and users also can halt the simulation or single step it one clock cycle at a time.

Models, statistical data, and screens showing the simulations at any desired point can be saved in disk files.

Setting up the routing

Overall operation of the package and control of the simulation display is exercised through a main menu, aided by use of Function keys for such matters as speeding up or slowing down the simulation.

To get the package into operation, the user has to write a routing file, with commands drawn from a set of over 40 directives and instructions. A typical simulation will require a routing file containing 10,000 to 20,000 characters, so a considerable amount of keyboarding is involved. The user must know how to use the commands, and, as with any computer language, there are formats and syntax rules to learn. One example simulation shown in the manual involves the movement of objects through two testing stations and one adjustment station. Setting up this simulation required the writing of over one hundred lines of statements in the PCModel language.

To use this package, one has to become a PCModel programmer and put in the study and practice time needed to attain this status. Although many engineers would find it helpful to run and use one of these models, it seems to me that few are likely to be willing to go through the apprenticeship required to be able to build their own models.

PSIM

Psim provides pieces of a package that computerwise users must put together if they wish to simulate. Only those systems whose operation can be described by state and output equations in required form can be simulated.

Psim runs on Apple II computers equipped with UCSD P-system software. It sells for $35 ($40 overseas) from Prof. A. Galip Ulsoy, College of Engineering, University of Michigan, 2250 G. G. Brown, Ann Arbor, MI 48109. (313) 764-3312.

Users must first set up the necessary equations that describe their system. The state equations must be of the form $dx/dt = f(x, u, t)$, where x and u are vectors of state variables and of input functions, and t is the time variable. The output equations must be of the form $y = g(x, u, t)$. After appropriate algebra to put equations into required forms, the user can turn to the software.

Users are expected to start by setting up two formatted procedures, which express the particular equations being used and which become part of their operative version of Psim. The Psim program must then be compiled and linked to two other programs, Psimpack (calculation routines) and Plotpack (graphical display routines), before the package is actually ready to run. Each time new equations are used, this setup, compilation, and linking procedure must be followed anew. The Psim manual assumes that users are familiar with all the necessary computer procedures and therefore gives no guidance in these matters.

When the operative version of Psim is executed, a menu is displayed that leads users to view or change parameter values or initial conditions, to run the simulation, view the results, and quit the package.

After setting parameter values from the state equations and setting initial conditions, the simulation run option can be taken. Users are then presented with another menu that allows them to set initial and final times and a display interval for the run. After this, the main menu run option is again taken, and the simulation is run.

To see results, users take the main menu display option. They can see or print a tabular display of T, $y(1)$, and $y(2)$ values. If they opt to see a graphical display, the package queries for the variables to be plotted. Then a menu with plotting feature choices is displayed. At this point the manual cops out and advises the user to look through the program listing of the Plotpack routine to find out how to make the appropriate choices.

There are some printing glitches in the plotting feature. Users are warned that getting a printout of a plot will take about 10 minutes. There is a choice of a smaller or larger display format, but a printout of the smaller will be blessed with some missing points.

Returning to the main menu, users can set new parameter values and initial conditions, and then rerun the simulation.

SLAM II/PC

SLAM II/PC is a personal computer adaptation of a simulation language originally designed for larger computers. The PC version of the language includes about 100 different functions and subroutines. Both continuous and discrete simulations are handled. Users have to learn the language and be proficient in the use of PC-DOS software; a knowledge of FORTRAN is also a requirement for serious use.

SLAM II/PC runs on IBM/PC/XT/AT and compatible computers. It sells for $975 ($200 to academic customers) from Pritsker and Associates, 1305 Cumberland Ave., PO Box 2413, West Lafayette, IN 47906. (317) 463-5557.

SLAM II/PC is a competent package for those who are willing to become programmers and learn this specialized programming language. It is open ended in that FORTRAN routines can be incorporated into almost any aspect of its operation. Competent FORTRAN programmers can enlarge the operations of this package, adding any features they wish, and there are many features that could and should be added to take advantage of the possibilities offered by personal computers.

Like many other translators of packages from large computers, the authors of the SLAM II/PC package have made relatively little use of the direct interactive capabilities of personal computers. There is no provision in the package for preparing the input files that describe the models. Users must turn to their own text editing or word processing packages for this. Nor is there any use of menu displays or interactive dialog between the user and computer to enhance the simulation process. No interaction is allowed during simulation itself. All the user can do is submit an input file, run it, and get an output file. If any changes are to be made, the input file is revised (outside the package) and resubmitted for another run, just as if the PC was a mainframe computer.

Using the package can be a time-consuming process. Before any simulation can begin, the user has to load the package, which takes about 15 minutes when floppy disks are used. Then the simulation itself is run, and the time depends on the complexity of the model. The package displays a trace report during execution that informs the user of the current simulation time.

SLAM II/PC has no provisions for storing or printing error messages. This means that it is difficult to interactively edit a model to iron out its imperfections. The trace reports shown during execution reveal errors that are connected with individual events in the model. Unfortunately, these reports move across the display screen too fast to allow the user to absorb what they contain.

Output reports can be displayed, but printing requires that users resort to their own computer systems (DOS) software. The displayed and printed output is where the user has ready access to reports of errors discovered in the input file and to the trace reports. In addition, output reports include statistics on variables and simulation activity.

A big book

There is a 600-page textbook that includes an introduction to simulation and a description of the SLAM II language. However, this textbook is not organized as a users manual, nor is it set up as a tutorial on the language. The result is that it is difficult to learn the language, even for those who set out to do so. Perhaps this creates an opening for tutorial courses on the language.

The package offers no online HELP facilities, no tutorial, no explanatory displays. It comes with some sample models recorded on disk, but these are not discussed in any printed material, nor is there any displayable explanation of their functions.

TUTSIM

Tutsim runs analog simulations on a personal computer. It makes a large number of different analog blocks available to the user and provides for hard copy output on a plotter or dot-matrix printer.

Tutsim runs on IBM/PC and CP/M systems. The package sells for $300 from Applied i, 200 California Ave., Palo Alto, CA 94306. (415) 325-4800

Before the word *computer* became completely taken over by the digital machines that so dominate the field today, there was another kind of computer that seemed, at least for a while, destined to share the future on equal terms. In these *analog* computers, functions such as integration are performed by circuit blocks plugged together to model processes that the engineer wished to simulate.

However, the integral and differential equations that are handled by an analog computer also can be solved digitally. The Tutsim package takes advantage of this capability to use a personal computer to simulate the operation of an analog computer.

The various analog functions, such as pulse formation, attenuation, and integration, are brought into play through Tutsim commands such as PLS, ATT and INT. The package allows the user to set up a number of such analog functions (blocks) and interconnect them into a model of some larger process.

A variety of function blocks are available in the package. These include addition, multiplication, division, cosine, sine, exponent, square root, logarithm, Boolean logic functions, and a block that will generate arbitrary piecewise-linear functions. There also are resistance and conductance blocks as well as blocks for constants, delays, gain, maximum and minimum, and white noise among others.

For each Tutsim block, the user is required to supply appropriate parameters. In the case of a pulse-forming block, for example, the on-time, off-time, and gain of the block must be supplied. Users also must specify the minimum and maximum values that the output of each block can assume as well as the step size and total simulation time for the model.

Each block is specified by a single-line Tutsim instruction. Once a simulation has begun, it may be interrupted at any time the user wishes to change the block or system parameters. Unfortunately, this editing process is laborious because it must be done on a line-by-line basis, rather than allowing the user to view and edit a screenful of simulation instructions.

Compared to the instantaneous response that analog computer users get, this package can grind rather slowly when presented with complex simulations. Simulation results can be preserved by outputting them on a printer or a Houston Instruments HiPlot plotter.

Chapter 9
Lighting Design and Analysis

A number of personal computer packages are available for indoor lighting design and analysis, some for daylighting and others for artifical illumination. An outdoor lighting package is also reviewed in this chapter.

CADLIGHT and DAYLITE are among the packages concerned with available daylight and the effect it can have on indoor lighting. CADLIGHT takes a room-at-a-time approach, allowing users to analyze simple rooms, consisting of no more than two adjacent rectangular spaces. It can take account of a number of specific features such as lightshelves, fins, overhangs, and, of course, fenestration, which can include skylights. The package is designed with the needs of the user as a foremost consideration. Procedures for entering data and viewing results are graphic and very easy to follow. The manual is exceptionally well written and helpful.

As for the results, CADLIGHT can display illuminance values on a plan view of the room. Users can choose the units (footcandles, lux, or daylight factor values) for the display. They also can opt to see only the direct component of illuminance, only the interreflected component, or the total illuminance. If desired, a graph can be displayed showing illuminance along a user-designated section of the workplane, or these results can be displayed in tabular form. Sunlight pattern analysis calculates where direct sunlight strikes the room and finds the window shading cutoff angles.

The DAYLITE package analyzes the effects of daylight on an entire building. Users assemble a building description by fitting together connecting items, such as surfaces, apertures, light shelves and overhangs, or fins, into "forms", such as walls, light wells, or monitors. These forms are then put together to make up the overall structure. The assembled building is checked by viewing 2D and 3D graphical images.

This package has a procedure for analyzing lighting power and control strategies. The user selects a control strategy (step or dimming) for electric lighting. The package calculates a monthly daylight utilization fraction for any standard work year. It also will produce monthly or yearly summaries of the extent to which daylighting alone covers required lighting loads.

Artificial lighting

On the side of artificial lighting, The Lighting Fixture Calculation package has a simple and well-defined task. It finds the number of lighting fixtures required for a given room. The zonal cavity calculation method, as described in the *IES Lighting Handbook,* is used. The package produces an energy economics report that compares initial, annual, four-year, and life-cycle costs of three alternative designs. There also are reports describing alternative fixtures and summaries of the lighting characteristics of the rooms under study with the alternative fixture choices.

Those who need a more detailed analysis can turn to the LUMEN-MICRO package, which calculates horizontal and vertical illuminance for an array of up to 20-by-20 points on a horizontal plane. It also will calculate equivalent spherical illuminance (ESI) for the same points, and an optional module will calculate visual comfort probability (VCP) for those points. An array of exitance and illuminance values can be calculated for any designated room surface. The luminaires involved can be point, line, or area sources. Lighting equipment can be direct, indirect, or a mixture of both.

Users can define their own grids of output data points, with equally or arbitrarily spaced rows and columns and with points at any desired height within the room. Separate, different grids may be defined for vertical illuminance, equivalent spherical illumination, and visual comfort probability results. A similar choice of grids, located on the appropriate surface, are available for calculations of room surface exitances and illuminances.

The LUMEN-MICRO package gives users considerable choice in selecting what will appear on printed results reports. Headings for the reports can be formatted and composed. Report material is available in ten modules, and users can select which of these will be printed.

Cost-benefit analysis

Although both the Lighting Fixture Calculation and LUMEN-MICRO packages take costs into account, economics is a more exclusive concern with the LUMEN$ package, which helps users perform a cost-benefit analysis for alternative indoor lighting systems. Users enter initial costs and system life estimates for up to four alternative lighting systems. They also enter the relevant economic data on such matters as interest rates, the cost of labor, electricity, equipment, and heating fuel.

LUMEN$ produces a cost summary for the alternative systems, a present worth summary, and shows the rate of return and payback period for the alternative systems compared to the first system. Yearly cost summaries also can be produced, comparing the alternatives over the period of the analysis.

Outdoor lighting, including roadway lighting, is addressed by the LUMEN-POINT package. Users can specify up to five analysis grids, having a total of 50-by-50 points, for a single calculation run, for example, one grid might be for a parking lot entrance, another for the main parking area, and a third for an annex parking area. The package prints out the results in tabular format with illuminance values shown at the grid points. A luminaire layout schedule gives location and aiming data.

Contour graphs

Some of the packages make use of two- and even three-dimensional graphics to exhibit their results. In the DAYLITE package, contour graphs are used to show illumination values at chosen points, along constant illumination contour lines, and in areas of direct sunlight. The package also can display an "illumin-net" 3D pattern of illumination contour lines between the floor of a room and a specified height above that floor. The pattern will be shown contained within the building outline.

The LUMEN-MICRO package produces contour plots that show repeated printed characters along lines of constant illuminance. The particular symbol denotes the value of illuminance on that contour. If desired, a contour plot can be printed with all areas above or below a specified value of illuminance covered by shading. The package also will display shaded renderings that use varying dot densities to represent different values of illuminance. High densities (dark areas) denote low illuminances. Users can select which parameters will appear on these graphical displays and also have some control over the use of shading.

Input convenience

Perhaps because of their involvement with shedding light, the authors of these packages have taken some care to give users convenient means of entering the sometimes voluminous data required before results can be produced.

Input data for the CADLIGHT package are entered, one item to a line on screen displayed lists. Users move an entry pointer up and down the lists, by pressing the keyboard arrow keys, to enter numeric values and word descriptions. At each input location, the user can press a Function key to get a display of HELP information on that item. Pressing other keys will move the pointer to the next item, cause a default entry to be inserted, erase individual entries, or restore the screen to the state it was at before the current entries began.

With the DAYLITE package, building items are each described by a few descriptive lines. For example, the descriptive lines for a surface will include width, height, and reflectance entries. Arrow keys are used to move the display cursor to the data positions on the descriptive lines. To edit an existing item, users type over the existing data.

The Lighting Fixture Calculation package uses a similar data entry scheme. Users key data into items on a list, pressing the Return key to move from one item to another. They also can move about the screen with the help of the arrow keys. Editing is done by using the Backspace key to erase characters and the space bar to move from one character to another. As many changes as desired can be made; they will take effect for package operation only after the data screen has been entered by pressing the Esc key.

The LUMEN$, LUMEN-MICRO, and LUMEN-POINT packages are all from the same vendor and all use very similar data entry procedures. Entry is onto a long form with as many as 500 lines. It is displayed, 24 lines at a time, on the computer screen. Various keys are pressed to maneuver the cursor around this display and to scroll or page to the rest of the form. Entries are filled in from the keyboard, and the form can be printed or stored at any stage.

Graphical input

The CADLIGHT and DAYLITE packages make good use of graphics, not just to display results but to guide input procedures. After they fill in a few preliminary information items, CADLIGHT users are presented with a screen-displayed room-layout drawing. This drawing is associated with a series of one-line item entries that specify the outline of the zone. As the user fills in these items (essentially the room dimensions), the displayed drawing will reflect the entered values, thus verifying that the room is taking shape as desired.

The DAYLITE package displays scaled, two-dimensional graphical representations of building rooms, walls, and other features once they have been numerically described. These drawings are reviewed, one by one. If not satisfied with their appearance, the user can return to the numerical entry procedures to make appropriate alterations.

The package also produces three-dimensional graphical displays of the buildings. After the appropriate menu choices, a 3D image of the entire building will be displayed, and the initial display will be followed by a sequence of rotated-position displays, so that all sides of the image can be checked.

Using stored data

Computer packages can include cataloglike data, and this can be an important tool in speeding analysis and design procedures. An obvious choice for lighting packages is to store data on available luminaires, and this is exactly what the LUMEN$ package does. Luminaire data can be entered from the keyboard or from previously stored files maintained by a Lumen-Data routine. The Lighting Fixture Calculation package comes with similar stored data on about 125 different types of fixtures.

Packages that do not deal with luminaires can still make use of prestored data. Thus, the DAYLITE package comes with stored daylight and sunlight availability data. Users choose appropriate climate data files from those available for the specific location of interest.

Built-in controls

Among the important input aids that computer software packages can provide are automatic controls that check for errors and inconsistencies in the data that users enter. One cannot expect to get correct computation results if the input data are faulty.

Unfortunately, the authors of these lighting packages seem to have paid only limited attention to such controls. Built-in CADLIGHT controls consist of screen-displayed warnings when users attempt to change data items, such as horizontal illuminance, that are calculated by the package from other user inputs. For some items, such as the Sky-type, there are only a few acceptable answers, and the package will reject all but these few.

The Lighting Fixture Calculation package has acceptable value ranges stored for some of the input parameters and will display a warning message if these ranges are exceeded. There are also some checks for completeness and correct numbering of input data items.

During input and editing processes, the LUMEN-MICRO package will not permit users to enter alphabetic characters where only numerals should be used. Nor will it allow users to key in an excessive number of characters for any item. Beyond these perfunctory prohibitions, there seem to be few built-in controls on the quality of the input data.

The LUMEN-POINT package has similar controls, and it also has a separate checking procedure that can be invoked, but there is no specific indication of exactly what is checked, nor is there a description of the error messages that result when flaws are found.

CADLIGHT 1

CADLIGHT 1 performs daylighting analysis for rectangular rooms and spaces or for spaces that can be described by two rectangular spaces that intersect along a single vertical wall or plane. Fenestration can be on any walls and can include skylights. Complex wall sections with lightshelves, fins, and overhangs can be included in the analysis.

CADLIGHT 1 runs on IBM/PC/XT computers. It sells for $595 from Wiley Professional Software, 605 Third Ave., New York, NY 10158. (212) 850-6788.

CADLIGHT 1 is designed with the needs of the user as a foremost consideration. Procedures for entering data and viewing results are graphic and easy to use. The manual is well written and helpful.

The package has its limitations; calculations are designed to handle unfurnished spaces and will give less accurate results for furnished spaces. It is recommended by its authors for use in comparing the relative performance of two or more different space design schemes, rather than giving very accurate predictions of actual lighting characteristics.

Full-screen data entry

Input data, to describe a design situation, consist of a Skydome description, plus Zone (room) and Site descriptions. All the entry items for the Skydome are listed, one to a line, on a single-screen display. Users can move an entry pointer up and down the list by pressing the keyboard arrow keys to enter numeric values and word descriptions.

At each input location, the user can press the F1 function key to get a display of HELP information on that item. Pressing the F2 key will cause a default entry supplied by the package to be inserted. When the Return key is pressed, the entry pointer will move to the next item on the list. Individual item entries can be erased by pressing the Esc key, or the user can press the F7 key and restore the screen to the state it was at before the current entries began.

When satisfied with the displayed item entries, the user can put them into effect by pressing another Function key. The information will then be stored in a file named by the user. If desired, the screen contents can be printed out using the IBM/PC PrtSc key.

Stored information can be recalled and edited, using the same full-screen procedures. This is useful not only for correcting errors and making needed revisions but also for keeping data entry to a minimum when similar designs are being analyzed.

Built-in controls include screen-displayed warnings when users attempt to change data items, such as horizontal illuminance, that are calculated by the package from other user inputs. Users can override this type of warning. For items such as Sky-type there are only a few acceptable answers, and the package will reject all but these few.

Visual verification

The procedures for entering Zone and Site information are similar to those for the Skydome information. However, with these data graphical displays are shown to guide the input process.

After filling in four general description items on the first Zone entry screen, the user is presented with a screen-displayed drawing of the Zone (room) layout. This drawing is associated with a series of one-line item entries that specify the outline of the zone. As the user fills in these items (essentially the room dimensions) the displayed drawing will reflect the entered values, thus verifying that the room is taking shape as desired.

A separate group of item entries is used to describe windows that are to be inserted. Complete window (and skylight) descriptions can be stored and invoked, with a minimum number of keystrokes, to fill in identical items. Other items to be entered include surface reflectances and partition openings.

During data entry, three item entries are displayed below the drawing. When these are filled in, the user can reveal further items by pressing the down arrow key. To view the entire list of fill-in items a Function key can be pressed, and the complete list will replace the drawing display. A second press of the same key will bring the drawing display back again.

Entry of Site information is done in a similar manner. The user is guided by a site drawing on which the Zone appears as one of the displayed objects. For each such object, the user fills in line items that set its height above the ground plane, distance from the left and bottom edges of the site plan, and orientation. In addition to the Zone, objects users can place on the site plan include trees, buildings, and special subsurfaces (used to represent objects such as swimming pools and parking lots).

Choosing the results

With the needed input data entered and approved by the user, a calculation run can be initiated. The user selects the desired type of analysis. Simple workplane illuminance analysis calculates the illuminance levels at three points along a line, specified by the user, through the Zone. Sunlight pattern analysis calculates where direct sunlight strikes the Zone (this can be displayed as shaded patches on a Zone plan) and finds the window shading cutoff angles. Surface illuminance, luminous exitance analysis produces data for a group of points on any surface of the Zone.

The calculations may take some time to complete. During processing a status screen keeps the user informed of progress. All results will be stored in a disk file. Users can select, from displayed menus of possible choices, which of these are to be displayed and printed.

For example, simple workplane illuminance values can be displayed on a plan view of the Zone. Users can choose the units (footcandles, lux, or daylight factor values) for the display. They also can opt to see only the direct component of illuminance, only the interreflected component, or the total illuminance. If desired, a graph can be displayed showing illuminance along a user-designated section of the workplane, or these results can be displayed in tabular form.

DAYLITE

DAYLITE is used to analyze the effects of daylight on a given building. Building descriptions are put together by connecting items, such as surfaces and apertures, into "forms", then assembling the forms to make up the overall structure. The structure is checked by viewing 2D and 3D graphical images. Detailed daylighting effects and their impact on electric power budgets can be calculated.

DAYLITE runs on IBM/PC, Apple III, and Lisa computers. It sells for $750 from Solarsoft, 1406 Burlingame Ave., #31, Burlingame, CA 94010. (415) 342-3338.

For input into this package, components that make up a building are divided into classes: surfaces, apertures, light shelves and overhangs, or fins. These classes are then assembled into forms such as walls, light wells, monitors, and these are, in turn, assembled into the building. Class and form items are stored in library files from which they can be retrieved as needed for individual or repeated use.

To enter a class item, the user first calls for a list of stored items, showing their names and numbers. Entering an item number calls up a description of that item with its name and descriptive data. For example, for a surface the descriptive lines will be width, height, reflectance, height of lower reflectance (to account for two-tone walls), and lower surface reflectance.

Each class has its own descriptive lines, with about four to eight lines needed to describe any class item.

Arrow keys are used to move the display cursor to the data positions on the descriptive lines. To edit an already-entered item, users type over the existing data. The item can then be saved under a name designated by the user.

Connecting class items into forms

From a form assembly menu, users can choose the type they desire. In response they get a list of existing forms of that type. Choose a number from that list and a short table of the class items that make up that particular form will be displayed, with one line for each class item.

For a wall form, the table gives the class code, name, and distance-left (distance from the left edge) for each class item. In the list heading, the class to be used as a basis for the form is shown (for a wall form this will be a specific surface class). At the bottom of the screen are listed the maximums allowed for that type of form. For a wall, for example, no more than six apertures, seven fins, and two light shelves can be used. Users can edit existing entries on the table or add new lines. When completed the form table is stored in a file named by the user. Similar tables are used to assemble class items into the remaining types of forms.

Two-dimensional and three-dimensional graphical checking

Having assembled all the needed forms, the user next calls for Building Wall Elevations, another menu choice. The package then displays scaled, two-dimensional graphical representations of the forms that have been numerically described. These drawings are reviewed one by one. If not satisfied with their appearance, the user can return to the forms construction procedures to make appropriate alterations.

The next step is to produce a three-dimensional graphical display (without hidden lines) of the building. After the appropriate menu choices, a 3D image of the entire building will be displayed, and the initial display will be followed by a sequence of rotated-position displays, so that all sides of the image can be checked. The axis of rotation, viewing distance, scale, and initial view can all be specified by the user.

Three-dimensional illumin-nets

A Daylight Gain calculation finds the exterior daylight available for the location, surface orientation, and slope of interest. The package comes with stored daylight and sunlight availability data. Users choose appropriate climate data files, from those available, for the specific location of interest. The calculation produces tables for global and diffuse illumination with azimuth figures for each hour of the day. Figures can be in either lux or footcandle units.

The user specifies the station points within the building that are to be used for the analysis, usually a grid of station points. The time frame for the analysis is also specified by the user. The package can then produce a table of illumination data that shows values for each specified station point.

Graphical display of the illumination data also can be obtained. Contour graphs can show illumination values at the chosen station points, constant illumination contour lines, and areas of direct sunlight.

The 3D capabilities of the package allow graphical display of daylight characteristics shown at a specified angle, height, and distance from the building. What is displayed is an "illumin-net" 3D pattern of illumination contour lines between the floor of a room and a specified height above that floor. The pattern will be shown contained within the building outline.

The package also provides calculations for a contrast ratio and glare index. For the index calculation the user selects an illuminance category from an IES table, and specifies the station point for the calculation, and the viewing eye height and angle and chooses either clear or overcast conditions. The package calculates the corresponding daylight glare index values. The calculated contrast is between a specified aperture in the room and the various station points.

Electric budget calculations

There also is a procedure for analyzing lighting power and control strategies. The user specifies the desired design and background illuminances and a cutoff level for dimming light fixtures. Then a control strategy (step or dimming) for electric lighting is selected. Calculations can determine monthly daylight utilization fraction for any standard work year for which station point illuminance has previously been calculated. The package will produce a summary of the portion of the hours during a specified month in which daylighting alone covers the required lighting loads. A similar yearly summary also is available.

Graphical displays include a chart of illumination versus time of day and a bar chart that compares supplemental electrical use under various control strategies.

Once suitable electric lighting parameters have been determined, the package can be used to calculate a lighting power budget. The tabular data produced show the total demand charges during each month of the year. A bar chart displays monthly kilowatt hours for entirely electric lighting and the portion that can be filled by daylighting.

LIGHTING FIXTURE CALCULATION

Lighting Fixture Calculation calculates the number of lighting fixtures required for a given room. The zonal cavity calculation method, as described in the IES Lighting Handbook, *is used. The package comes with stored data on about 125 different types of fixtures.*

Lighting Fixture Calculation runs on IBM/PC, TRS-80, and CP/M computers. It sells for $495 from Elite Software Development, PO Drawer 1194, Bryan, TX 77806. (409) 846-2340.

Users are advised to organize their input data, before entering it into the computer, by manually filling out data sheets. Blank forms are supplied as part of the user manual. The items on these forms correspond, more or less, to the display screens that will have to be filled in during data entry.

When a data entry screen is displayed, the cursor is initially located at the first item. Users key in the information for that item, then press the Return key and the cursor moves to the second item. While that screen is displayed, users can also move about from item to item with the help of the up and down arrow keys. Editing of the entries is done by using the Backspace key to erase characters and the space bar to move from one character to another. As many changes as desired can be made; they will take effect for package operation only after the data screen has been entered by pressing the Esc key.

There are a few built-in controls designed to maintain the accuracy and validity of the inputs. The package has acceptable value ranges stored for some of the input parameters and will display a warning message if these ranges are exceeded. There also are some checks for completeness and correct numbering of input data items.

Alternative designs

The package is set up to take input for three alternative fixture types with each design to allow a comparison of costs for the various choices. To include these alternative fixtures, they must be named on the General Project Information entry screen, and they must each be selected from available stored fixture descriptions by a Fixture Schedule Data screen. Data on their use will be calculated by the package and entered on the Room Data screen.

If the user wishes, the package will make use of an internally calculated figure for the number of fixtures needed to achieve the foot-candle level specified for a room. Otherwise, the package will operate with a user-supplied number.

In addition to the stored fixture descriptions provided with the package, users can enter and store data on other fixtures of their choice. Detailed manufacturers' data, including a table of coefficients of utilization, are needed for each such fixture. A separate procedure is provided by the package for this type of entry.

Fixed format reports

The package produces three main reports, which can be displayed or printed. An energy economics report compares initial, annual, four-year, and life-cycle costs of the three alternative designs. A fixture requirements report summarizes the lighting characteristics of the rooms under study and the results produced by the three alternative fixture choices for each room. A fixture schedule report gives descriptive information on each of the fixture types proposed for use in any of the rooms.

The formats of these reports are fixed, but users can select which of them will be displayed or printed. Selections also can be made to limit the scope of the reports to specified groups of rooms.

LUMEN-MICRO

LUMEN-MICRO calculates horizontal and vertical illuminance for an array of up to 20-by-20 points on a horizontal plane. It also will calculate equivalent spherical illuminance (ESI) for the same points. An optional module will calculate visual comfort probability (VCP) for those points. A similar array of exitance and illuminance values can be calculated for any designated room surface. The luminaires involved can be point, line, or area sources. Lighting equipment can be direct, indirect, or a mixture of both.

LUMEN-MICRO runs on IBM/PC/XT/AT computers. It sells for $2,660. An optional visual comfort probability (VCP) module sells for $500. The vendor is Lighting Technologies, 3060 Walnut, #203, Boulder, CO 80301. (303) 449-5791.

The package is opulent, using five different floppy disks to carry the standard routines, plus one more for the VCP calculations.

Four hundred input lines

Data input to this package is to a form that is 400 lines long. Included are room and luminaire data, information on the desired types of calculations, and on the output the users wish to be produced. During this input process, data on the different types of luminaires used in the design can be obtained from files stored by a companion package called LUMEN-DATA.

Only 24 input data lines are displayed on the screen at any given time. The data are keyed into spaces enclosed by square brackets. Users move the screen cursor forward from one item space to another with the help of the Enter key or backward using the Tab key. They also can move the cursor one line at a time with the up and down arrow keys. The PgUp and PgDn keys move the display to the next 24 lines of the form. Scrolling up or down, one line at a time, can be done with the Scroll Lock and up–down arrow keys.

This is truly a full-screen, full-form entry process. Until the entire table is saved on a disk file, none of this information becomes available for calculations. If for any reason the computer should lose power during the entry process, before disk storage takes place, the just-entered data will be lost. Users nervous about this possibility can store a partially completed input table, then call a copy from storage and add items to it or change existing items, and then once again store the amended table. In fact, this is the normal method of editing the input data.

If they wish, users can get a printed copy of the input table or print a blank input table for use as a worksheet.

During the input and editing process, the package will not permit users to enter alphabetic characters where only numerals should be used. Nor will it allow users to key in an excessive number of characters for any item. Beyond these perfunctory prohibitions, there seem to be few built-in controls on the quality of the input data.

Specifying output arrays

Unless the user requests another format, output data will be given for a grid with 19 columns and rows spaced uniformly around the room being analyzed. The grid points will be located 2.5 feet above the floor. Users can, however, define their own grid, with equally or arbitrarily spaced rows and columns (but no more than 19 of each), and with points at any desired height within the room.

Such a grid must be specified for horizontal illuminance results, and separate, different grids may be defined for vertical illuminance, equivalent sphere illumination, and visual comfort probability results. A similar choice of grids, located on the appropriate surface, are available for calculations of room surface exitances and illuminances.

The package gives users considerable choice in selecting what will appear on printed results reports. Headings for the reports can be formatted and composed. Report material is available in ten modules, and users can select which of these will be printed.

Beyond the printed reports, the package will produce graphical results in the form of (numerical symbol) contour plots and (dot-matrix) shaded renderings. Users can select which parameters will appear on these graphical displays and also have some control over the use of shading. Plots also can be rotated 90 degrees, if desired.

LUMEN-POINT

LUMEN-POINT is for design of outdoor lighting, including roadway lighting. Illuminances are calculated for up to 50-by-50 points on a grid.

LUMEN-POINT runs on IBM/PC/XT/AT computers. It sells for $995. An optional digitizer module sells for an added $995. The vendor is Lighting Technologies, 3060 Walnut, #203, Boulder, CO 80301. (303) 449-5791.

A lot of data may have to be input into this package. The grid of calculated points can be horizontal, vertical, or tilted. Illuminances on the grid can be calculated at any orientation the user specifies. Ability to locate, group, and aim luminaires has to be described in detail. Altogether, input data for an analysis may consist of up to five hundred lines of information.

The long worksheet that contains this information is displayed to the user, 24 lines at a time. The data are keyed into spaces enclosed by square brackets. Users move the screen cursor forward and from one item space to another with the help of the arrow keys. They also can move the cursor one line at a time with the up and down arrow keys. The PgUp and PgDn keys move the display to the next 24 lines of the form. Scrolling up or down, one line at a time, can be done with the Scroll Lock and up–down arrow keys. Alternatively, IBM/PC Function keys can be used for scrolling and paging and for other functions as well. In fact there are so many of these command key functions that the ? key is reserved to produce a list of such available keyboard actions.

In some worksheet areas, such as those for luminaire types, analysis grids, and output heading lines, there are input locations that can represent whole groups of input lines. If desired, the user can enter these points to expand the size of the worksheet and expose the full details of the input data.

This is truly a full-screen, full-form entry process. Until the entire table is saved on a disk file, none of this information becomes available for calculations. If for any reason the computer should lose power during the entry process, before disk storage takes place, the just-entered data will be lost. Users nervous about this possibility can store a partially completed input table, then call a copy from storage and add items to it or change existing items, and then once again store the amended table. In fact, this is the normal method of editing the input data.

A separate program floppy disk is devoted to the input process, and the input data are kept on another disk supplied by the user. If that data disk gets filled, a warning will be displayed and the data disk can be replaced by a new one. If they wish, users can get a printed copy of the input table, or print a blank input table for use as a worksheet.

Users can choose the units in which input, calculation, and output will be expressed. The package will produce American standard dimensions to scale in the output but is not able to properly scale metric dimensions.

To assist in the entry of data on luminaires, files of such data can be maintained by a companion package, LUMEN-DATA, and up to eight different luminaire types from these files (or from lamp descriptions entered by the user) can be used in any analysis run.

Error controls and runs

During the input and editing process, the package will not permit users to enter alphabetic characters where only numerals should be used. Nor will it allow users to key in an excessive number of characters for any item. Beyond these prohibitions, there seem to be few built-in controls on the quality of the input data.

A separate checking procedure can be invoked. The manual assures us that this process is very thorough, but there is no specific indication of exactly what is checked, nor is there a description of the error messages that result when flaws are found.

Numerical and graphic output

Since the input data include specifications for the desired calculation and output, the calculation can be started, by an appropriate menu choice, once the input data table has been completed and stored. Users can specify up to five analysis grids for a single calculation run, for example, one for a parking lot entrance, another for the main parking area, a third for an annex parking area.

The analysis grids are printed out in tabular format with illuminance values shown at the grid points. A luminaire layout schedule lists location and aiming data for each luminaire in a tabular format. Detailed tabular and other data on each luminaire type used are also printed.

Different types of plots are available. Ordinary contour plots show repeated printed characters along lines of constant illuminance. The particular symbol denotes the value of illuminance on that contour. If desired, a contour plot can be printed with all areas above or below a specified value of illuminance covered by shading.

Instead of printed characters, shaded contour plots use varying dot densities to represent different values of illuminance. High densities (dark areas) denote low illuminances.

Design examples

Included in the manual are samples of output for four design problems. The first is a single luminaire in the center of a 100-by-100-foot area. The second has 16 luminaires of a single type in a large parking area that contains a traffic island (an excluded area; the package is equipped to handle such areas). The third example is a parking lot with entrance road and guardhouse, lit by two different luminaire types. The final example is of tennis court lighting, using four floodlights with point aiming.

LUMEN$

LUMEN$ does zonal-cavity and life-cycle cost-benefit analysis for alternative lighting systems.

LUMEN$ runs on IBM/PC/XT/AT computers. It sells for $495. The vendor is Lighting Technologies, 3060 Walnut, #203, Boulder, CO 80301. (303) 449-5791.

Like other packages from this vendor, the general procedure for LUMEN$ is to enter data on a long form that is displayed, 24 lines at a time, on the computer screen. Arrow, Pg, and other keys are pressed to maneuver the cursor around this display and to scroll or page to the rest of the form. Entries are filled in from the keyboard, and the form can be printed or stored at any stage.

Zonal cavity analysis

For the zonal cavity analysis, users input data on room geometry. Luminaire data can be entered from the keyboard or from previously stored files maintained by a LUMEN-DATA routine. The user specifies luminaire suspension lengths, number of luminaires, or footcandle level desired, a light loss factor (or detailed maintenance and other factors from which the package can calculate such an LLF).

The output summarizes these data for one to four alternative lighting systems and produces a table that includes number of luminaires required and used, average illuminance, and watts per square foot.

Economic analysis

For the economic analysis, users enter initial costs and system life estimates for up to four alternative lighting systems. They also enter an interest rate, operating periods and electricity costs, maintenance costs, depreciation data, data on air conditioning and heating for the room, and estimated escalation rates for labor, electricity, equipment, and heating fuel.

Package output produces a cost summary table for the alternative systems, a present worth summary, and shows the rate of return and payback period for the alternative systems compared to the first system. Yearly cost summaries also can be produced, comparing the alternatives over the period of the analysis.

Chapter 10
Control Packages

Several different aspects of control systems design and use are addressed by currently available personal computer packages.

Handling control system mathematical functions and equations is the aim of packages such as LINSYS and SYSPAK1. LINSYS is for analysis and design of linear single-input, single-output systems represented by transfer functions. The package can be used to design compensators to satisfy overall frequency- or time-domain specifications.

LINSYS can produce several types of graphical displays. Time-domain step or impulse response can be displayed at the touch of a key. To produce Bode or Nyquist response plots, the user answers short sequences of prompts. A similar procedure is followed to generate a root locus plot. The data in each of these cases can be displayed on the screen in graphic or tabular form and printed, if desired.

After a plot has been displayed, the package allows the user to change the overall transfer function and then superimpose a plot for the new transfer function over the original plot. This provides a means to do design by repeated trial changes in the transfer function.

SYSPAK1 includes four routines that allow users to put control system equations into state-space format, calculate transfer function coefficients, tabulate and graph the frequency response of a system transfer function, and calculate the coefficients of a control unit that will provide desired poles in a closed-loop system.

Both packages require the user to do some preliminary transformations before entering system descriptions into the computer. LINSYS requires that systems, and the blocks within them, be represented as simple, cascade, two-branch parallel, or single-loop feedback arrangements. Systems that are not in these forms must be rearranged so they conform. Even after the block diagram is in the proper format, users next must come up with a binary-tree representation of the system. The user manual for the package suggests several block diagram transformations and explains how to meet the binary-tree requirement.

With SYSPAK1 the equations to be entered must be either algebraic or linear-differential and have constant coefficients. Equations with higher-order derivatives must be transformed by the user into two or more first-order equations before they are entered into the computer.

Programmable controllers

If you think about the benefits of using a personal computer together with programmable controllers, several possible improvements come to mind: simplified ways to prepare ladder logic, clearer logic diagrams with helpful notes, computer-aided operation of the controllers, and reports that clearly present the parameters of concern.

There seems to be no single software package that performs all these functions, certainly not for all the popular progammable controllers. However, each package has its own particular combination of functions.

Making programmable controllers easier to set up and use is the aim of the LADDERDOCTOR packages. There is an L484 package used to create and document ladder logic for Gould Modicon 484 programmable controllers; the M484 package is for monitoring those controllers. Similar L584 and M584 packages serve Modicon 584 controllers, and a companion package can be used to translate 484 logic for use with 584 controllers.

Monitoring and reporting are the areas covered by the DMC-5 package, which includes three programs for Modicon programmable controllers. DMC-2 displays and prints coil, input, and register data. DMC-3 monitors current values and change of state for register and discrete data, tags and logs events, and provides alarms. The DMC-4 package is for preparing customized reports. To construct a DMC-4 report, users fill out a table that specifies the row and column location at which each item in the report will appear. Items can be notes and other text items prepared by the user as well as the variable data collected from the controllers. Time and date notations also can be defined as report items.

Logic annotation and documentation are the functions performed by The Soft Complement. This package is designed to download ladder logic from a Gould Modicon Micro 84 programmable controller, to annotate the logic, and to produce annotated ladder listings, reference listings, and cross references.

The Indelec Ladder Diagram Documentor has similar functions but performs them for a wider variety of controllers. This package is used to document, annotate, and provide comments to ladder diagrams for programmable controllers. Versions are available for Allen Bradley PLC-2, PLC-3, and for Gould 384, 484, 584, 884, and 984 controllers. The Indelec output procedures are flexible. They offer options for including or omitting rung and global comments and cross references. Diagrams can be shown from beginning to end or can be restricted to the portions between specified rung numbers. Printouts can be continuous or divided into pages.

Flexible monitoring

A computer display screen can hold only a limited amount of data, far less than would be needed to give a complete picture of typical ladder logic function. The DMC package seems to have chosen 16 lines of display data as the proper amount, with each line indicating the status of a particular point in the ladder logic.

In DMC-2 displays, users can specify freely what controller input, coil, or register is to be represented by each row of data. Users call for a PC Address (the Modicon address of the programmable controller that is to be polled) and a Reference Number (the Modicon designation of the input, coil, or register to be queried) for each data row. They also can choose whether the data will be shown in decimal, octal, hexidecimal, binary, or two-character decimal representations.

To get beyond the 16 displayed points, users can scroll or page the list up or down to reveal additional points. A displayed screen of data can be printed using the IBM/PC Shift-PrtSc keyboard function. The package also allows a list of current point data selected by point type and state to be printed.

DMC-2 has a parameter-setting feature as well. To change a controller value, the user simply moves to the data position on the desired displayed line and enters the new value. To prevent unauthorized persons from changing controller parameters, the package has a password feature. No password is needed merely to read a parameter, but to change one, the package asks the user to enter the password.

With the DMC-3 package, points can be defined as Alarm, Status, or Warning types. For each point, the user also provides high- and low-limit values and a message to be displayed when the data from that point exceed these limits. Users can specify a series of points by giving the starting Reference number and telling the computer how many sequential points are to be polled.

The LADDERDOCTOR package packs values for as many as 64 points on a single-screen display. Contents of up to four groups of registers can be displayed with as many as 16 registers in each group. In each case the current value of the register is displayed next to a register reference number. Similarly, the current state of 64 coils or inputs can be displayed with current values shown next to the reference number for each item. To select the items to be displayed, users enter a reference number to get that item plus the 15 following items. Selections can be made only in contiguous groups of 16 reference numbers. Only four such groups will be accepted for display at a one time.

Editing ease

Some of the packages make it quite simple for users to annotate and edit ladder logic diagrams, but the techniques used differ from one package to another.

The Indelec package uses the ten IBM/PC Function keys to provide quick access to editing functions. Brief labels for these Function keys are displayed on the screen during the editing process. To edit a displayed diagram, the user moves the cursor to the desired location on the screen (arrow keys control this movement), then a Function key is pressed. For example, key F3 will delete the line on which the cursor has been placed. Comments for the diagram are inserted, using separate procedures for ladder-rung comments and for global and contact comments, again with the help of the Function keys.

Selecting the LADDERDOCTOR menu options to create new ladder logic causes a 7-row-by-11-column network grid to be displayed. Users move the screen display cursor around on this grid by means of the IBM/PC arrow keys. Numerical keys and some alphabetical keys are used to insert desired elements at cursor locations. For example, a coil is inserted by pressing the 1 key; a relay, normally open, by pressing the 7 key. Editing is done in a similar manner. LADDERDOCTOR comment generation features can be applied to a diagram or to a specific element within a given diagram. By pressing the right keys, the user can move or jump from one diagram or element to another, inserting or deleting as desired.

Entries and editing of notations and comments for the Soft Complement package are done one at a time, without reference to any displayed picture of the ladder logic. The result is a file of notations and comments that will later be merged with the actual ladder diagram. Users also are allowed to formulate up to five comments, each up to 250 characters in length, for each ladder network. To reach a particular comment that is to be edited, the user has to start at the first network then step through each comment on that and subsequent networks until the desired comment is reached.

DMC-5

There are three packages here. All are designed to work with Modicon programmable controllers. DMC-2 displays and prints coil, input, and register data from controllers. DMC-3 monitors current values and change of state for register and discrete data, tags and logs events, and provides alarms. The DMC-4 package is for preparing customized reports.

DMC 5 packages run on IBM/PC computers. The price for the three packages is $1,295. They also are available at $595 each. The vendor is The J. D. Truba Agency, 2091 15 Mile Rd., Sterling Heights, MI 48077. (313) 826-8520.

Individual Modicon controllers can be connected to an IBM/PC through a standard (RS232) communications port. Connections to more than one controller are accomplished with the help of modem equipment. The packages will handle either kind of connection.

Display package

The DMC display package presents a screen that displays 16 rows of information. Each row represents a specific controller input, coil, or register that is to be queried for data.

Users may specify the desired parameters by making entries in the PC Address, Reference Number, and Mode columns. PC Address is the Modicon address of the programmable controller that is to be polled. The Reference Number is the Modicon designation of the input, coil, or register to be queried. Mode refers to the kind of display desired for that parameter. A choice of decimal, octal, hexidecimal, binary, or two-character decimal representations is available. The manual lists the codes used to designate these choices.

Alternatively, this specification can be made with an entry in the Mnemonic column; for this type of access, the user must create a stored table of mnemonic references that the package can access. Each mnemonic represents a particular combination of PC Address, Reference Number, and Mode.

The Data column on the screen is the location where the polled parameters are displayed. To change a controller value, the user moves to the Data position on the desired line and enters the new value. The screen will show the old value until the next poll operation is complete, then will display a new value. This may not match up with the user's desired value because of some error or the effects of controller logic. An error message column on the screen helps account for such differences.

To prevent unauthorized persons from changing controller parameters, the package has a password feature. No password is needed merely to read a parameter, but to change one, the package asks the user to enter the password. Since this password can be changed through the same package, not only the current password but the means of changing it must be protected.

The displayed table can be printed by simply pressing the P key. Timed-interval printing is also available. The user sets the desired interval and the parameter on the table to be monitored. The package will collect and print the desired parameter readings.

In addition to the user manual, this package has a HELP feature. However the HELP screens, which do not contain as much information as the printed manual, are intended only as an aid.

Monitor-alarm package

The DMC-3 package is designed to handle up to 16 programmable controllers with over 999 points. The last 200 changes in polled readings are recorded in a log file.

Displayed on the screen during monitoring are 16 points. The user can scroll or page this list up or down to reveal additional points. Selection of this display or an alternative log file display is made by pressing individual numerical keys or combinations of these keys. The needed keys and combinations are listed at the bottom of the monitoring screen.

A displayed screen of data can be printed using the IBM/PC Shift-PrtSc keyboard function. The package also allows a list of current point data selected by point type and state to be printed.

Setup of monitoring is performed through a Design mode. As with the DMC-2 display package, points to be monitored are defined by PC Addresses, and point Reference numbers. Users can specify a series of points by giving the starting Reference number and telling the computer how many sequential points are to be polled. Users also must specify at what location on the display list (1–999) the initial point in such a series shall be located.

Points can be defined as Alarm, Status, or Warning types. For each point, the user also provides high- and low-limit values and a message to be displayed when the data from that point exceed these limits. The package has a routine for setting communication parameters.

Report generation

To prepare a report, the package takes current data from the programmable controllers that are to be included in the report. With the help of the package, the user can then arrange these data to form the desired report.

Data to be collected are defined by the times at which they are to be taken. Up to four times can be set for each report. These might correspond to three shifts plus one summary. Controllers and points within them, from which data are to be taken, are also specified. The collected data are kept in user-defined files.

To construct a report, users fill out a table that specifies the row and column location at which each item in the report will appear. Items can be notes and other text items prepared by the user as well as the variable data collected from the controllers. Time and date notations also can be defined as report items. Variable data items are specified by the files in which they are stored. Users must indicate the format (number of digits and location of decimal point) in which the variable data will appear on the report.

The package will put together up to 31 reports at a time. The construction tables for these reports are each stored in a file and displayed when the user wishes to add to a table, delete a report, or make changes. Entries for a report construction table are made a line at a time. Lines may be inserted, deleted, and edited.

Calculations can be performed on the collected data, if desired. For example, two counts may be converted in a percentage change figure. Any calculation that can be defined in the version of the BASIC computer language used by the package (BASICA) can be invoked. Of course, to do such calculations users must have some familiarity with BASIC. A calculation can be defined for each line in the report construction table.

Some trial and error practice is needed to put together reports that end up with the desired format. Users make up the construction tables but do not see the resultant reports until after table entries have been completed.

INDELEC LADDER DIAGRAM DOCUMENTOR

The Indelec Ladder Diagram Documentor is used to document, annotate, and provide comments to ladder diagrams for programmable controllers. Versions are available for Allen Bradley PLC-2, PLC-3, and for Gould 384, 484, 584, 884, and 984 controllers.

The Indelec Ladder Diagram Documentor package consists of software and a plug-in communications board for IBM/PC computers. The package sells for $2,245 from Holmor Associates, 169 Route 206, Flanders, NJ. (201) 584-2500

The manual includes detailed instructions for installing the communications board in an IBM/PC. Users are expected to perform this installation themselves, a simple task for anyone who feels fairly confident in dealing with circuit boards and dip switches.

Software installation is another matter. Some of these installation tasks "may not be considered user friendly," according to the manual. Users are expected to be familiar with IBM/PC file naming and manipulation procedures. For each ladder diagram stored in a file, there will be a file named by the user, plus nine other auxiliary and backup files created by the package.

When the software is loaded, a master menu appears, offering the user a choice of the various functions the package will perform. The plug-in board includes a battery-backed clock–calendar that keeps track of real time even when the computer is shut off. One of the menu choices is a procedure to initially set the time and date.

Ladder listings

Data describing the ladder logic that is to be documented can be acquired from an industrial terminal. This involves setting up a prescribed search sequence on the terminal so it will transmit the desired data to the computer.

Aquisition can also take place from a STARLINK 1770-SB unit. After setting up the unit and its connection to the IBM/PC, the user initiates a computer procedure to automatically transfer the data and then convert them to the form needed by the Indelec package. Conversion may take a minute or more to complete.

A third way to acquire the ladder logic data is from a PLC2 Data Highway connected to the IBM/PC. To access a particular Allen Bradley PLC2 controller connected to the highway, the package will prompt the user for a coded controller address. Again, the procedure involves a conversion process that will last a minute or more.

Editing functions

The package allows users to display and edit a stored ladder diagram. To aid in this process, the ten IBM/PC Function keys provide quick access to editing functions. Brief labels for these Function keys are displayed on the screen during the editing process.

To edit a displayed diagram, the user moves the cursor to the desired location on the screen. The IBM/PC arrow keys control this movement. Then a Function key is pressed. For example, key F3 will delete the line on which the cursor has been placed; key F5 is used to select the desired color for text entries; key F10 is used to initiate a search for a specific contact, rung, or text entry. This is a rather unforgiving search for exact text, and the search order must include all details such as upper and lower case, spaces, and punctuation exactly as they appear on the diagram.

Other Function keys allow the user to expand space allotted to a ladder rung, to recover deleted material, to move to the top or bottom of a diagram, to page up or down on a diagram, and to exit the editing session. If the exit key should be inadvertantly pressed while a previous editing step is in process, all the editing during that session will be lost.

Comments for the diagram are kept in a separate file, with each comment associated with a given rung number. The comment editing procedure is somewhat different from that for editing diagrams. Function keys are again used, but some of them have different meanings during comment editing.

Global comments and contact comments are inserted and edited with still another procedure. The Function keys are assigned a unique set of meanings for this procedure.

After editing, users may wish to examine or print out the annotated ladder diagram. The output procedure includes options for including or omitting rung and global comments and cross references. Diagrams can be shown from beginning to end or can be restricted to the portions between specified rung numbers. Printouts can be continuous or divided into pages. If desired, a table of input–output usage throughout the diagram also can be produced.

LADDERDOCTOR

LADDERDOCTOR packages are for creating and documenting ladder logic for Modicon 484 and 584 programmable controllers and for monitoring those controllers. The L484 programmer and documenter package sells for $249; the similar L584 packages sells for $349. The M484 monitor package costs $129; the M584 monitor package sells for $149. A 484 to 584 translation package sells for $99. All five packages can be purchased together for $795. The packages run on IBM/PCs and Z-80 based personal and microcomputers with CP/M. The vendor is Datablend, PO Box 1095, Wodinville, WA 98072. (206) 481-4030.

Communications between the computer and the programmable controllers take place through standard (RS-232) interface connections. Each package includes procedures to set up and conduct these communications so ladder diagram programs can be up- and downloaded from the controllers to the computer.

Setting up ladder logic

L484 can generate software to operate the Modicon 484 programmable controller. It also can read existing Modicon 484 software and generate appropriate documentation.

Selecting the menu options to create new ladder logic causes a 7-row-by-11-column network grid to be displayed. Users move the screen display cursor around on this grid by means of the IBM/PC arrow keys. Numerical keys and some alphabetical keys are used to insert desired elements at cursor locations. For example, a coil is inserted by pressing the 1 key; a relay, normally open, by pressing the 7 key.

Whenever an element that requires a reference number is inserted, the package will display a list of legal reference numbers, and the user can choose from these. Users also are prompted for added information, as needed, for other types of elements, for example, to specify timers and arithmetic elements.

Editing is done in a similar manner by moving the cursor to the desired location on a ladder logic network, then pressing keys to delete or add elements as desired. Other keyboard actions allow users to move from one network to another and to save or discard created or edited versions.

Existing ladder diagrams can be read from a Modicon 484 controller and uploaded to the computer for editing and annotation. Communications capabilities also include initializing, starting, and stopping the controller from the computer keyboard.

The comment generation features of the package can be applied to either a network or a specific element within a given network. Network comments can be made on three lines with up to 60 characters per line. Most elements can be annotated by up to three lines but are limited to eight characters per line; some elements, such as arithmetics, timers, and counters are limited to just six comment characters per line.

By pressing the right keys, the user can move or jump from one network or element to another, and insert or delete characters or networks. The annotated diagrams can be printed and stored in files. The package also will produce a cross-reference list that relates reference numbers to the networks in which they appear.

Counting coils

The L584 package seems to be an earlier product. Users have to supply counts of coils, discretes, input and holding registers, and some other values before the 7-by-11 network grid will be displayed. This means that users must have access to a completed ladder diagram before they can begin to design the desired ladder diagram. In practice, this may be the way many designs are made up, but it places an added step (getting counts and values) into the procedure for those who want to make designs from scratch.

Element insertion and editing use essentially the same methods as the L484 package. Starting with initial element counts does not seem to restrict the user's ability to make as many changes as desired to the logic design. In addition to the regular keyboard keys, L584 makes use of the IBM/PC Function keys to implement ladder programming and comment annotation functions.

Comments for the networks and elements within them are handled in the same way as in the L484 package. The same is true of printouts, cross references, and communications.

According to the vendor, it takes just 352 bytes to store a 7-by-11 network, and 284 such networks can be kept in 100,000 bytes of disk storage.

Monitoring and logging

The M484 package is for those users who want to monitor and log the operation of a Modicon 484 programmable controller. The data display produced by this package can show the contents of up to four groups of registers. Contents of as many as 16 registers in each group can be displayed on the screen. In each case the current value of the register is displayed next to the register reference number.

Similarly, the current state of 64 discrete items (coils or inputs) can be displayed, with current values shown next to the reference number for each item.

In addition, users can specify up to four alarm conditions. These can consist of any combination of register or discrete item values. Alarm relationships can be =, <, >, < =, or = >. The display will show the set and current values for each of these four alarm conditions. When an alarm occurs, the computer will give an audible signal and a visual alert symbol (< <) will be shown by the appropriate alarm value display.

Sixteen at a time

To select the registers and discrete items to be displayed, users enter a reference number to get that item or register plus the 15 following items or registers. Selections can only be made in such contiguous groups of 16 reference numbers. Only four such groups will be accepted for display at one time.

Users may elect to monitor all displayed groups or only one group at a time. The selections are made from a displayed menu. Alarm conditions also are specified from a menu display. The package seems to insist that four alarm conditions be set up, though less than four can be chosen for monitoring. Lists of legal reference numbers for coils and registers are displayed, and the user must validly choose from these before proceeding. Then the package prompts for the set value of the alarm and the relationship between the current and set values needed to cause an alarm.

The package can produce a printout but only as a snapshot of what is displayed on the screen. There is no continuous printer logging capability.

Operation of the M584 package is similar to that of the M484, except that the M584 is designed to monitor Modicon 584 programmable controllers.

Translation package

The function of the Translator package is to convert Modicon 484 programs so they will run on a Modicon 584 programmable controller. The essential procedure is to use the L484 package to read a program from a Modicon 484. Then the Translator package is used to do the translation and store the result in a disk file. Finally the L584 package is used to modify and correct the disk-stored program, as needed, and to install it in the Modicon 584 controller.

Four types of elements or references cannot be handled by the Translator package. When these are encountered, the package will leave a gap in the translated network. Filling these gaps is a major reason for employing the L584 editing capabilities. The translator does not locate the missing elements, leaving that up to the user, but it does keep a count of the number of elements that were not translated.

Users also must supply configuration data for the targeted 584 controller. These data include counts of coils, discretes, and registers, and some other data items. These data can be read from the 584 controller using the L584 package.

LINSYS

LINSYS is for analysis and design of linear single-input, single-output systems represented by transfer functions. The package designs compensators to satisfy overall frequency or time-domain specifications.

LINSYS runs on IBM/PC ($650), Apple II and II+ ($325), HP-85 ($325), *and HP 9000 Series 200 computers ($500). The vendor is Parametrics, Inc., 1129 West Oak, Fort Collins, CO 80521. (303) 221-3163.*

LINSYS requires that systems, and the blocks within them, be represented as simple, cascade, two-branch parallel, or single-loop feedback arrangements. Systems that are not in these forms must be rearranged, without disturbing the overall input–output relationships, so they conform. The user manual suggests several simple block diagram transformations that meet this requirement.

When the block diagram is in the proper format, users next must come up with a binary-tree representation of the system. The manual explains how this is done. Nodes on the tree represent transfer functions and connections in the block diagram. In the words of the manual: "This procedure may seem somewhat cumbersome; however, after a little practice you should be able to construct the binary tree representation by inspection."

An example in the manual shows the transformation of a system diagram containing seven blocks and two summing points into a form with four summing points. Then the manual shows a 15-node binary-tree representation that corresponds to the transformed block diagram.

A series of prompts leads the user to enter the required input information, working from the binary-tree representation of the system. The package will then calculate an overall system transfer function, store, and display the binary-tree information in tabular form. An editing procedure allows the user to add and delete nodes and alter node attributes without reentering the entire set of input information.

Generating plots

Once the system description has been entered, the package can produce several types of graphical displays. Time-domain step or impulse response can be displayed at the touch of a key. To produce Bode or Nyquist response plots, the user answers short sequences of prompts. A similar procedure is followed to generate a root locus plot. The data in each of these cases can be displayed on the screen in graphic or tabular form and printed, if desired.

After a plot has been displayed, the package allows the user to change the overall transfer function and then superimpose a plot for the new transfer function over the original plot, providing a means to do design by repeated trial changes in the transfer function. The user manual suggests adjusting parameters such as the transfer function damping coefficient to get plots that show the effects of such adjustment.

Package limitations

Transfer function polynomials can be no larger than degree 15. Binary-tree representations of system block diagrams can have no more than 15 nodes. There also are limitations on the size of time intervals and step sizes for the impulse and step responses that can be handled. The package will show incorrect responses if these limits are exceeded, but there are no fixed rules for avoiding such erroneous output. Experimentation by the user will be needed to test these limits for any particular case.

THE SOFT COMPLEMENT

The Soft Complement package is designed to download ladder logic from a Gould Modicon Micro 84 programmable controller, to annotate the logic, and to produce annotated ladder listings, reference listings, and cross references.

The Soft Complement runs on IBM/PC, NEC 8021, and TRS 80 Mod 100 computers. It sells for "less than" $600 from the J. D. Truba Agency, 2091 15 Mile Rd., Sterling Heights, MI 48077.(313) 826-8520.

Ladder logic data enter the computer through any standard (RS 232) input connection. A J375 Modbus interface is required to connect the Micro 84 to the computer input. Connection is via a Gould "null modem" cable. Installation involves setting dip switches on the J375 unit, a relatively simple procedure for those generally familiar with electronic equipment.

In some cases it may be necessary to edit the line of code in the program that sets up the computer to correctly receive the incoming signals. This one-time procedure requires a working knowledge of the BASIC computer language but may also be accomplished by nonprogrammers who follow explicit instructions from the vendor via telephone.

The user assigns a file name for the ladder data. When the data transfer begins, lamps on the J375 unit will blink.

Annotating the items

Ladder diagram reference notations written with the help of this package are limited to five or seven characters each. With the IBM/PC and NEC versions of the package, existing notations can be edited in place on the diagram. With the TRS-80 version, whole new notation entries must be made when changes are desired.

Entries and editing of notations and comments are done one at a time, without reference to any displayed picture of the ladder logic. The result is a file of notations and comments that will later be merged with the actual ladder diagram.

Entering a ladder logic reference number causes the notation file entry for that number to be displayed, and corrections or changes to the notation can then be made. The package will follow the first entry with the notation for the next sequential ladder reference number and continue that sequence after each entry. Users can stop the sequence by keyboarding a special entry.

Users are allowed to formulate up to five comments, each up to 250 characters in length, for each ladder network. To reach a particular comment that is to be edited, the user has to start at the first network then step through each comment on that and subsequent networks until the desired comment is reached. This seems to be a clumsy procedure.

Merging the files

To apply notations and comments to a ladder diagram, the user selects the ladder listing option on the package main menu, enters the ladder data file name and the name of the comment and notation file that is to be used. The user also provides a title for the merged diagram and specifies which networks should be included, since the entire diagram need not be merged.

The package than produces a listing of the merged diagram. In addition, if desired, a separate listing of notations and reference numbers can be printed.

There seems to be no internal check on the extent to which the notation and comments file matches up with the actual ladder diagram. Apparently, users must examine displays or printouts of the merged listings before they can be confident that the annotations are properly placed.

The package does include a procedure that creates cross-reference listings for all or portions of the ladder diagram.

Minimal manual

The user manual for this package seems to have been written with a minimum of attention on the part of the vendor. Explanations of package operation are skimpy. The text is full of grammatical and spelling errors. At one point, users are advised to "experiment" with the package to find out how editing is performed on their particular type of computer.

SYSPAK1

With the four routines contained in SYSPAK1, users can put control system equations into state-space format, calculate transfer function coefficients, tabulate and graph the frequency response of a system transfer function, and calculate the coefficients of a control unit that will provide desired poles in a closed-loop system.

SYSPAK1 runs on IBM/PC/XT/AT computers and sells for $750 from Engineering Software Co., Three Northpark E., #901, 8800 North Central Expressway, Dallas, TX 75231. (214) 361-2431.

The STATE1 routine takes process and interaction equations and reduces them to a state-space format. The equations to be entered must be either algebraic or linear-differential and have constant coefficients. Equations with higher-order derivatives must be transformed by the user into two or more first-order equations before they are entered into the computer.

Formats of equations to be entered are prescribed by the package manual, and inputs must be arranged by the user in these formats before entry. What the user actually enters are seven matrices; the sizes of the matrices depend on the number of input equations used.

After the user keyboards a description to identify the problem to be run, the package cues for the number of each type of variable used. This defines the sizes of the input matrices. The package then asks for each of the elements of the first matrix, A. When these are entered, the A matrix is displayed. The user can either accept it as correct or cycle once again through the element entry process. A similar input and checking procedure is followed for the remaining six matrices.

An alternative input method allows the user to enter only non-zero elements, convenient when matrices are sparse.

When the user accepts the last input matrix, the package displays the seven input matrices, along with the calculated output. This consists of two matrices for the state-space form of the equations and four additional matrices that can be used to define the input compactly. If desired, the output can be saved on disk, under a file name defined by the user, for further use.

To exemplify the use of STATE1, the manual describes preparation of input equations and shows computed output for a simple mass-spring-damper system, an RLC circuit, and a circuit with four resistance elements, a capacitor, an inductor, and a voltage source. These examples give some idea of the systems that can be handled by this routine. Input can consist of up to about 15 equations.

The SISO1 routine takes the state-space equation coefficients of a system, along with the coefficients of an output of that system, and calculates transfer function coefficients.

After entering a description of the problem, the user is cued for the number of state variables in the state-space equation, then for the elements of the coefficient matrices. If desired, the output of the STATE1 routine can be used as input to SISO1.

SISO1 then produces a table showing the coefficients of the transfer function. The original coefficients also are listed to allow users to exercise their judgment in dropping terms with very small coefficients. If desired, the results can be stored in a disk file, named by the user, for input to other routines in this package.

The FREQRESP routine is used to evaluate, tabulate, and plot the frequency response of a system transfer function. Data from SISO1 can be used as input to this routine, or the input data (coefficients from the numerator and denominator of the transfer function) can be entered in response to displayed cues. The routine can handle polynomials of order up to 15.

For calculated results and plots, the user can choose either Hertz or Rad/Sec for the frequency units. A table of results is produced with columns showing frequency, amplitude ratio, and phase angle. Ordinarily, plots will be shown with a frequency spacing of 20 points per decade but the user can specify any desired number of points. Plots of amplitude and/or phase can be displayed or printed. The plots are crude, using O and * characters to indicate points

SIGAINS1 generates the coefficients of a control system that will provide desired closed-loop poles. This calculation assumes there is an overall System consisting of a Plant that operates into a Control unit, with negative feedback. The Plant, the Control unit, and the overall closed-loop system are each described by polynomials.

Input to the routine consists of the coefficients of the Plant polynomials and of the prescribed System characteristic polynomial. The routine produces the coefficients of appropriate Control unit polynomials and the gains of the System transfer function. It will accommodate input Plant polynomials of order 15 and 14.

Chapter 11
Symbol and Object Manipulation Graphics Packages

The graphics packages described in this chapter allow users to construct drawings by positioning prepared symbols or objects on the screen. These packages generally incorporate the ability to do some drawing of lines as well as figures such as polygons, circles, and ellipses, and they also allow users to annotate the drawings with text.

However, the packages fall short of providing drafting capability in that they are not equipped to efficiently produce standard engineering drawings. For this, the ability to automatically, or at least conveniently, handle drafting necessities such as dimensioning and to construct highly complex drawings with rapid manipulation of drawing processes would be needed. The packages described here do not meet those requirements.

Nevertheless, even with their limited capabilities, symbol and object manipulation packages have a wide range of possible applications. They can be used to draw layouts useful in conceptual and planning activities and to produce the many types of diagrams and charts that help engineers to put their projects into operation.

If they are to be more than mere toys, the implicit question that packages of this type must answer is: Can they produce results that are better, more useful, or less expensive to achieve than what can be done by manual drawing methods?

Store and draw

The basic idea of these packages is to use the storage capabilities of the personal computer to eliminate the need to constantly redraw often-used objects or symbols. To allow creation of useful drawings, there also must be some means to connect these objects together. And, like other symbols, images of alphabetic letters and numerals should be able to be drawn from storage to form text entries on the drawings.

PC-Draw provides those basic capabilities in pretty much the expected way but that is about all it provides. For most of the other packages, the vendors apparently felt they had to offer something extra.

Benchmark, for example, tries to be as universal as possible. This package includes, along with its graphics capabilities, a data manager, word processor, spelling checker, and spreadsheet. Benchmark is integrated in that the graphics can be used together with the data manager and spreadsheet. However, there seems to be no useful interaction between graphics and the word processor.

In addition to symbol manipulation, the EnerGraphics package does line, bar, and pie charts, and it even has some 3D capabilities. Objects made up of planar panels can be constructed, and perspective, isometric, and oblique drawings can be displayed.

Multiple text fonts are offered by the QUICK-DRAFT package. Twenty different character fonts come with the package, and these include many symbols useful in making drawings as well as different types of numerals and alphabet letters. There is a Japanese Katakana font, a Greek font, and an Esperanto font. Most of the other packages provide only a few fonts. For example, PC-Draw gives a choice of only two different text fonts for drawing annotation.

The extra feature of the TGS package is its ability to animate. TGS can put display frames containing simple shapes composed of lines and circles, along with text, into apparent motion. However, the drawings are simple and somewhat crude, unsuitable for most engineering animation needs.

MGI/Schematic Drafter has a unique approach. It divides the drawing area into cells. There is no overall ability to draw connecting lines, but symbols in the cells can be prepared and placed so that lines that are part of one symbol match up with those in an adjacent cell.

Symbol creation and use

Current versions of PC-Draw come with symbols for software flowcharting, logic diagrams, and electrical drawings. Users can create their own stored symbols as desired, and this can be done while a drawing is in progress. Stored symbols are selected from displays, using cursor or light pen pointing, then placed on the current drawing. An excellent feature of this package is the split-screen display, with the symbol selection menu on the right and the drawing shown on the left as it takes shape.

Most of the other packages use similar approaches, providing stored libraries of symbols and allowing users to add their own symbol creations to those libraries.

The MGI/Schematic Drafter is again unique in that it devotes most of the computer keyboard to symbol selection. The user refers to a printed template guide for the particular symbol library in use. There the available symbols are shown in positions corresponding to those of the keyboard. The user moves the cursor to a cell in the displayed drawing space, then presses the appropriate key to call a copy of the associated symbol into that cell. The user can move the cursor from cell to cell, inserting symbols in each. An undesired symbol can be erased from a cell by moving the cursor to it, then pressing the space bar. Or it can be replaced by another symbol by a similar key action. If desired, two symbols can be placed within one cell.

Some packages provide only limited symbol or object storage. Thus, QUICK-DRAFT is able to store and repeat just 20 sequences of keystrokes and cursor motions. In other words, it can automatically redraw up to 20 user-created objects on command. The object definition process is simply the sequence of keyboard actions used to draw the object. The package automatically numbers the objects as they are created, and they have to be recalled by number. Therefore, it is left up to the user to keep some record of which stored object is which.

Complexity of drawings

For engineering applications, drawings often have to show considerable detail, and this is an area where these packages tend to be deficient.

EnerGraphics drawings, for example, are limited in complexity by two restrictions. Overall, a 2D drawing may not contain more than 800 line segments and 200 arcs and symbols. In addition, up to ten other "overlay" drawings can be added from storage to the working drawing. However, they are simply put into place, and no editing or alteration is allowed.

The complexity of an MGI/Schematic Drafter drawing is set at the outset by the user. To start a drawing, the user must select a mesh size for the background grid of the drawing. The grid may contain up to 50 cells across its width, or as few as 6 cells. A large number of cells may be needed to portray complex drawings since only one or two symbols can be placed in a cell.

The quality of the displays and printed output produced by the QUICK-DRAFT package is limited by the coordinate matrix on which the package is based. This has a width of 280 positions and a height of 192 positions. Lines and figures are always made up of straight orthogonal segments from one coordinate position to another. The result is that vertical and horizontal lines are smooth in appearance, while all lines and curves that deviate from these two directions have a discomfortingly jagged appearance.

A similar effect is seen in the drawings produced by the PC-Draw package. All lines that are not horizontal or vertical suffer visibly from the "jaggies." Although these jagged angled lines are quite easy to draw, they are difficult to erase and may make the drawings unacceptable for some uses.

Manipulation of symbols

To properly place and orient symbols taken from storage, users need considerable flexibility in manipulating them. Most packages provide several different symbol manipulation modes. For example, EnerGraphics symbols can be rotated, reduced, and enlarged, as well as translated to any position.

With the MGI/Schematic Drafter, the manipulation is more discrete. Symbols can be rotated into eight possible orientations: four 90-degree rotations, plus four more 90-degree rotations of a mirror image of the symbol. All these are obtained from one of the Function keys. Each time it is pressed, the symbol moves to the next in the sequence of eight orientations. If desired, these orientation changes can be imposed simultaneously on all the symbols in the current drawing. A symbol can be shifted in position within a cell, up, down, or sideways, but only in increments of 1/4 of the cell dimension.

Because the MGI package treats line segments only as symbols and has no separate line drawing ability, it gives users an extra symbol manipulation feature. Any given symbol can be rapidly copied into a series of cells. The user types the number of cells that are to receive copies, then presses the desired symbol key, and cells are filled, following the direction of last movement of the cursor. This feature can be used to create long continuous lines from individual cell line segments.

In most of the packages, text characters and chunks of text can be manipulated in much the same way as any other symbols. However, text positioning is sometimes restricted. In the PC-Draw package, for instance, text annotations must be either horizontal or vertical.

Manipulation of drawings

Not only symbols but entire drawings must be freely manipulated in order to produce the variety of results that these packages promise. These capabilities vary from package to package.

Thus, with the EnerGraphics package, 2D drawings can be scaled, and the scale of a drawing can be changed as needed. There is a zoom feature that allows portions of a drawing to be enlarged up to 1,024 times. Sections of a drawing can be defined, and these sections can then be rotated, translated, enlarged, and reduced.

The user can define a rectangular block of cells with the MGI/Schematic Drafter package, then copy, move, or delete that block. And the block also can be placed in, and recalled from, disk storage. However, MGI scrolling is a rather clumsy operation. The user has to key an entry to the scroll mode, then the scroll direction, and then the number of cells that are to be scrolled, before the actual scrolling can take place. Zooming is controlled by defining the block that is to be zoomed, then specifying the width of the zoomed view in number of cells. Other zoom choices include a return to the last previous view, to the original grid view, or to a view that will display all the cells in the drawing.

A QUICK-DRAFT drawing can occupy a space equivalent to four display screen areas. The package has no zooming or panning features, but it does have a scrolling capability to allow the user to move the display about, over this four-screen area, in half-screen increments. QUICK-DRAFT also allows full-screen drawings to be stored in the computer memory. The number of such storage positions, up to 33 with a main memory of 256K, depends on the available memory space. The current contents of the drawing screen can be stored in one of these memory positions at any time, and stored drawings can be summoned to the display for viewing or alteration.

Printing and plotting the results

For most of the packages, a printout is simply a hard-copy version of what is displayed on the computer screen. PC-Draw adds a bit to that capability. It allows drawings to be divided up into pages, so that each screen display can be a component page of a larger drawing. Printouts allow two screens to be shown on a single printed page; with condensed printing, four screens can be shown.

The only package reviewed here that makes use of plotting equipment is the MGI/Schematic Drafter. Drawings created with this package can be reproduced using a Houston Instruments DMP series plotter. To use the Hewlett-Packard 7580 plotter, special added software is needed. To get plotted output, the user specifies the desired paper size (A–E) and the package automatically scales the output to place the drawing on that size paper. Drawings also can be reproduced, as displayed, using a dot-matrix printer.

Special features

Something close to an itemized bill of materials, but not quite, can be produced by the MGI/Schematic Drafter package. The user can call for a symbol listing that will show all the symbols used in a drawing and the number of times each symbol appears, and these listings can be edited to remove unwanted items.

Some of the packages offer color to users. Thus with QUICK-DRAFT, any region of the drawing that is surrounded by lines or arcs can be quickly filled with color. The user moves the cursor into the selected region, then calls for any of 22 available colors. This package also can operate in a Wash mode that allows the user to fill in the entire background of a drawing with a choice of eight colors. This same mode can be used to draw rays from a fixed point.

BENCHMARK: GRAPHICS INTEGRATED WITH OTHER FUNCTIONS

Reviewer: Shen C. Lee, University of Missouri-Rolla

Benchmark is an integrated software package that includes some graphics capabilities along with a data manager, word processor, spelling checker, and spreadsheet. The strength of the package is that the graphics can be used together with the data manager and spreadsheet. However, the word processor is not able to combine graphs with text. No operational problems were encountered with the package during this review.

Benchmark runs on IBM/PC and compatible computers with minimum 256K main memory. Use of color display, with high resolution color adapter, and of color-supported printer is optional. The package sells for $795 from Metasoft Corp., 6905 W. Frye Rd., #12, Chandler, AZ 85224. (800) 621-1908, (602) 961-0003.

This reviewer found the graphics capability of the Benchmark package most interesting. It has the ability to move and rotate the images, to zoom, and also to add text to the graphics. Since no high-resolution color display or color-supported printer was available, the output was viewed on a monochrome monitor and the reviewer was favorably impressed with the results.

The package can generate simple engineering design drawings using 18 primitive commands, such as circle, arc, ellipse, line, and rectangle. Drawings can be stored for future use, revision, or modification.

Using data from the package's spreadsheet or data manager, presentation charts and graphs can be produced. These include pie charts, clustered bar charts, horizontal bar charts, line graphs, and *x, y* plots. These displays are controlled by 11 presentation graphics commands.

Handling multiple functions

An overall directory program called the Benchmark Administrator provides access to the various functions the package provides. It also performs some disk housekeeping functions.

The data manager is similar to many conventional data base packages for personal computers. It allows the user to design formats for data retrieval and storage. A very useful feature is the ability to transfer stored data (through a DIF file format) for use with the package's graphics capabilities.

The spreadsheet feature includes a three-dimensional capability allowing the user to store, retrieve, and manage data in the columns and rows of one or of several spreadsheet pages at the same time. Spreadsheet data also can be transferred for use with graphics display.

The word processor presents the user with more operational procedures than most simple similar packages. This reviewer expected that it would provide a choice of fonts and allow graphs to be included in the text. However, the word processor seems poorly integrated with the other features of this package. The vendor has stated that an advanced word processor feature, able to work in an integrated manner with the rest of the package, was being completed at the time of this review.

For use with the word processor, there is a spelling checker with an authority file of words that will be compared to those in text. The user can add to or subtract from this authority list as desired.

Limited help for the user

The vendor has a toll-free phone number through which users can seek answers to a limited number of questions they may have about the package. For the rest, they must rely on the documentation supplied with the package.

This written material seems to be directed mainly to experienced users. There are clear instructions for configuration of the package so it can be used with the many different computers and printers for which it is designed. However, the documentation does not provide clearly illustrated procedures for tutoring an inexperienced user to become proficient with the package. During the review process, it became clear that the user manual should be expanded to include more examples to minimize the amount of guesswork on the part of the user in figuring out necessary procedures. The manual also needs an index for quick reference.

ENERGRAPHICS

EnerGraphics is a multipurpose graphics package for creating charts, simple 2D drawings, and displays of some 3D objects. It is not suitable for drafting since there is no ability to use essential tools such as graphics tablets, nor is attention paid to providing dimensioning.

EnerGraphics runs on the IBM/PC and compatible computers. It sells for $250. A plotting option costs an additional $100. It is available from Enertronics Research, 150 North Meramec, #207, St. Louis, MO 63105. (314) 725-5566.

Extensive use is made of the ten Function keys on the IBM/PC keyboard. These keys are used to implement ten different modes of operation. Some attempt was made to make this use consistent across all the modes, but this is only partially successful, so the same key may have several different uses. However, users are reminded of the key functions by menu displays.

Charts and slide shows

There are separate but similar procedures for creating bar, pie, and line charts. In each case, after the chart is set up, it can be edited to meet user requirements. Data inputs to these chart creation procedures can be in equation form, or in the form of (DIF) files derived from spreadsheet packages as well as from direct keyboard entry.

Completed charts can be displayed one at a time, and they also can be shown in sequence, presenting a sort of slide show. In addition to its displays, the package can produce printed and plotted outputs.

Simple two-dimensional drawings

The package provides for 2D symbols, made up of line segment, arc, and dot elements, to be created by the user, stored and then placed, as needed, on drawings. The symbols can be rotated, reduced, and enlarged as well as translated to any position. When creating these 2D drawings, the same elements can be used along with the symbols, and text can be added from the keyboard as desired, with a choice of font styles. Like elements and symbols, text can be rotated, translated, reduced, and enlarged.

These drawings are limited in complexity by two restrictions. Overall, a 2D drawing may not contain more than 800 line segments and 200 arcs and symbols. In addition, other (up to ten) ''overlay'' drawings can be added from storage to the working drawing. However, once on the working drawing the overlays cannot be altered.

Drawings in 2D can be scaled, and the scale of a drawing can be changed as needed. There is a zoom feature that allows portions of a drawing to be enlarged up to 1,024 times. Sections of a drawing can be defined, and these sections can then be rotated, translated, enlarged, and reduced.

Limited three-dimensional capabilities

EnerGraphics allows users to create three-dimensional surfaces that represent mathematical functions or sets of points in three-dimensional space. Given the function, the package will calculate the data points to be displayed.

Three-dimensional objects made up of planar panels also can be constructed. Points and lines can be copied and rotated to quickly construct these panels in a variety of shapes. The panels can then be oriented in 3D space, copied and rotated to form the 3D objects. In turn, these 3D objects also can be rotated.

Perspective, isometric, and oblique drawings of a 3D object can be displayed, and the zoom feature can be used with these displays.

There is an automatic hidden-line removal feature that functions with some restrictions. If 3D objects are not constructed with careful attention to the hidden-line removal rules, this feature will not operate satisfactorily.

No three-dimensional hard copy

Graphs and 2D drawings can be printed and, with the help of a separate optional program, can be plotted as well. However,there seems to be no way to print or plot the 3D creations of EnerGraphics. Three-dimensional drawings can be stored, and they can be included in the slide shows the package provides for, but the package provides no way to produce hard-copy versions.

MGI/SCHEMATIC DRAFTER

MGI/Schematic Drafter allows the user to place and interconnect prepared symbols on a grid and to annotate drawings produced in this way with text. Symbol libraries are available for a number of engineering applications.

This package runs on IBM/PC/XT and compatibles and on Tandy 2000 computers. Use of the 8087 math chip is recommended to increase the speed of package operation. It sells for $2,495, and that price includes a choice of three of the symbol libraries offered by the vendor, Microcomputer Graphics, 13468 Washington Blvd., PO Box 10819, Marina del Rey, CA 90295. (213) 822-5258.

To start a drawing, the user must select a mesh size for the background grid of the drawing. The grid may contain up to 50 cells across its width, or may have as few as 6 cells across. A large number of cells may be needed to portray complex drawings since only one or two symbols can be placed in any cell. During the subsequent drawing process, it is possible to insert blank rows and columns of cells into the drawing, and rows or columns also may be deleted.

The package uses keyboard Function keys to control cursor movement, rather than the customary arrow keys provided on the keyboard. Users are advised to paste arrow labels over the appropriate Function keys to remind them of each key's function.

Placing symbols

Other keys are used to transfer stored symbols to the drawing cells. To do this, the user refers to a printed template guide for the particular symbol library in use. There the available symbols are shown in positions corresponding to those of the keyboard. The user moves the cursor to a cell, then presses a symbol key to call a copy of the associated symbol into that cell. The user can move the cursor from cell to cell, inserting symbols in each. An undesired symbol can be erased from a cell by moving the cursor to it, then pressing the space bar, or it can be replaced by another symbol by a similar key action. If desired, two symbols can be placed within one cell.

Manipulating symbols and drawings

Symbols can be rotated into eight possible orientations: four 90-degree rotations, plus four more 90-degree rotations of a mirror image of the symbol. All these are obtained from one of the Function keys. Each time it is pressed, the symbol moves to the next in the sequence of eight orientations. If desired, these orientation changes can be imposed simultaneously on all the symbols in the current drawing. A symbol can be shifted in position within a cell, up, down, or sideways but only in increments of 1/4 of the cell dimension.

A given symbol also can be rapidly copied into a series of cells. The user types the number of cells that are to receive copies, then presses the desired symbol key and cells are filled, following the direction of last movement of the cursor. This feature is used to create long continuous lines from individual cell line segments.

Block operations also are provided. The user can define a rectangular block of cells, then copy, move, or delete that block, and the block also can be placed in and recalled from disk storage.

Scrolling is a rather clumsy operation. The user has to key an entry to the scroll mode, then the scroll direction, and then the number of cells that are to be scrolled, before the actual scrolling can take place. Zooming is controlled by defining the block that is to be zoomed, then specifying the width of the zoomed view in number of cells. Other zoom choices include a return to the last previous view, to the original grid view, or to a view that will display all the cells in the drawing.

Text insertion

Text can be added to the drawing by locating the cursor at the desired start of text, going into the text mode, then keyboarding the text characters using the normal keyboard numeral and letter keys. Like symbols, text entries can be rotated in 90-degree increments. The entries also can be dragged using the cursor arrow keys, erased, and replaced.

Using and creating symbols

Library files of stored drawing symbols are provided by the vendor. Among the available libraries are those with symbols for project scheduling, space planning, pneumatic and hydraulic drawings, P & ID designs, electrical and electronic circuits.

Users also may create their own symbols and add them to existing symbol files or to newly created symbol library files. Like the overall drawings, the symbols are created on grids, but the procedures are somewhat different from those previously described. Straight lines, circles, and arcs can be drawn and erased. Text also can be included as an integral part of a symbol. Existing symbols can be recalled and modified, and they also can be selected and rearranged into new symbol library files. Printed guides, for use when calling up the symbols from the keyboard, can be produced.

Plots, printouts, and listings

Drawings created with this package can be reproduced using a Houston Instruments DMP series plotter. To use the Hewlett-Packard 7580 plotter, special added software is needed. To get plotted output, the user specifies the desired paper size (A–E) and the package automatically scales the output to place the drawing on that size paper. The drawing also can be reproduced, as displayed, using a dot-matrix printer.

Something close to an itemized bill of materials, but not quite, can be produced by this package. The user can call for a symbol listing that shows all the symbols used in a drawing and the number of times each symbol appears. These listings can be edited to remove unwanted items.

PC-DRAW

PC-Draw supplies a number of tools for drawing but not a full-fledged drafting capability. Shapes can be created, stored, and placed as desired in the drawing area.

The package runs on IBM/PC and some compatible computers. Use of a light pen is optional. Price of the package is $395. It is available from Micrographx, 8526 Vista View Dr., Dallas, TX 75243. (214) 343-4338

PC-Draw provides for using stored sets of prepared symbols and drawing elements. Current versions come with symbols for software flowcharting, logic diagrams, and electrical drawings. Users can create their own stored symbols as desired, and this can be done while a drawing is in progress. When selected from the symbol display, with cursor or light pen pointing, a symbol can be placed at any location on a drawing, expanded or compressed, and rotated using keyboard commands.

For most of the manipulations, a single-key command is used, but there also are a number of multiple-key commands that have to be remembered. A reference card, supplied with the package, lists these keyboard commands. It will take most users considerable practice to get them down pat.

Users work from menu displays to create new symbols or to update old ones and to create, edit, and print out drawings. The keyboard commands are used in carrying out menu-selected procedures. The optional light pen can be used to make selections from any PC-Draw displayed menu, to control cursor movement, to produce free-hand line drawings, to locate circles, and to perform many combination-key functions.

An excellent feature of the package is the split-screen display, with the symbol selection menu on the right and the drawing shown on the left as it takes shape.

Jagged lines

Resolution of the drawings is poor, and all lines that are not horizontal or vertical suffer visibly from the "jaggies." Although these jagged angled lines are quite easy to draw, they are difficult to erase.

Drawings can be divided into pages, where several screen displays can be component pages of a larger drawing. Printouts allow two screens to be shown on a single printed page; with condensed printing, four pages can be squeezed onto one printed page. Larger drawings must be made up by pasting together several printed pages. A choice of two different text fonts for drawing annotation is available, and the annotations may be placed either horizontally or vertically.

The user manual explains how to use the package, but it lacks an index and therefore is not too useful as a reference source. There is some indexing in the online HELP material, but no facility for calling up the specific instructions pertinent to the current package operations being used.

PC-Draw can be used with a variety of dot-matrix printers or with Hewlett-Packard 7470A, 7475A, 7550A, Houston Instruments, DMP, or Roland plotters.

QUICK-DRAFT

QUICK-DRAFT is a versatile symbol manipulation and drawing package. Twenty different character fonts come with the package. Up to 20 objects may be drawn, defined, and stored, then recalled and positioned on a display. A variety of drawing modes are provided. However, the quality of drawing displays and printouts is limited, producing jagged lines, and is unsuitable for showing accurate detail.

QUICK-DRAFT runs on Apple II computers. It sells for $50 from Interactive Microware, PO Box 139, State College, PA 16804. (814) 238-8294.

Line segments are drawn by defining their beginning and end points while the package is in its Line mode of operation. When a line is drawn, the cursor automatically moves to the end point, allowing a series of connected segments to be easily constructed. Arrowheads can be added to the ends of lines by simply pressing the A key.

Circles are drawn by defining a center point, then moving the cursor away to define another point through which the circle should pass. Ellipses are similarly started by locating a center point, then moving the cursor to define the width and height of the figure. Arcs of circles and ellipses are defined by angles from the centers of these figures.

To draw rectangles, squares, and diamonds, the user moves the cursor to define two opposite vertices, then keys a command for a rectangle or diamond shape. Equilateral triangles and polygons are drawn by defining the locations of a center point and one vertice, then specifying the desired number of sides (3–9).

Movement of the display cursor can be controlled by key actions, game paddles, a joystick, or a Koala pad. The distance the cursor will move at each keystroke is adjustable. To aid in keeping track of the cursor position, its *x, y* coordinates can be shown at the bottom of the display. It can be made to jump, on command, to any *x, y* position on the drawing.

To more accurately locate items in a drawing, a background grid of dots can be called for. There is a slight problem here. Items drawn after the grid is put in place will have grid-dot holes in them if the grid is removed, and, if the user wishes to fill these holes, this will have to done one at a time.

Objects can be defined and used

QUICK-DRAFT is able to store and repeat up to 20 sequences of keystrokes and cursor motions; in other words, to automatically redraw up to 20 user-created objects on command. The object definition process is simply the sequence of keyboard actions used to draw the object, but that sequence must be preceded by pressing the O key immediately followed by the $ key to define it as a stored object sequence.

The package automatically numbers the objects as they are created, and they have to be recalled by number. Therefore, it is left up to the user to keep some record of which stored object is which.

This feature has a procrustean peculiarity. If the user tries to place an object recalled from memory in a position where it would run over the boundaries of the screen, the object will be drawn in a distorted manner so it will fit within those boundaries.

Drawings in memory

A QUICK-DRAFT drawing can occupy a space equivalent to four display screen areas. The package has no zoom or panning features, but it does have a scrolling capability to allow the user to move the display about over this four-screen area in half-screen increments. QUICK-DRAFT also allows full-screen drawings to be stored in the computer memory. The number of such storage positions, up to 33 with a main memory of 256K, depends on the available memory space. The current contents of the drawing screen can be stored in one of these memory positions at any time, and stored drawings can be summoned to the display for viewing or alteration.

To help users keep track of where they are in the four-screen drawing space, row and column numbers can be displayed at the bottom of the screen. These indicate a division of the drawing space into three horizontal and three vertical regions. However, this notation is hard to grasp and has to be rehearsed before it can be helpful.

Color filling

Any region of the drawing surrounded by lines or arcs can be quickly filled with color. The user moves the cursor into the selected region, then calls for any of 22 available colors to be used.

The package also can operate in a Wash mode that allows the user to fill in the entire background of a drawing with a choice of eight colors. This same mode can be used to draw rays from a fixed point. These are constructed much like line segments, except that the cursor returns to the same beginning point each time the user defines another endpoint. With the help of a joystick or Koala pad, this feature can be used to fill areas of the screen with rays.

Selectable cursor symbol

Text for this package is made up of dot characters, and the standard symbol set contains 122 characters, including about two dozen squares, circles, crosses and such, that may be useful for drawing figures. Nineteen additional text fonts, including a Katakana Japanese and a Greek font, can be called up from storage, and tables of other symbols can be stored as well.

Many of these characters can be chosen as the cursor symbol. In the Paint mode of the package, the chosen symbol can then be continuously replicated along the path over which the cursor is moved. Further freedom in this mode is offered by the ability to rotate the position of the cursor character, to scale it in one of nine different sizes, and to have different colors as well.

To remind the user of how cursor use is currently arranged, there can be a display of the last-chosen scale factor and rotation angle at the bottom of the screen.

A Text mode of operation enters text onto a drawing from the keyboard. When in this mode, the available symbols are split into three fonts, and the user must be aware of which symbols are in each font, which font is currently in use, and also be able to manipulate a few special keyboard text commands.

Output on a dot-matrix printer can be obtained at any desired point. The printout simply replicates what is shown on the screen. No plotting equipment is supported.

Limited graphics quality

The quality of the displays and printed output is limited by the coordinate matrix on which the package is based. This has a width of 280 positions and a height of 192 positions. Lines and figures are always made up of straight orthogonal segments from one coordinate position to another. The result is that vertical and horizontal lines are smooth in appearance, while all lines and curves that deviate from these two directions have a discomfortingly jagged appearance.

The QUICK-DRAFT manual includes explanations of each of the package's commands, and there is a HELP facility that calls up a menu of 26 chunks of explanation text that the user can choose to display.

TGS: ANIMATED GRAPHICS

Simple shapes composed of lines and circles, along with text, can be created and put into apparent motion with TGS. It is not designed for animation of engineering drawings.

TGS runs on Apple II computers. It sells for $149.95 from Accent Software, 3750 Wright Place, Palo Alto, CA 94306. (415) 856-6505

Drawing is done by pointing the display cursor in the desired direction. It can then move off in that direction until the user stops it, or changes its direction, or until it reaches the edge of the drawing area on the screen. The movement can be used to relocate the cursor or, at the user's command, to leave a drawing trace behind it. To remove what has been drawn, the cursor can be run over existing traces in an erase mode. Alternatively, an entire screenful of drawings can be erased at a keystroke.

In addition to lines, circles and text can be created and located on the displays.

Drawing can take place at two levels. At the LO level a drawing can be composed on the screen. Drawings made in this way can then be dropped into small rectangles on the HI level display. To alter the LO level drawings, they can be elongated or compressed on command, scrolled to move them about, and mirror imaged.

These windows can then be moved about to place them as desired on the HI level display area, and they can be replicated. For example, a square can be replicated and moved to form a cube. New images, formed on the HI display, can then be transferred back to the LO display to make changes, such as removing the hidden lines in the cube.

A sequence of operations allowed by the package can be easily combined into a single macro instruction. The user simply gives a keyboard command that means "I am starting the macro," then keyboards the sequence itself, followed by instructions that say "I am ending the macro." Thereafter, pressing R will automatically reexecute the whole sequence.

Animation technique

HI level displays can be stored and retrieved quickly. To make a square move across the screen, the user would create and store a series of HI level display frames, placing an image of a square at successive positions in each frame. Then, when the frames are rapidly displayed one after another, a moving square will be seen.

The package gives the user control over the speed at which such frames will be displayed. It also gives the user control over the sequence in which frames will be shown, allowing a sequence to be created, then frames to be removed from that sequence and other frames inserted in any desired position.

To see what is in a sequence, the user can step through it frame by frame. Objects in the frames can be colored and backgrounds for the images created and changed.

Chapter 12
Curve and Chart Graphics Packages

Charts that display data points, curves, bars, and pie sections are familiar tools to engineers who use them in reports, conference papers, and other presentations. In the past, preparation of such curves and charts, beyond those for personal recordkeeping, have usually required application of some drafting skills.

The packages discussed in this chapter provide a way to prepare curves and charts in acceptable presentation form with the help of a personal computer. The idea is that, with the help of these packages, such charts can be quickly set up, then displayed, printed, or output on a computer-connected plotter.

Plotting data points

Some of the packages are devoted entirely to producing plots of data points. For example, the SCIENTIFIC PLOTTER package produces point plots, with the option of having the points interconnected by straight-line segments.

Despite its title, there is no actual use of plotters with PCPLOT. This little bargain-priced package will display or print plots of x, y data points. Up to four data sets can be shown on one plot, but it can be fully used only by those with some knowledge of programming. The PLOTPRO package produces similar plots, operates from displayed menus, and sells for even less than PCPLOT.

For those who program, the ASYST package provides a set of computer commands specifically designed to produce charts of data points.

Most of the point-curve packages have few built in user conveniences. Apparently they were put together with the idea that engineers are so handy with computers that they can afford to spend a good deal of time doing little programming tricks. In actuality, these packages are in competition with manual methods of drawing data plots. An engineer cannot be expected to turn to a software package if it will be easier and faster to turn out the same graph on the drafting board.

It is by no means clear that, for the occasional user, point-charting packages represent an improvement in the state of the art of graphmaking. In terms of convenience for the occasional user, SCIENTIFIC PLOTTER is probably the best of those discussed in this chapter.

Bars, pies, text, and more

A large number of graphics packages capable of producing presentable line, bar, pie, and other charts, using personal computers, have recently become available. Most of these packages are sold for use in business applications, but they can be very helpful in preparing materials for engineering presentations as well. Vendors of these packages seem to have given careful attention to making them easy to learn and easy to use in preparing a wide variety of professional-looking charts.

For example, users of the Graphwriter package have a choice of 23 different graph and chart formats. Ten of these are variations on the bar chart, with bars vertical or horizontal, solid or segmented, single, clustered, paired, or ornamented. Then there are pie charts, line, surface-line, and scatter charts, text display formats, including tables, Gantt charts and range-bar charts, flow charts and pie-bar, bar-line combinations.

The KeyChart package produces a similar variety of charts by cleverly combining common characteristics of the different formats. The CURVE II CRT package gives the user the option of employing standard line, bar, and pie formats or of custom designing their own charts.

In addition to these more or less general-purpose curve and chart packages, there are some with much more specific applications. Thus, for those who need contour plots, the GEOCONTOUR is available. Specifically designed for this purpose, this is truly a bare bones package, put together with little thought for convenience of the user, and with very few options of which the user can take advantage.

Giving graphics instructions

To give an idea of the difference between packages that cater to user needs and those that ignore the user, let us take a brief look at some of the ways these packages provide for users to express what they want.

Perhaps the best arrangement is the one provided by the Keychart package. Users input information for setting up their charts directly onto data displays. For example, there is a screen for naming, sizing, and labeling charts. Altogether it has spaces for 24 user-supplied inputs. Pressing the Return key moves the screen cursor from one data entry item to another, and the keyboard arrow keys can be used to move back and forth between items. When entry items are keyboarded, they show up in bright or inverse-video characters. Users also can move easily from this input screen to others, back and forth if needed, by pressing the PgUp and PgDn keys.

To attain this high level of operational simplicity, yet still provide for complex options, the designers of this package found they had to combine keyboarded commands with display entry. Altogether, the user has to remember about 22 special key actions to master full control of the data entry process. The manual includes a handy reference sheet listing these actions, but it will still take some practice to become familiar enough with them to make convenient use possible.

Graphwriter uses a menu display approach to present the many choices users can make. To navigate from one item and one menu to another, the user presses appropriate keys. This package has an alternative approach that should be well suited to experienced users. It is not necessary to actually wait to see a menu before making selections. If users are familiar enough with what is on the menus, they can type ahead, entering their choices in advance. The package will then move immediately through one typed option after another to bring the user to the desired point in the package's operations.

If these are the best of the lot, from the point of view of ease of learning and use, what do the others look like?

The regular procedures offered by the SCIENTIFIC PLOTTER package are spelled out by a lengthy and fixed sequence of screen-displayed prompts followed by user answers. The sequence includes setting up a graph format, inputting and scaling the data, plotting the data, labeling the graph, and storing the results of these steps. There is, however, some flexibility. At any point in the sequence, the user can call for a current view of the graph, manipulate the files, go into a BASIC language programming mode, or return to the display of prompts. And if users are dissatisfied with one of the steps, they can return directly to it through a redraw menu. The package will then erase the graph, automatically redraw it to the indicated step, and the user can proceed from there.

The CURVE II CRT package also uses long sequences of prompts and replies. Thus, a sequence of more than 60 prompts leads the user through construction of bar charts and includes the options of horizontal or vertical bars. Another 28 prompts provide for placing text on the chart. The procedures for putting pie charts together are similar.

The prompt sequences are rather rigid. If errors are made and not immediately caught, a whole sequence must be replayed to make a correction, and it is not easy for users to keep track of the choices they have made. Experienced, constant users of this package will find little difficulty in visualizing their results as their instructions are entered, but novices and occasional users may find they have to cycle again and again through the procedures to get the results they want.

The ASYST package assumes that the user is ready to do programming. Although, from a programming standpoint, ASYST commands are very powerful (that is, one command will cause a lot to happen), all that power does little good for those who are not programmers.

Data entry

Point plotting can often involve entering a lot of point data. That creates a special problem in terms of presenting the data in a form that the software package can accept while imposing minimum work on the user. Some packages handle this problem well; others seem to give it little consideration.

With KeyChart, the input data are assumed to be arranged in a row and column format, and the package receives one column of data at a time. The entry action is arranged so the cursor skips from position to position as the user methodically enters a column title, row name, and value for each item of data in the first column. When the screen for that column is completed, it is entered and the package presents another column entry screen, and so forth, until all the input data have been entered.

In lieu of keyboard entry, row and column data can be fed directly into KeyChart from packages such as Lotus 1·2·3, Symphony, Supercalc, Multiplan, and several other popular spreadsheet programs.

CURVE II CRT will accept data from the keyboard or from stored files in the DIF format used by VisiCalc spreadsheets and similar packages. In either case, the data to be used can be reviewed, ten points at a time, and edited to correct errors or make desired changes.

PCPLOT draws on data files to trace out its plots. However, users have to rely on their own programming skills to provide and enter those files. If that sounds inconvenient, consider the GEOCONTOUR package where the input information is required to be formatted as if it were being read into the computer by a data-processing card reader. Thus, the first line of element data gives, in positions 1–5, the total number of elements involved. The second line shows, in positions 1–5, a number identifying an element. Then in positions 6–10, 11–15, and 16–20 it shows the three node numbers for that element. Other input must be keyboarded in similar rigidly prescribed formats.

Graph Construction

Packages that handle formats for a number of different types of graphs need to have a way to make those formats available to users. For instance, Graphwriter comes with a total of ten different floppy disks. Two of these are concerned with the main program and with setting up for specific kinds of plotters and printers. The other eight contain the software needed to set up particular chart and graph formats. To get a desired format the user has to insert the correct disk into the computer's disk drive.

The Keychart package takes a different approach. In this nicely integrated package, once the user creates an input description file for a set of data, that same file can be used to create bar, line, and pie charts with minimum added effort. To adopt the stored chart data for presentation of other types of display, such as *x, y* graphs, stacked bars, or exploded pies, added input has to be entered. For this, the user works through an Options Screen. Here, with surprisingly few entries, the necessary added instructions are given.

See and do

One of the main problems in using this kind of package is to form a clear picture of what effect various choices will have on the actual appearance of the graph that is to be produced. Most of the packages make the user wait until virtually all instructions and data have been entered before allowing any view of an actual graphics display.

CURVE II CRT has the advantage of letting the user see a displayed version of the partially completed plot. This works particularly well in the line chart procedures, since the display can be viewed quite early in the chart construction process. For bar and pie charts, there is a lot more input before a displayed chart can be seen.

SCIENTIFIC PLOTTER is something of an exception. At any point in the input sequence, this package allows the user to call for a current view of the plot. However, other packages are much less generous. With Graphwriter, it is only after all the entry material is completed that the user gets a look at what the visual results will be. That is done by calling for a Preview on Screen from the plot menu. If that display seems acceptable, the user can go on and produce plotted or printed output.

Keychart also requires users to complete the input process before getting any screen display of graphs, but then this package relents and actually allows editing of the graph while it is being displayed. At this time, the size of the chart can be expanded or reduced, and it can be moved about within the plotting space. While observing the chart display, users can also edit titles to change their size and location. Charted variables may be edited to show them in alternative formats. For example, a variable shown in bar form can be changed so it will be displayed in line format. Types of crosshatching also can be changed.

The ideal method of constructing charts and curves would be to give instructions and enter data while using a split or windowed screen to simultaneously watch the desired display take shape. None of the packages reviewed in this chapter provide this capability.

Editing procedures

No user, even the most experienced, can expect to flawlessly enter exactly the right instructions and specifications to produce a perfect graph. That is one reason why editing procedures are needed to provide for changes in what has been entered. Another reason is that a new chart is often very similar to one that may have been previously constructed and stored. In this situation, a lot of effort can be saved by simply recalling the previous chart and editing it to meet the new requirements.

With the completion of the last data entry screen, the Keychart package stores the whole set of chart data. The package includes a cataloging arrangement in which stored charts are listed by name for later recall and use. When a chart is to be updated or another chart very similar to one already stored is to be created, the user can call up a copy of the stored chart. The displayed input screens will then appear with all the previously entered data in place, and the user need only change items that need updating. To aid in keeping track of the stored data, the package provides for printing out the contents of any selected file of chart data.

SCIENTIFIC PLOTTER uses three files for each graph. Format, Data, and Picture information are each stored in their own named file. These files can be recalled to speed the construction of similar future graphs, since the user is allowed to skip quickly through the entire graph construction sequence, accepting previously installed values and altering only those that need changing.

Graphwriter has a similar arrangement. Users can go through the entire sequence of prompts, keyboarding the answers that are requested, or they can call up a previous chart from a storage file and make changes only in selected segments to alter that chart to meet current needs.

ASYST: GRAPHICS LANGUAGE

ASYST is a computer language in the same sense that BASIC and FORTRAN are computer languages. It was designed with engineering and scientific problem solving in mind and has a variety of capabilities, including those in the areas of equation solving and instrumentation. Here we will discuss only the graphics abilities of this language.

ASYST runs on IBM/PC and compatible computers. The Systems mod-

ule, which sells for $795, includes graphics and statistics. The vendor is Macmillan Software Co., 866 Third Ave., New York, NY 10022. (212) 702-3241.

ASYST is not designed to do curves or charts conveniently. It provides the user with no menus of available functions, nor does it have predesigned formats to organize its use. Instead, like the BASIC language, it simply presents the user with an OK prompt and waits for commands to be keyboarded. As with any other language, these commands must reside primarily in the user's head, which means that the ASYST language must be learned before it can be used.

As with any other computer language, regardless of its intrinsic merits, the success or failure of ASYST will rest on the extent to which it is adopted and used by programming-minded engineers and scientists. If and when these users write programs that can, in turn, be put to use by the vast majority of engineers who have little or no interest in doing programming, then ASYST will indeed become a significant engineering tool.

The graphics capabilities of ASYST are directed specifically to making plots of mathematical functions and of sets of data. Some of the commands are very powerful. For example, plotting can be called into action by a single command, Y.AUTO.PLOT, that scales the *x* and *y* axes, sets up tick marks, and provides a background grid of dots. The same command goes on to plot a data set that the user has previously put into place.

Of course, a command that does so much leaves the user with little power to control the details of the graphs it produces. However, other commands give the user more detailed control of the appearance, scaling, and labeling of plots of functions and data points. There are also commands for controlling Hewlett-Packard 7470 and 7475 plotters.

CURVE II CRT

CURVE II CRT is used to create, print, and plot line, bar, and pie charts. Users able to do BASIC programming can produce custom-designed graphs from programming subroutines provided with the package. Other users can work from standard chart formats that do not require any programming. An added feature is the ability to produce plots of three types of equations.

CURVE II CRT runs on IBM/PC, Victor 9000, and Apple II and IIe computers. The package sells for $325 from West Coast Consultants, 4049 First St., #234, Livermore, CA 94550. (415) 449-0900.

Operation of this package proceeds from a main menu choice such as Line Chart to a sequence of prompts and user replies. For a line chart, the user first replies to 12 such prompts to specify the scaling, placement, and appearance of the chart. Then the data to be charted are entered.

CURVE II CRT will accept data from the keyboard or from stored files in the DIF format used by VisiCalc spreadsheets and similar packages. In either case, the data to be used can be reviewed, ten points at a time, and edited to correct errors or make desired changes.

Sequences of prompts

Following data entry and editing, the user responds to another set of nine prompts that provide for fitting a plotted curve to the data points and establish further details of the appearance of the plot.

At this point, the user can take a first look at a display of the plot itself. While viewing this display, the user gives further instructions, again in response to prompts from the package. These concern the makeup and position of labels on the chart. The labels and title appear on the display as the user creates them. When they have been completed, the package permits the user to order plotted output.

A sequence of file storage and data editing prompts follows. The file storage provisions for the chart data are flexible. Users can save x and y data in separate files. All the data or only selected portions can be saved, and data points can be edited. Then new plots can be constructed.

A similar sequence of more than 60 prompts leads the user through construction of bar charts and includes the options of horizontal or vertical bars. Another 28 prompts provide for placing text on the chart. The procedures for putting pie charts together are similar.

Still another sequence of prompts controls the creation and placement of plotted text and numbers. These can be added to already-completed charts with a choice of 15 character sizes, pen color, and standard or bold characters. Block characters also can be produced.

CURVE II CRT supports the use of several plotters, including the HP 7470A and the HIPLOT DMP -3, -4, -6, -7, and -29.

Equation plotting

Aside from the fact that users must enter their equations in BASIC language format, the equation plotting features of this package operate on much the same basis as chart construction. Three types of equations can be plotted: Cartesian equations of the form $y = f(x)$; parametric equations $y = f(t)$, $x = g(t)$; and polar equations $r = f(s)$.

Limits on usefulness

There are drawbacks to the incessant use of prompts in this package. The prompt sequences are rather rigid; if errors are made and not immediately caught, a whole sequence must be replayed to make a correction, and it is not easy for users to keep track of the choices they have made. Users might be better served if full-screen displays were used on which entries could be made, compared, and edited before being processed.

One of the main problems in using this kind of package is to form a clear picture of what the effects of the various choices are on the actual plot that is to be produced. CURVE II CRT has the advantage of letting the user see a displayed version of the partially completed plot. This works particularly well in the line-chart procedures, since the display can be viewed quite early in the chart construction process. For bar and pie charts, there is a lot more input before a displayed chart can be seen.

Experienced, constant users of this package will find little difficulty in visualizing their results as their instructions are entered, but novices and occasional users will probably find they have to cycle again and again through the procedures to get the results they want.

In the end, the usefulness of this kind of package rests on its ability to provide an improvement over manual composition and drafting of charts. To fully succeed, it must be fast and convenient for an engineer, even one who only occasionally uses the package, to turn out the needed results.

GEOCONTOUR

GEOCONTOUR is designed specifically to make contour plots. It is truly a bare bones package, designed with little thought for convenience of the user and with very few options of which the user can take advantage.

GEOCONTOUR runs on IBM/PCs. It sells for $500 from GEOCOMP Corp., 342 Sudbury Rd., Concord, MA 01742. (617) 369-8304.

To produce results with this package, three sets of input information are needed: the first describes the data points (elevation, or temperature, or whatever is being contour plotted) involved, the second contains *x, y* node coordinates for each data point, and the third (optional information that will speed the plotting process if supplied) describes how the plot can be made up from triangular elements, each containing three node points.

Data processing input formats

This input information is required to be formatted as if it were being read into the computer by a data-processing card reader. Thus, the first line of element data gives, in positions 1–5, the total number of elements involved. The second line shows, in positions 1–5, a number identifying an element. Then in positions 6–10, 11–15, and 16–20, it shows the three node numbers for that element. Other input must be given in similar rigidly prescribed formats.

The information for controlling the appearance of the plotted output is input in a similarly rigid format. Included here are such items as the number of contours to draw, scale factor, height of letters in the legend, location of legend, specification of any circles or rectangles that are to appear. When the actual run is made, the package will prompt the user for the file names in which node coordinates, contouring information, and data are stored. The user is asked if node locations are to be plotted, if contours are to be plotted, andif element information is being supplied. Users also are given the choice of plotting on the display screen or on the plotter.

Plotters supported by this package include the HP 7470A, 7475A, IBM XY/749, 750, DMP-4, -5, -6, -7, -29, -40, -41, and Calcomp M84.

Graphwriter

Graphwriter puts together display charts and graphs. It allows a person with little or no knowledge of typeface specification to produce a wide variety of lettered graphics, suitable for use at professional meetings and presentations.

Graphwriter runs on IBM/PC/XT computers. It sells for $595 from Graphic Communications, 200 Fifth Ave., Waltham, MA 02254. (617) 890-8778.

Users of this package have a choice of 23 different graph and chart formats. Ten of these are variations on the bar chart, with bars vertical or horizontal, solid or segmented, single, clustered, paired, or ornamented. Then there are pie charts, line, surface-line, and scatter charts, text display formats, including tables, Gantt charts and range-bar charts, flow charts, and pie-bar, bar-line combinations.

The first step in using the package is to select which of these choices is appropriate for the use at hand. Once that choice is made, the user is led through a series of menu choices to set up the desired graphic display and can produce printed or plotted output.

Along the way there are lots of choices to be made. The displays have to be scaled, colors and fill patterns have to be chosen, headings and titles have to be composed. These choices are made through menu displays. To navigate from one item to another, the user presses appropriate keys. Experienced users do not actually have to wait to see a menu before making their selection. If they are familiar enough with what is on the menu, they can type ahead, entering their choices in advance. The package will then move immediately through one typed option to another to bring the user to the desired point in the package's operations.

A total of ten floppy disks comes with the package. Two of these are concerned with the main program and with setting up for specific kinds of plotters and printers. The other eight contain the software needed to use particular types of charts and graphs, and to do so the correct disk must be inserted into the computer's disk drive.

Entering the graph description

The prompts that ask for answers from the user are arranged in a number of series or segments: chart headings, with up to three 48-character lines of text usually allowed; notes; axis titles and scales; text for legends; color and fill; and, finally, the data that are to be represented.

To allow data to be fed in from existing files, prepared with other software, the package can read the DIF files used by some spreadsheet packages.

Once a block of text has been added to a graph, it can be moved (up, down, left, right). The user follows prompts that ask for the movement distances in millimeters.

Users can go through the entire sequence of prompts, keyboarding the answers that are requested, or they can call up a previous chart from a storage file and make changes only in selected segments to alter that chart to meet current needs.

Within the limitations of the available graph and chart formats, the package offers considerable freedom in selecting line types, colors, and bar widths, as well as fonts, character sizes, color, and placement of legends and comments, and other stylistic treatments, such as rotating the pie chart orientations.

These choices can be quite detailed, bringing the user into design territory usually occupied by graphic artists. For example, one page of the user manual spells out the relationship between length of text lines and number of characters used for various choices of font size. The bulk of the user manual is actually devoted to explaining the various available text and graphics alternatives.

Choice of plotters

Some menu-choice controls are also provided for the plotted output. The size of the plot area can be chosen from a list of seven possibilities. There is a choice of output for plain paper, transparencies, or coated paper; paper sizes A or B can be selected. Slides, sized for 35-mm projectors, also can be produced with the help of Polaroid Palette Computer Image Recorder equipment.

It is only after all this entry material is completed that the user gets a look at what the visual results will be. That is done by calling for a Preview on Screen from the plot menu. If that display seems acceptable, the user can go on and produce plotted or printed output.

An impressive list of plotting devices is supported by the package. These include the HP 7220C, HP 7470A, HP 7475A, HP 7550A, IBM XY/749, 750, Mannesman Tally PIXY3, and Calcomp M84.

KEYCHART

KeyChart is designed to produce charts easily and conveniently in such formats as line, bar, pie, x, y *and scatter, as well as text for display and presentation. Single charts, or several at a time, can be plotted on a single sheet. The package will work with a variety of different types of plotting equipment.*

KeyChart runs on IBM/PC and compatible computers and also on Kaypro and Epson personal computers. It sells for $375 from SoftKey, Inc., 2727 Walsh Ave., Santa Clara, CA 95051. (408) 986-8148.

One of the key features of this nicely integrated package is that once the user creates an input description file for a set of data, that same file can be used to create bar, line, and pie charts with minimum added effort.

Data input screens

Users input information, for designing a chart, onto data displays. For example, there is a screen for naming, sizing, and labeling charts. Altogether, it has spaces for 24 user-supplied inputs. When these are keyboarded, they show up in bright or inverse-video characters. Pressing the Return key moves the screen cursor from one data entry item to another. The PC arrow keys can be used to move back and forth between items. Users also can move easily from one input screen to another, back and forth if needed, by pressing the PgUp and PgDn keys.

Altogether, the user has to remember about 22 special key actions to master full control of the data entry process. The manual includes a handy reference sheet listing these actions, but it will still take some practice to become familiar enough with them to make convenient use possible.

The input data are assumed to be arranged in a row and column format, and the package receives one column of data at a time. On the naming screen, the user enters the number of columns that are to be charted. When that screen has been completed and entered, the package will next show the user a screen for the first column of data. Here the way that column is to be handled graphically is spelled out by filling in several input items.

The user then fills in the column data. The entry action is arranged so the cursor skips from position to position as the user methodically enters a column title, row name, and value for each item of data in the first column. When the screen for that column is completed, it is entered and the package presents another column entry screen, and so forth, until all the input data have been entered.

In lieu of keyboard entry, row and column data can be fed directly into KeyChart from packages such as Lotus 1·2·3, Symphony, Supercalc, Multiplan, and several other popular spreadsheet programs.

The final data entry screen is concerned with the scaling of the chart. Using the already-received data, the package calculates and inserts provisional scaling values on this screen. The user can either accept these or change them.

Chart storage and retrieval

With the entry of the last data entry screen, the package stores the completed set of chart data. The package includes a cataloging arrangement in which stored charts are listed by name for later recall and use.

When a chart is to be updated or another chart similar to one already stored is to be created, it may pay for the user to call up a copy of the stored chart. The displayed input screens will then appear with all the previously entered data in place, and the user need only change items that need updating.

To aid in keeping track of the stored data, the package provides for printing out the contents of any selected file of chart data.

Shaping specific chart types

Selection of the desired type of display ordinarily is made from the main menu, where the choices are pie, bar and/or line, or text charts.

To adapt stored chart data for presentation of other types of display, such as *x, y* graphs, stacked bars, or exploded pies, added input has to be entered. To do this, the user retrieves the desired chart data and then calls for the Options Screen. Here, with surprisingly few entries, the necessary added instructions are given.

Choice of each specific type of chart also involves giving plotter instructions for pen speeds, use, and changes. Text charts require the use of an added input screen to define and locate the lines text that are to be used.

Editing what is displayed

Before actually plotting, the user can call for a displayed version of completed charts. These can be viewed one at a time, or several can be viewed on the same display.

At this time the appearance of the charts can be edited. The size of the chart can be expanded or reduced, and it can be moved about within the plotting space. The same thing can be done when several charts are to be plotted on one sheet.

While observing the chart display, users also can edit titles to change their size and location. Charted variables may be edited to show them in alternative formats. For example, a variable shown in bar form can be changed so it will be displayed in line format. Type of crosshatching can also be changed.

The package supports the following plotters: HP 7074A, 7475A; Calcomp M84; DMP-29,-40; Zeta 8; MP 1000; 6-Shooter; Sweet P; Strobe 200, M260; and Roland DG DXY 800.

PCPLOT

Despite the title, PCPLOT uses no actual plotters. This little package will display or print plots of x, y *data points. Up to four data sets can be shown on one plot. It can be fully used only by those with some knowledge of BASIC and FORTRAN programming.*

PCPLOT runs on IBM/PC, Compaq, and Columbia personal computers, It sells for $62.95 from BV Engineering, 2200 Business Way, #20-7, Riverside, CA. (714)781-0252.

The user replies to a few screen-displayed prompts and questions, and this package produces a plot. The procedure is very simple because the choices available are very limited. Nevertheless, the plots can be linear, log-log, or semilog (x or y).

The locations of the title and labels for the vertical and horizontal axes are fixed, but the user can decide what characters to use in these labels and what the axis scaling should be. Once a chart of a given type has been formatted, that format can be stored and reused.

The trick with this package is to make up the data files that are to be drawn on to trace out the plot. The only clue the manual gives to the needed file format is to list for the user a short, 17-line BASIC program that can be run to set up the needed data file from a given function with specified increments of the variable. The equation used in the example is $y = (\sin x)/x$. It will be helpful to those familiar with programming to know that the variables are to be stored in string variable format, with data points each entered as separate lines. However, this explanation will hardly be of interest to the nonprogramming user.

PLOTPRO

PLOTPRO produces point plots from data entered through the keyboard or from a previously prepared data file. The plots can be displayed or printed on a dot-matrix printer. The package also can produce continuous plots limited in length only by the length of available printing paper. The package works from menu displays. There is no provision for output to plotting equipment.

The price is a bargain. What the user gets is a package with some obvious, though correctable, faults and a manual that could stand considerable improvement.

PLOTPRO runs on IBM/PC, Apple II+, TRS-80 Models I, III, IV, and Victor 9000 computers. It sells for $52.95 from BV Engineering, 2200 Business Way, #207, Riverside, CA 92501. (714) 781-0252.

This package operates from two menus. The main menu allows the user to choose whether to make a normal or a quick graph, enter data, create a graph template, ask for HELP, or exit the package.

If no graph templates (formats) exist, the user must start putting one together by calling up the second menu display. Here a set of format parameters is shown, including titles, type style, and scaling (automatic or manual), along with choice options for changing these parameters. Each menu choice brings on a series of screen-displayed prompts that call for keyboarded answers from the user. As these responses are completed, the display returns to the second menu, until the Done with Changes option is selected. Then the first menu display reappears.

Once created in this way, a graph template is automatically stored for further use. The storage file will have the name assigned to the template by the user. This name must conform to the rules for file naming required by the particular computer system on which the package is being run. The manual assumes that users are familair with these requirements.

Data files for programmers

For the user who does no programming, the only realistic way to get data into this package is through the keyboard. The manual sketchily describes the file formats needed for input and shows some sample lines in BASIC and FORTRAN for generating such files, but these will be meaningless to nonprogrammers.

Some cautionary rules must be observed when keyboarding data. For example, the user who enters the point 0,0 in the middle of a data set will find that entry is abruptly terminated. The solution to this defect in the package is to substitute a very small number for one of the zeros.

Curve creation and cautions

With data in place and a plotting template ready, the user is ready to call for a graph to be produced. When one of the appropriate menu choices is made, the series of prompts that follow will ask whether grid lines are desired and what character is to be used to trace out the curve. If several curves are called for, this last question is repeated for each one.

A number of package peculiarities require caution on the part of the user at this point.

Grid lines, made up of strings of the period character, can be invoked by the user. However, there is a defect here. Because the period character is used, these lines show up about a half character space below their true position. As a result, curves that should pass through grid intersections will miss.

As many curves as desired can be plotted on one set of axes. However, the automatic scaling of the axes will be controlled by the first curve that is plotted. If the option of two sets of axes is used, automatic scaling of the second set will be controlled by the first curve entered after that set is introduced.

When a quick graph is drawn, up to ten curves can be plotted on it. However, the user must open a special buffer file for each such curve. This requires a knowledge of the particular version of the BASIC language in use on the computer on which the package is being run.

The package has an automatic phase-plotting feature that is invoked by properly labeling a file of input data. The manual assumes that users have the minimum programming skills needed to create such a file.

The PLOTPRO manual is a sketchy affair. Its 25 pages do not contain a complete and detailed explanation of the various keyboard commands and menu choices available to the user, so it cannot be employed as a reference guide. In addition, it fails to meet the information needs of those who have no programming background. The vendor might find it worthwhile to clear up some of the package's foibles and provide an improved manual, thus justifying an increase in the current bargain price of this package.

SCIENTIFIC PLOTTER

Construction of point and line charts follows a rigid, compulsory prompt-and-reply sequence in this bargain-priced package. Users can store their data and charts and modify an existing chart by rapidly skipping through the sequence. Output can be displayed, printed, or plotted.

SCIENTIFIC PLOTTER Version II runs on Apple II computers. It sells for $25. There also is a version called Scientific Plotter-PC for the IBM/PC. Adaptation software packages for use of Hewlett-Packard, Houston Instruments, or Apple pen plotting equipment sell for $25 each. The vendor is Interactive Microware, PO Box 139, State College, PA 16804. (814) 238-8294.

Users are led into operation of this package through a series of screen-displayed prompts. In addition, about 12 special keyboard commands have to be kept in mind, since these control viewing of graphs, use of a cursor, and other package activities.

Fixed procedures

The regular procedures offered by the package are spelled out by a lengthy and fixed sequence of screen-displayed prompts, followed by user answers. The sequence includes setting up a graph format, inputting and scaling the data, plotting the data, labeling the graph, and storing the results of these steps. If users are dissatisfied with one of the steps, they can return directly to it through a redraw menu. The package will then erase the graph, automatically redraw it to the indicated step, and the user can proceed from there.

At any point in the sequence, the user can call for a current view of the graph, manipulate the files, go into a BASIC language programming mode, or return to the display of prompts.

Formatting and data procedures

After defining the graph name, the first step in the sequence is to set up the way a graph will appear. This is done by answering a series of questions that define a format for the graph. Thus, for the x axis, the user is asked to set the coordinates of the ends of the axis, assign numeric values to the endpoints, specify if logarithmic scaling is to be used, what the interval between numeric labels will be, how many digits will appear to the right of their decimal points, and at what interval tick marks should appear. A similar series of prompts and answers sets up the y axis. Users can select to have a background grid of dots over the graph, at tick mark intervals, and a rectangular frame that surrounds the graph.

The next step is to provide data for the graphs, which can be done by keyboarding the data, by reading previously prepared data from disk storage, or by a short user-composed program that calculates the desired data from a formula or equation. The SCIENTIFIC PLOTTER manual shows a five-line BASIC language program for storing data points on disk and assumes that users are able to do programming. If the data are keyboarded, they can be stored on disk by a keyboard command.

Scaling and labeling

To properly display a set of data on an existing set of graph axes, it may be necessary to scale the data, which is the next step in the graph construction procedure. The package allows users to multiply x and y values by scale factors and to add offset values to them as well.

When whatever scaling is desired has been completed, the next step is to actually plot the data. Several different sets of data can be plotted using the same set of axes. SCIENTIFIC PLOTTER also allows the use of a second set of axes on a graph. If there is room on the display, it even allows users to show several different graphs simultaneously.

Points on the graphs can be shown with a choice of five plotting symbols, each available in four different sizes. In addition, the larger sizes can be filled if desired. The points can be interconnected with straight-line segments in a choice of seven colors.

Graphs can be labeled with text. Upper and lower case letters plus 32 plotting and mathematical symbols can be used. With the graph displayed on the screen, the user moves a cursor to the desired position and types the label.

Storing the graphs

The final step in the SCIENTIFIC PLOTTER procedures is to store Format, Data, and Picture information, each in its own named file. These files can be recalled to speed the construction of similar future graphs, since the user is allowed to skip quickly through the entire graph construction sequence, accepting previously installed values and altering only those that need changing.

Chapter 13
Drafting Packages

A close relationship exists between personal computer packages that provide for graphics display and those that can be usefully employed to perform drafting tasks.

Basic graphics capabilities may include the ability to create drawings by freehand sketching, to replicate and mirror image drawings or symbols that can be used to make up drawings, to move symbols and sections of drawings, and to join items together to make up an overall drawing. Many graphics packages also provide for creating text annotations and moving or manipulating them to appropriate positions on drawings.

Basic drafting features

Personal computer drafting packages build on these features offered by graphics packages. However, computer-based drafting could offer little or no improvement over manual drafting methods unless it also allowed engineering drawings to be created conveniently and rapidly. Thus, a capable drafting package such as VersaCAD provides lines, arcs, circles, rectangles, fillets, and regular polygons as primitive drawing elements. These can be placed on the screen at any location in any desired size. There also is a cross-hatch function that works automatically to fill closed areas.

Dimensioning is another essential feature of conventional engineering drawings. To make the use of drafting packages efficient, varying degrees of automatic dimensioning are offered. For example, a package such as RoboCAD will automatically annotate lines and other features with dimension text, but dimension lines, arrows, and reference marks have to be drawn by the user. In the CADKEY package, once the distance between two designated points is computed, the user may choose to dimension the horizontal, vertical, or parallel distance. After the location of the dimension text is selected, the package automatically creates the text, arrowheads, and associated lines. Size and font of the text, arrowhead size, and other parameters can be adjusted as desired. Dimensioning of arcs, circles, and angles is carried out in a similar manner.

The ability to handle both English and metric units, when calculating dimensions, is a useful one offered by several drafting packages, as is the ability to translate automatically from one system of units to the other. Some drafting packages make dimensioning an optional feature. Thus, CADPlan offers an optional automatic dimensioning program, but it costs an added $250.

In personal computer drafting packages, control over text can be quite flexible. For example, with the Drawing Processor II package, the user can define the size, slant, and ratio of width to height of the text, as well as its location on the drawing. And in 3D drawings, the text will remain attached to the proper plane when the image is moved or tilted.

The ability to handle drawings sufficiently complex to represent the actual objects with which engineers have to deal is another essential capability for drafting packages. For example, the EnerGraphics package is a very capable graphics tool that comes close to being useful for general drafting applications. However, the package limits drawings to 800 line segments and 200 arcs and symbols. At that level of complexity, many simple layout drawings can be put together but detailed and complex objects cannot be properly represented by engineering drawings. In contrast, capable drafting packages are able to handle thousands of elements in a single drawing.

Speed of operation is another very important characteristic of capable drafting packages. Most of the presently available packages are slow, particularly when they redraw displayed objects. The panning process, in which the user shifts the displayed view of a drawing that is larger than the computer screen, involves time-consuming redrawing. For a complex drawing, a single panning move may take over a minute to complete, and that is a large enough time-delay to try the patience of most users.

One way of speeding up the redrawing process is to simplify the calculations that it involves. Thus, the CADPlan package attains its relatively speedy operation by using integer arithmetic instead of the more accurate floating point arithmetic used by most other drafting packages. For many applications, the slight distortion introduced by this approach may not be significant. For example, a rectangle 10 feet long will expand about a quarter of an inch when it is rotated 30 degrees by CADPlan, and it will continue to expand each time it is rotated. This growth is due to the arithmetic rounding that takes place.

Ability to create drawings in layers is an important feature for many applications of drafting packages. The layers can be used to portray such items as overlapping plumbing, electrical wiring, and other details in a building design or to plan the levels of a layered printed circuit board. Some packages provide for only a few layered drawing levels, others for many. Thus, the VersCAD package allows each drawing to include up to 250 layers of material, and each layer can be shown in a different background color to help keep track of layer location.

A variety of features

Personal computer drafting packages offer users many different kinds of features that may be more or less important, depending on the uses to which the package will be put.

Most packages will offer the user choice of a number of different types of lines for drawing. These can include solid lines of several widths as well as dashed, dotted, and combination lines. Similarly, a variety of different text styles and fonts may be available.

Special editing features can be helpful. For example, the CADKEY package provides some very specific selection functions. When editing a drawing, the user can choose to affect only points, or lines, or arcs, or other specific types of items. If, for example, the user chooses to edit and delete points, the package will not allow adjacent lines to be inadvertently deleted.

Being able to create easily lines that are accurately horizontal or vertical is a helpful feature. In the AutoCAD package, an orthogonal mode of operation converts all lines at less than 45 degrees to the horizontal, and all those at more than 45 degrees to the vertical. If they wish, users can define the angle from the vertical and horizontal at which this orthogonal "snapping" will take place.

Another useful drawing aid is a displayed background grid. Many packages provide the ability to "snap" points on a drawing into alignment with the crosspoints of this kind of grid, and they also allow the user to specify the spacing between grid dots.

A powerful feature, which seems to be offered only by the AutoCAD package, is to provide nonprogrammers with the ability to customize package operations. Users who want tailored package functions will appreciate the ability to define their own display menu choices and also the ability to create macros, special command sequences to get the package to perform customized tasks.

Stretching and distorting basic drawing shapes can be a valuable drafting asset. The CADPlan package, for example, allows rectangles to be pulled into squares or long thin shapes, with size changes as desired. With the Design Board package, a circle can be stretched into a keyhole shape, using a function that causes parallel motion of selected points.

Three-dimensional capabilities

The beauty of 3D packages is their ability to create a computer-stored representation of a three-dimensional object. The stored image then can be turned about and examined on the display screen, and conventional engineering drawings can be readily created from two-dimensional views of the stored data.

Some 3D packages, such as CADKEY and MicroCAD, provide both object-creation and 2D drawing capability. Others, such as Design Board 3D, just allow the user to set up and store 3D objects, leaving the production of engineering drawings to other, compatible packages.

For viewing 3D objects, a desirable feature is automatic hidden-line removal. Without it, displayed views of 3D objects can often be confusing, and it is difficult to recognize whether objects are complete in all details. Unfortunately, automatic hidden-line removal is a feature still under development and is not always completely satisfactory. Thus, the MicroCAD package has a hidden-line removal feature that works on only one object at a time. If one object is in front of another, the package will not remove all the lines in the area where the two overlap. The vendor of this package believes that, at present, the most efficient way to remove hidden lines is to have the user do it manually. The CADKEY 3D package provides no automatic hidden-line removal at all. Manual editing functions can be used to remove such lines. However, if the display of the object is then rotated, it will have to be reedited to correct these lines.

Drawing elements, such as lines and curves, are plentiful for 2D drawing, but few vendors offer any similarly convenient 3D drawing elements. One exception is the optional 3D program offered with the VersaCAD package, which includes a 3D "box" as a primitive element. The Design Board 3D package provides meshlike spherical and elliptical dome elements and allows mesh-formed 3D surfaces to be defined flexibly.

Curve-fitting is made available by some 2D drafting packages. The VersaCAD package has a Bezier curve-fitting feature that allows drawing segments to be connected by smooth curves. Drawing Processor II allows curves to be fitted to any selected points, then stored and used as defined drawing elements.

Curve-fitting capability is extended into three dimensions by the Design Board 3D package. Here, surfaces are defined by first fixing four corner points, which the package uses to set up a rectangular mesh. The user then sets the position of some of the desired points of the surface above or below the plane of the rectangle; using a Bezier curve-fitting procedure, the package then forms a curved surface that goes through these points. Depending on the fineness of the mesh used, quite complex curved surfaces can be formed.

The EnerGraphics package, which has been mentioned as having near-drafting capabilities, also provides an example of partial 3D capabilities. It creates 3D surfaces but only as representations of mathematical functions or sets of points in three-dimensional space. Three-dimensional objects also may be created, but these can be made up only of planar panels that are copied and moved about to form the 3D forms. In addition, there is no ability to use views of the 3D images as 2D drawings.

Viewing images and drawings

Most personal computer drafting packages provide a variety of viewing capabilities. As has been mentioned, panning across a drawing tends to be slow because of the massive calculation and manipulation involved in the redrawing process. Zooming features usually work more rapidly, and, to some extent, these can be used in place of panning.

A feature artfully employed by some packages is the ability to display several different images, views, or items at the same time by dividing the screen up into display windows. For example, on the CADKEY package display, in addition to an area or window where the current part is shown, there is a status window that indicates such parameters as the current level, color, line type, pen number, working scale, and depth, while a third window displays the currently available menu choices. Other information on the screen display includes the current coordinates of the drawing cursor and prompts to aid the user in choosing menu options.

In a similar manner, the Design Board 3D package presents the user with a display screen divided into four windows that show a perspective view, along with top, bottom, and right side views of the object being modified.

Storage features

More than any other feature, the ability of personal computer drafting packages to retrieve predrawn symbols and sections of drawings from storage has contributed to their productivity. For example, users of the VersaCAD package can make up parts that then are stored in pages containing up to 100 parts each. The stored parts can be called up and placed where desired in the currently active drawing. The user simply points to the desired part with a graphics tablet stylus and presses the Enter key. The vendor makes about five thousand predrawn part symbols available.

Some packages allow entire drawings to be retrieved from storage and added to the currently active drawing. A point to watch for here is that, although the retrieved drawings can be displayed and printed or plotted, there may be no capability to modify them or to connect them to lines on the currently active drawing.

As stored files of drawings and symbols continue to grow, locating the desired item from all those that are stored can become a problem. Packages that allow drawings to be filed under descriptive names can help to ease this problem, but it is mainly up to the users to take care to keep proper track of what they have stored. Some packages allow objects within drawings, as well as the drawings themselves, to be named for retrieval from storage.

Computer-stored information can include data about the items that are drawn as well as the graphical data. Such information can be very useful when it comes to compiling bills of material, cost estimates, and other documents related to drawings. Some packages allow written specifications, costs, and other information to be filed for each object in a drawing. To create a bill of materials, the package may automatically identify all the appropriate objects in a drawing and print a summary showing such information as the vendor, order number, and cost for each item. Compiling a bill of materials in this way is generally a slow process but can nevertheless be a convenient one.

One storage problem that can be most annoying arises when the capacity limits of the computer's disk files are reached. What is needed is a procedure that can avoid any loss of information entered into the computer but not yet stored on the disk. This problem is not too well handled in some packages. For example, when an AutoCAD drawing gets too large to be contained on a single storage disk, the display is turned off and the user is asked whether the drawing should be retained or abandoned. However, either answer produces the disheartening message "AutoCAD gives up."

Generating computer output

When the user has finished creating objects and drawings with one of these packages, the next order of business generally is to produce plotted output or some other form of useful output.

One of the main problems encountered with plotters is that each drafting package is able to utilize only certain models of plotting equipment. If a prospective purchaser of a package already has a plotter, it is important to find out if the package being considered is designed to accommodate that plotter and, if not, what it will cost to get an appropriate interface.

It can be useful to exchange drafting information between one computer and another. Often a personal computer may be used to prepare drawings that are to become part of a larger set of drawings stored on a larger computer. AutoLINK is a package designed to transfer information between personal computers using AutoCAD software and larger computers using INTERGRAPH software. In the near future more packages performing this kind of communication function can be expected to appear.

Some packages make it convenient for the user to display objects and drawings one after another as in a slide show. An extension of this, providing animation of drawings, is achieved in the AutoCAD package by a feature that saves images in a special file, then allows them to be played back in rapid sequence.

An active art

Developing drafting packages for personal computers is an active art. What is available today is only a prelude to the tools that will be available in the near future. This is evident in the continuing changes and updates to the best of the existing packages. That means that the best choice is probably a package that is open to such improvements, makes them frequently to keep up with the state of the art, and passes the improved versions on to their users at reasonable cost.

AUTOCAD: A SEASONED AND VERSATILE DRAFTING PACKAGE

Among 2D drafting packages that run on popular personal computers, AutoCAD is widely regarded as a standard of comparison. It is certainly the best-selling package. AutoCAD supports a wide variety of input and output equipment, has most of the features available in such packages, and is well designed. Storage operations are particularly well handled.

AutoCAD is designed to run on several computers, including the IBM/PC XT, Victor 9000, Zenith Z100, NEC APC, Colombia, Eagle PC, Digital Microsystems, TI Professional, DEC Rainbow, Compaq, and many CP/M computers. Price of the package is $1,000. Semiautomatic dimensioning costs an added $500. It is available from AutoDesk Inc., 150 Shoreline Highway, #B20, Mill Valley, CA 94941. (415) 331-0356.

AutoCAD provides a choice of easily invoked drawing elements. In addition to straight lines, there are circles, arcs, trace lines, and solid, filled-in areas.

Once created, objects can be nested within other objects. For example, identical doors in a drawing of a building can be produced by calling for the door shape and specifying the locations at which it is to appear. Shapes can be placed, stretched, shrunk, scaled, rotated, and repeated. Materials within a screen-display window, defined by its four corner points, can be moved as a group.

Flexible drawing features

An orthogonal mode of operation converts all lines at less than 45 degrees to the horizontal and all those at more than 45 degrees to the vertical, or the user can define the angle from the vertical and horizontal at which this orthogonal "snapping" will take place. As an optional drawing aid, an alignment grid may be displayed, and the user can specify the spacing between the grid dots.

For dimensioning, the package will compute the distance between any two points and also the area enclosed by a set of points. An optional semiautomatic dimensioning feature is available.

A zoom feature allows a selected small area to be enlarged to fill the screen or a portion of a drawing to be reduced to fit into a larger context. A portion that is to be viewed in this way can be defined by simply pointing to the upper-right and lower-left corners of the rectangular area that is desired.

Internal data are maintained in floating-point format. The package allows a ratio of a million to one or better between the largest and smallest objects that are represented.

Drawings can be stored in overlays, a convenient feature for multiple-layer designs. With a 640K main memory in the computer, AutoCAD can handle up to 127 such layers. However, once a layer is created it remains an active part of the drawing. Material displayed on all such layers will continue to be scanned, eating up processing time, even when they are not being used.

Animation of drawings can be achieved by a feature that saves images in a special file so they can be played back in rapid sequence.

Other features include: partial deletion of drawing elements, such as breaking an existing wall to insert a window; a cross-hatch command that can draw from a library of 38 predefined patterns or invoke user-defined patterns; a freehand sketch mode of operation; automatic creation of fillets; automatic closure of polygons; the use of circular or radial arrays; alternate text fonts; units that can be decimal or in feet and inches.

Speed of operation

Like other personal computer drafting packages, AutoCAD is slow, particularly when it is redrawing. The package takes time scanning all drawing layers in the redrawing process. Panning a displayed drawing requires redrawing; for a complex display, a panning move may take over a minute to complete.

As the drawing in progress gets larger, program operation gets slower.

However, use of the 8087 coprocessor can help speed up zooming, editing, and panning. Another feature used to speed up operation is the ability to temporarily eliminate filled areas, thus speeding up image regeneration after panning or zooming.

When a drawing gets too large to be contained on a single storage disk, the display is turned off and the user is asked whether the drawing should be retained or abandoned. However, either answer produces the disheartening message "AutoCAD gives up." To avoid this situation, the floppy disk user must keep in mind that, because of work and backup files, only about a third of the disk space is available for storage of the actual drawing file. In general, it is wise to assign only one drawing to a floppy disk.

Wide choice of peripherals

The package offers excellent peripheral support. A light pen, touch pen, or a mouse as well as the keyboard or a digitizer pad can be used for input.

AutoCAD is designed to run using two display screens simultaneously, one for graphics, the other for text. If only one screen is used (this can be done when it is connected to the IBM Color/Graphics Adapter), the user has to constantly switch back and forth from text to graphic mode. If higher resolution than that provided by IBM displays is desired, the package can be used with a Hercules Graphics card that provides 640-by-400 pixel resolution.

The package provides for use of electronic tablet input. In tablet mode, however, no reference cursor is displayed on the screen to indicate the user's position. A nice tablet input feature is the ability to make menu choices by pointing to tablet areas, instead of having to refer to the display screen.

Use of a plotter for output is required by the package. Maximum plotting resolution is 0.025 millimeters. Paper from size A to E may be used, provided the chosen plotter will accommodate it. Multiple plotting pens with different colors and line types can be supported.

Customizing and communication

Users who wish to customize AutoCAD operation, or have it customized for them, will appreciate the user-defined menus and the ability to create macros, special command sequences to get the package to perform customized tasks. The package is designed to be customized by nonprogrammers. For example, previously created shapes or drawings can be recalled by pointing to a single item on a menu display. Drawing sessions can be initiated from a command file that sets up the custom commands that are to be used in that session. This kind of customizing may also include special prompts to remind the package user of how to proceed, and these can be conveniently set up using a text editor or word processing program.

For those who use digitizers with multiple buttons, menus can be customized to show button numbers, so menu choices can be conveniently made from the digitizer.

Communication with central drawing files or other computers is handled by storing drawings in a standard ASCII format for transmission or reception. This allows drawings to be sent easily over telephone line connections.

The package is well supported. The vendor claims that it now has over fifteen hundred users. It is written in the C computer language, which can be moved easily from one type of computer to another, so improved performance on more-capable machines than the IBM/PC can be expected in the future.

AUTOLINK: TRANSLATES BETWEEN AUTOCAD AND INTERGRAPH

It is sometimes desirable to transfer drafting work done with a personal computer over to a larger computer or to move work in the other direction. This package allows such transfers to take place between a personal computer using the AutoCAD package and a minicomputer or mainframe using INTERGRAPH software. These transfers take place over wire or telephone connecting lines.

The AutoLINK package resides on the larger computer and includes communications software. Both computers must be equipped with serial asynchronous communications equipment. AutoLINK costs $10,000 and is available from Interactive Graphic Services Co., 200 South Meridian St., Indianapolis, IN 46225. (317) 638-5057.

CADAPPLE: SEE REVIEW OF VERSACAD

CADKEY: DRAWINGS OF WIREFRAME PARTS

CADKEY is a well-organized two- and three-dimensional drafting package. The user creates three-dimensional objects or parts, which appear in wireframe form. Drawings then can be displayed or plotted to represent projected or isometric views of the part.

The package runs on IBM/PCs and compatible computers. It sells for $1,895; a yearly upgrade package is available for $200; the vendor is Micro Control Systems, 27 Hartford Turnpike, Vernon, CT 06066. (203) 647-0220.

CADKEY users choose the operations they wish to perform from a well-organized sequence of screen-displayed menus. The menu options can be chosen from the keyboard or from the user's digitizer board. Experienced users can bypass the menu sequences, invoking CADKEY functions directly by entering command codes.

The screen display is also well organized. In addition to an area or window where the current part is displayed, there is a status window that indicates such parameters as the current level, color, line type, pen number, working scale, and depth. These parameters can be changed immediately, as desired, by entering special keyboard commands. A third window displays the currently available menu choices. Other information on the screen display includes the current coordinates of the drawing cursor and prompts to aid the user in choosing menu options.

Creating lines

Drawing can be done by making a string of lines, in which the next line begins where the last one left off, or lines can be defined by making them parallel to, or perpendicular to, existing lines. In a more general way, a line can be created by defining its endpoints.

Because of the difficulties that can arise in accurately placing lines on a three-dimensional drawing space, CADKEY provides nine different ways to define the endpoints of a line. These range from positioning the cursor to indicate the desired endpoints, to locating them at endpoints or center-points of existing lines or arcs, to entering the exact endpoint coordinates on the keyboard.

In addition to lines, the package provides for the creation of points, arcs, circles, rectangles, and fillets.

Transforming shapes

Once a basic shape such as a square has been created, it can be transformed by translating, rotating, shrinking, or expanding it. Shapes also can be replicated or copied. The material to be copied can be defined by surrounding it with a box, or the type of item to be copied can be chosen and then the particular item to be copied can be spotted by placing the cursor on it.

Created items can be viewed from the top, front, back, bottom, left, or right sides, and there also are two isometric views that can be chosen. Each of these views can be used as drawings to be displayed or plotted.

Viewing manipulations provided by the package also include panning and zooming. A two-dimensional grid can be set up, with spacing selected by the user. Drawing points can be forced to align with the grid to a snap resolution selected by the user.

Specific and wholesale editing

Editing functions allow the user to trim or extend lines and arcs and to delete items that have been drawn. The package provides no automatic hidden-line removal. The editing functions can be used to remove such lines. However, if the part is then rotated, it will have to be reedited to correct these lines.

To aid in handling complex drawings, the package provides very specific selection functions. Thus, the user can choose to select only points, lines, arcs, or other specific types of items. If, for example, the user chooses to edit and delete points, the package will not allow adjacent lines to be deleted inadvertently.

Sections of a drawing also can be deleted in a wholesale manner by first surrounding the desired material with a defining box, then deleting it. If desired, the specific selection of types of items to be deleted can be combined with this box deletion technique.

Dimensioning automatically

Automatic dimensioning is accomplished by having the package compute the distance between two designated points. The user may choose to dimension the horizontal, vertical, or parallel distance. After the location of the dimension text is selected, the package automatically creates the text, arrowheads, and associated lines. Size and font of the text, arrowhead size, and other parameters can be adjusted as desired. Dimensioning of arcs, circles, and angles is carried out in a similar manner. In addition, labels, with witness lines, leaders, and arrows, can be created and positioned as desired.

Filing parts, shapes, and drawings

When work on an object has been completed, it can be stored as a part under a file name selected by the user. Other stored files may contain information on shapes and notes that can be retrieved whenever needed to be added to the part that is being created or modified. Views (drawings) that are to be plotted are stored in special plot files.

Plotting itself is carried out as a separate operation from the rest of the package functions with the advantage that, provided the computer is capable, plotting can proceed while the user turns to other CADKEY tasks.

CADPLAN: FAST AND FLEXIBLE

CADPlan operates rapidly, faster than many other personal-computer-based CAD packages. It has a good supply of primitives, good editing features, and good support of peripherals. CADDraft is a junior version with smaller drawing-size capability.

Written for the IBM/PC, it requires 320K main memory, two floppy disk drives, IBM Color/Graphics Adapter, and plotter or printer. List price of the package is $1,200. An optional data base feature sells for $350. An autodimensioning feature costs $250. An abbreviated version of the package, called CADDraft, is available for $95. The vendor is Personal CAD Systems, 981 University Ave., Los Gatos, CA 95030. (408) 354-7193.

Package operation and documentation are designed to make CADPlan easy to learn, and they do their job well. The prompts, menus, and other displayed information items are easily and quickly understood.

An initial menu lists 18 commands. These are supplemented by added subcommands. Drawings are filed and can be retrieved by name; no provision is made to allow added displayed notes in case the user should fail to remember the name assigned to a given drawing.

Primitive elements available for drawing include lines, arcs, circles, and rectangles as well as predefined symbols and text. Available line types include solid, dashed, and dotted in three different colors. Line width and text size can be defined by the user. Symbol libraries can be created for frequently used drawing items.

Initial use of the package is quite easy, but to take full advantage of its capabilities, the user must employ the package commands with considerable skill and knowledge.

The package allows shapes to be stretched as desired to modify or create a drawing. Rectangles, for example, can be pulled into squares or long thin shapes, with size changes as desired. There is, however, no provision for automatically changing the scale of an entire drawing.

Editing can be performed on any stored drawing. This is somewhat complicated by the fact that shapes created by different techniques have to be edited differently. For example, if a rectangle was originally created by constructing four separate lines, each can be separately edited. However, if it was created as a program-generated rectangle, it must be edited by overall commands. There is no way to tell, simply by looking at a drawing, just how its shapes were created.

Like other similar packages, CADPlan makes libraries of ready-to-use symbols available.

The price of speed

CADPlan attains its relatively fast operating speed by using integer arithmetic instead of the more accurate floating-point arithmetic used by most other drafting packages. For most applications, the distortion introduced by this approach may not be significant. For example, a rectangle 10 feet long will expand about a quarter of an inch when it is rotated 30 degrees by CADPlan, and it will continue to expand each time it is rotated. This growth is due to the arithmetic rounding that takes place.

Layering, zooming, windowing

Drawings can be layered, with up to 10 layers allowed on an IBM/PC with a 320K main memory, or up to 65 layers with a 640K memory. The layers can be used to show overlapping plumbing, electrical wiring, or other detail in a building design, or they can be used to plan the levels of a layered printed circuit board or any other type of drafting that lends itself to layering.

With several layers in simultaneous use, the operation of CADplan may slow down. However, the user has the option of switching off layers not in current use to speed things up again.

A zoom command works together with a pan command that allows the user to move a screen display over the entire surface of a drawing larger than the screen. The package must redraw the screen each time the pan moves outside the presently viewed area, greatly slowing down the panning process. Zoom operates in steps of 2X magnification.

The windowing feature is a helpful one. Up to five separate windows, each showing a different drawing area, can be displayed simultaneously, and the user can quickly switch back and forth between these areas, thus avoiding the need to use the slower panning feature.

Two helpful options

An optional automatic-dimensioning program ($250) measures the distance between any two selected points and draws a dimensioning line with accompanying dimension text. This feature is quite flexible. Text can be placed as desired.

Another optional program module is a data base designed to ease the process of creating a bill of materials from a finished drawing. This feature allows a library of useful information related to shapes and objects to be maintained. For example, written specifications, costs, and other information can be filed under each object, a very important feature for many types of design.

To create a bill of materials, the package counts the number of objects in a drawing and prints a summary showing the number of each object along with the vendor, order number, and cost, automatically taking these items of information from the data base. Compiling a bill of materials in this way is a slow process, nevertheless a convenient one.

The package is designed to work with a three-button mouse input device, but it also accommodates at least three different brands of input tablet devices. Input via digitizer tends to be slow.

About a dozen different plotters can be used with the package, as well as IBM and Epson dot-matrix printers. A recent addition is the eight-pen Model 84 plotter from Calcomp in Anaheim, California.

CADDraft limitations

The low-cost version of CADPlan, called CADDraft, comes with considerably smaller drawing capacity. It can handle only ten drawing layers and a 500-point data base, while CADPlan can handle up to 65 layers and 67,500 points. Features of CADPlan missing in CADDraft include: copy, step, and repeat commands; vertex highlighting; absolute and relative coordinate input; view recall and fit commands; and area and volume calculations.

DESIGN BOARD 3D: CREATING THREE-DIMENSIONAL OBJECTS

Design Board 3D is a capable package for creating computer-stored representations of three-dimensional objects, including hidden-line perspective views. Although it can produce plotted versions of various views of the created objects, it is not a full-fledged drafting package since it lacks the ability to dimension and annotate the drawings. The package vendor provides optional software to convert Design Board views into input for the AutoCad drafting package.

Design Board 3D runs on IBM/PCs and compatible computers equipped with a math chip and mouse input device. It sells for $750. Software to *convert its output for use with AutoCAD sells for $295. The vendor is MEGA CADD, 401 Second Ave. So., Seattle, WA 98104. (206) 623-6245.*

This package is remarkable, if only for the fact that the keyboard goes dead as soon as the software is powered up. Users had better get familiar with using a mouse for input, for it, along the display screen, will be the only means of communication they have with Design Board 3D.

This means that when words, such as file names, are to be entered, they have to be picked out one letter at a time by moving crosshairs around on an alphabet display; the same is true for numbers. Since the Design Board package includes no capability to add annotation or dimensions to drawings, this limited means of alphanumeric communication is adequate to the task at hand. When it comes to entering lines and points, the mouse is far easier to use than a keyboard. Nevertheless, why the keyboard had to be completely dead is something of a mystery.

Polygons plus

The basic shape created by this package is the polygon. Any polygonal shape with up to 17 sides can be formed in a plane and then extended in the third dimension to create a building block shape. These blocks then can be combined, as desired, to form more complex 3D objects. In addition to polygons, the user can also play with three net-type forms: a spherical dome, an elliptical dome, and a saddle-shaped curved surface. Each of these are formed by an open mesh of intersecting curved lines.

While the user is creating an object, the screen shows a large creation space complete with background grid, a smaller perspective view space, and small menu rectangles that contain the various commands that can be mouse chosen. As the object is created, its perspective view will take shape automatically. Included on the perspective space is a little x, y, z-axis orientation figure and a numerical reading of the current location of the drawing cursor along these axes.

To create a square, the user defines the endpoints of one side by moving the crosshairs with the mouse, then places one more point to show on which side of the line the square will lie. The package does the rest. To form a rectangle, a similar procedure is followed, but the distance from the line to the opposite side must be specified. Arbitrary plane shapes can be formed by specifying their vertex points. To turn these plane figures into parallelopipeds, all the user has to do is specify a top and a bottom point in the third dimension. The principle of this package seems to be to have the user do only the minimum necessary to achieve the objective, an admirable approach.

Spherical domes are created by defining a center and radius points. In addition, the user must indicate the degrees of the dome, that is, the angle from the center of the sphere that defines the extent of the dome. For a hemisphere, that angle would be 180 degrees. A similar angle must be chosen for elliptical domes, along with major and minor axis dimensions.

Surfaces are defined by first fixing four corner points; the package uses these points to set up a rectangular mesh. The user then sets the position of some of the desired points of the surface above or below the plane of the rectangle; using a Bezier curve-fitting procedure, the package then forms a curved surface that goes through these points. Depending on the fineness of the mesh that is used, quite complex curved surfaces can be formed. However, the package will not display grids if more than 1,800 grid points are to appear on the screen.

The user has a lot of control over the object creation process, including the ability to rotate, move, or scale the objects, repeat parts or whole objects, mirror image, snap to the grid or turn off the grid, connect one object to another, and pan and zoom.

Perspective viewing

If the creation-mode views of the objects are not sufficiently enlightening, the user can view the model in perspective and move the viewing around to expose all sides. Creation produces a wireframe version of the object, but automatic hidden-line removal can be invoked for better viewing. Curved surfaces and domes are not affected by this hidden-line removal process. They continue to appear in mesh form. Some hidden-line imperfections may show up where different objects intersect.

Pulling and twisting

Once objects have been created using Design Board, they can be modified to take many different forms. For example, one face of a cylinder can be collapsed into a single point to pull it into a conical shape. Or a circle can be stretched into a keyhole shape, using a function that causes parallel motion of selected points. In a similar fashion, an entire face of an object can be moved in parallel fashion, stretching all the lines that connect it to the rest of the object as it moves.

During these modification procedures, the user is presented with a display screen divided into four windows that show a perspective view, along with top, bottom, and right side views of the object being modified. This allows added insight into the results achieved by the various changes the user makes.

Printing and plotting

Views displayed on the screen can be printed on a dot-matrix printer or plotted, provided your plotter is a Hewlett-Packard or Houston Instruments model supported by the Design Board package. The simplest way to get hard copy is to dump the screen, that is, have the printer reproduce exactly what is displayed on the screen. If the user wants to get a scaled drawing, including a perspective drawing with hidden lines removed, these too can be produced.

DRAWING PROCESSOR II: SOME NICE DRAFTING FEATURES

Drawing Processor II is designed for digitizer pad input. It has a curve-fitting feature and easily generates ellipses, but line handling can sometimes be awkward. There is no panning.

The package is for the IBM/PC and compatible computers. It requires 256K of main memory and the use of the 8087 math coprocessor chip is recommended. It also requires use of a color graphics adapter. The IBM color adapter or graphics cards from several other vendors may be used. If a second display screen is available, the package will support two-screen operation.

List price of the package is $995. A version that lacks overlays, computer-aided dimensioning, command file input, and component menu extension is available for $495. Program revisions for a period of one year cost $100 extra. It is available from BG Graphics Systems, 824 Stetson Ave., Kent, WA 98031. (206) 852-2736.

Unlike most other drafting packages, this one does not operate from screen-displayed menus. A digitizer pad is the preferred form of input. Drawing commands are selected from the digitizer pad by spotting them on a menu strip (at the top of the digitizer) with the drawing stylus or puck. Editing of on-screen drawings can be done in a similar manner so, if desired, the computer keyboard can be ignored and all drawing tasks, including ordering of output plots, can be conducted from the digitizer pad. However, commands also can be keyboarded.

Unless instructed otherwise, the package assumes the user is drawing straight lines. Pressing the stylus to define a point, then moving elsewhere and pressing the stylus again will result in a line being drawn between these two points. If the user does not want this to happen, the PEN UP command must be given. If not, the user runs the danger of drawing unintended lines while moving from one part of the drawing to another.

Handling lines sometimes can be awkward. The package does not include rubberbanding of lines or automatic creation of orthogonal lines. When the stylus is moved across the digitizer, the cursor on the display screen does not follow it.

Control over text is quite flexible. When text is incorporated in a drawing, the user can define the size, slant, and ratio of width to height of the text, as well as its location on the drawing. In 3D drawings, the text will remain attached to the proper plane when the image is moved or tilted.

The Drawing Processor has a useful curve-fitting feature. In addition to creation of lines, arcs, circles, and ellipses, curves can be fitted to any selected points, then stored and used as defined drawing elements. Components, such as these curves, can be manipulated by stretching, shrinking, or rotation and can be added as choices on the digitizer menu strip.

Lines drawn close to the horizontal or vertical can be automatically adjusted so they become orthogonal. The package allows a tolerance to be set for this correction process to take place. The package also will report on distances between points, areas enclosed by lines, and angles between lines.

Pan by zoom

There is no panning facility provided with the Drawing Processor. To get a view of portions of a drawing beyond the immediately displayed area, the user has to zoom out for a wider view, then zoom back in again on the desired area.

Operation of this package depends on use of the 8087 math coprocessor chip, which is capable of dramatically speeding up calculation. Nevertheless, the package tends to run rather slowly compared to some of the other CAD packages for personal computers.

Like most packages, this one is supported by telephone-based technical support. Unfortunately, there seem to be only a few support people available, and users have found the support number is often continuously busy. The vendor also has been found to be slow in delivering replacement disks when they are needed.

ENERGRAPHICS: CHARTS AND DRAWINGS WITH LIMITED THREE-DIMENSIONAL CAPABILITIES

EnerGraphics is a multipurpose graphics package for creating charts, simple 2D drawings, and displays of some 3D objects. It is almost but not quite suitable as a drafting tool since there is no ability to use essential tools such as graphics tablets nor is attention paid to providing dimensioning.

EnerGraphics runs on the IBM/PC and compatible computers. It sells for $250. A plotting option costs an additional $100. It is available from Enertronics Research, 150 North Meramec, #207, St. Louis, MO 63105. (314) 725-5566.

Extensive use is made of the ten Function keys on the IBM/PC keyboard. These keys are used to implement ten different modes of operation. Some attempt was made to make this use consistent across all the modes, but this is only partially successful, so the same key may have several different uses. However, users are reminded of the key functions by menu displays.

Charts and slide shows

There are separate but similar procedures for creating bar, pie, and line charts. In each case, after the chart is set up, it can be edited to meet user requirements. Data inputs to these chart creation procedures can be in equation form or in the form of (DIF) files derived from spreadsheet packages as well as from direct keyboard entry.

Completed charts can be displayed one at a time, and they also can be shown in sequence, presenting a sort of slide show. In addition to its displays, the package can produce printed and plotted outputs.

Simple two-dimensional drawings

The package provides for 2D symbols, made up of line segment, arc, and dot elements, to be created by the user, stored and then placed as needed on drawings. The symbols can be rotated, reduced, and enlarged as well as translated to any position. When creating these 2D drawings, the same elements can be used along with the symbols, and text can be added from the keyboard as desired, with a choice of font styles. LIke elements and symbols, text can be rotated, translated, reduced, and enlarged.

The drawings are limited in complexity by two restrictions. Overall, a 2D drawing may not contain more than 800 line segments and 200 arcs and symbols. In addition, overlay drawings can be added from storage to the working drawing. Up to ten such overlays can be added. However, once added they cannot be altered.

Two-dimensional drawings can be scaled, and the scale of a drawing can be changed as needed. A zoom feature allows portions of a drawing to be enlarged up to 1,024 times. Sections of a drawing can be defined, and these sections can then be rotated, translated, enlarged, and reduced.

Limited three-dimensional capabilities

Energraphics allows users to create 3D surfaces that represent mathematical functions or sets of points in 3D space. Given the function, the package will calculate the data points to be displayed.

Three-dimensional objects made up of planar panels also can be constructed. Points and lines can be copied and rotated to construct quickly these panels in a variety of shapes. The panels can then be oriented in 3D space and copied and rotated to form the 3D objects. In turn, these 3D objects also can be rotated.

Perspective, isometric, and oblique drawings of a 3D object can be displayed, and the zoom feature can be used with these displays.

An automatic hidden-line removal feature functions with some restrictions. If 3D objects are not constructed with careful attention to the hidden-line removal rules, this feature will not operate satisfactorily.

No three-dimensional hard copy

Graphs and 2D drawings can be printed and, with the help of a separate optional program, can be plotted as well. However, there seems to be no way to print or plot the 3D creations of EnerGraphics. These drawings can be stored, and they can be included in the slide shows the package provides for, but the package provides no way to produce hard copy versions.

ENTRY LEVEL: SEE VERSACAD REVIEW

MICROCAD: THREE-DIMENSIONAL DESIGN

MicroCAD is a 3D package, directed more to design and modeling than to drafting. It runs on the IBM/PC and requires 128K main memory, two floppy disk drives, and a color monitor board. List price is $500 from Computer-Aided Design, 764 24th Ave., San Francisco, CA 94121. (415) 387-0263.

With MicroCAD, wireframe shapes can be created, rotated, and viewed in isometric or perspective. Shapes defined by mathematical functions can be plotted. The package draws lines and arcs, inserts text, edits drawings, and stores symbols that can be retrieved and inserted into the drawings.

A particularly pleasing feature is the ability to use the display cursor to represent the observer's eye when producing varying views of an object. The cursor can be moved to the desired viewing position, then the new perspective can be generated at the push of a button.

A hidden-line removal feature works on only one object at a time. Thus, if one object is in front of another, the package will not remove all the lines in the area where the two objects overlap. There are other restrictions as well. If lines connecting planes are not drawn twice, they will disappear in some views. The vendor believes that at present the most efficient way to remove hidden lines is to have the user do it manually.

The main weakness of the package is its slowness in redrawing, which becomes more obvious when large or complex drawings are being handled. Text and curved lines are easily added to drawings, but they add measurably to the time needed to redraw. To reduce the need to redraw, a zoom function can be used to get close-up or distance views.

Input and output devices

The package can be used with keyboard, display, and dot-matrix printer. Use of a plotter is not required by the software, but the HP 7470A plotter is supported. To supplement the keyboard, a light pen also can be used, and several digitizers will work with the package. Operation is through a series of menus. Points can be entered numerically through a coordinate menu. There is also a main menu, an edit menu, and a plot menu for getting hard copy output.

Other features

Plan and elevation views are available for all drawings. Both are ordinarily isometric views, but perspective views are available at the press of a key. A quick rotation feature allows a 3D object to be viewed from each of four sides (90 degrees apart) by successively pressing a two-key combination.

Component shapes can be stored, rotated, and scaled. Although components also can be joined to each other, attaching lines must be moved separately when the components are moved.

An optional rotatable character set feature ($150) allows text to be attached to any of the three planes, so that it will move or rotate as the images are moved. Another option is an automatic-dimensioning feature. Callouts, arrowheads, dimension lines, and strings are generated. There also is an optional bill of materials module that reports on quantity, unit, and extended cost for assemblies of parts that have been appropriately prepared using MicroCAD.

The package will find the area size, center of gravity, and moment of inertia for any selected area of a displayed object. It also will find the volume of solid objects. Circles and arcs can be generated. The user must define a point on the arc, the center of the arc or circle, and the angle of the arc.

An obtuse manual

The manual is somewhat peculiar. It reads like a calendar rather than a book, and the introduction section begins on page 56. This section is preceded by a discussion of software and equipment requirements phrased in computer talk that may be unfamiliar to many engineers, and it is followed by a listing of a BASIC routine for reading MicroCAD files. In short, the manual seems to have been written for the convenience of the vendor rather than of the user. However, once it begins to describe actual package operation (on page 84), it is quite easy to follow.

PC-DRAW: DRAWING BUT NOT FULL-FLEDGED DRAFTING

PC-Draw supplies a number of tools for drawing but not a full-fledged drafting capability. Shapes can be created, stored, and placed as desired in the drawing area. The documentation is good.

The package runs on an IBM/PC with 128K main memory, two disk drives, and a color graphics adapter. Use of a light pen is optional. Price of the package is $250. It is available from Micrographx, 8526 Vista View Dr., Dallas, TX 75243. (214) 343-4338.

PC-Draw provides for using stored sets of prepared icons or drawing elements. Current versions come with symbols for software flowcharting, logic diagrams, and electrical drawings. Users can create their own stored symbols as desired, and this can be done while a drawing is in progress. When selected from the symbol display, with cursor or light pen pointing, a symbol can be placed at any location on a drawing, expanded or compressed, and rotated using keyboard commands. For most of the manipulations a single-key command is used.

The optional light pen can be used to make selections from any PC-Draw displayed menu, to control cursor movement, to produce freehand line drawings, to locate circles, and to perform many combination-key functions.

Resolution of the drawings is poor, and all lines that are not horizontal or vertical suffer visibly from the "jaggies." Although these jagged angled lines are quite easy to draw, they are difficult to erase.

Drawings can be divided up into pages, where several screen displays can be component pages of a larger drawing. Printouts allow two screens to be shown on a single printed page; with condensed printing, four pages can be squeezed onto one printed page. Larger drawings must be made up by pasting together several printed pages. A choice of two different text fonts for drawing annotation is available, and the annotations may be placed either horizontally or vertically.

ROBOCAD-2: FOR DRAWINGS OF LIMITED COMPLEXITY

RoboCAD-2 is relatively inexpensive and very straightforward and simple to use. It handles 2D drawings of limited complexity quickly and easily. However, changing drawings is an awkward process, and some features such as automatic dimensioning are only partially effective.

RoboCAD-2 runs on Apple II+, and IIe computers. It sells for $1,495, and the price includes a controller device with joystick, dial, and pushbuttons. It is available from Chessel-Robocom Corp., 111 Pheasant Run, Newtown, PA 18940. (215) 968-4422.

With this package users can keep their eyes on the computer display screen, maneuvering and creating the drawings with the help of a joystick-dial controller and making very little use of the keyboard.

Memory countdown

Drawings are created a screenful at a time. To help users keep track of the computer memory space currently available for drawing, RoboCAD displays a countdown counter. The correct procedure is to avoid reaching the lower memory limit (10K) by storing drawings on disk as they begin to grow too large.

Disk-stored drawings can be retrieved as needed to be incorporated in larger, overall drawings. These retrieved drawings can be rotated, reduced, flipped about vertical or horizontal axes, stretched, or squeezed, but their details cannot be altered.

To modify a drawing, it can be put through an editing process. This process is inconvenient since the user has to step through the lines of the drawing, one at a time, to find the one that is to be altered. For $600 an accelerator card can be plugged into the computer that will speed up, but not shortcut, this process.

To find their way through partial or component stored drawings that will be used to make up a larger overall drawing, users create a skeleton or base drawing that shows how key reference points on the partial drawings relate to the overall picture. If a drawing is created in the package's scale drawing mode, it will be automatically rescaled to the correct size when incorporated into a scaled work drawing.

Screen-displayed features

Displayed around the edges of the screen are symbols and codes representing various choices available to the user. These are invoked by pointing to them via the controller joystick. Choices include straight line or several curved drawing elements, selectable pitch cross-hatching, six line colors, four line types, and a number of other drawing choices.

The package makes symbol libraries available. These are displayed on the screen, and symbols are selected from the screen, rather than by entering a code on the keyboard.

RoboCAD will calculate line lengths, measure angles, arcs, and radii. It uses decimal quantities for all dimensions and will automatically annotate lines and other features with their dimensions. English and metric dimensions can be used on the same drawing and will be correctly scaled. Dimension lines, arrows, and reference marks must be drawn by the user. Blocks of text can be rotated as needed but are limited to 40 characters each.

VERSACAD: IMPRESSIVE DRAFTING FEATURES

VersaCAD has an impressive array of useful features and options. It provides for a huge number of drawing layers, and it makes up, to some extent, for speed limitations by offsetting features. In support, ease of learning, and use, VersaCAD is very much oriented to user needs.

VersaCAD runs on IBM/PCs and HP Series 200 computers; it sells for $1,995. An Apple computer version called CADAPPLE is also $1,995. ENTRY-LEVEL versions with more limited capabilities are available for $495. The vendor is T&W Systems, 7372 Prince Dr., #106, Huntington Beach, CA 92647. (714) 847-9960.

Available with VersaCAD as drawing primitives are lines, arcs, circles, rectangles, fillets, and regular polygons that can be placed on the screen at any location in any desired size. The package also has a Bezier curve-fitting feature that allows drawing segments to be connected by smooth curves and the ability to create ellipses by specifying their major and minor axes. A hatch function works automatically to fill closed areas.

Users set up their own overall coordinate system for each drawing. Polar coordinates can be used. Drawings may be larger than the display screen. A screen readout informs the user of the location of the current display on the overall drawing.

Two monitors are used for display. One shows the drawing itself; the other is a command screen used to select drawing procedures and features. For simultaneous work on different sections of a drawing, the package allows up to five drawing windows to be set up and used. These can be used to provide the equivalent of zooming in and out.

Input may be by means of a graphics tablet or a joystick or by entering coordinates, angles, and so forth through the computer keyboard.

Each drawing may include up to 250 layers of material, and each layer can be shown in a different background color, provided a color display is used, to help keep track of layer location.

Many convenient features

The user controls the size, proportion, and rotation of text used with a drawing. Drawings can be modified conveniently by spotting the object to be changed on the screen and giving an appropriate change command.

Drawing primitives may be used to make up parts that can, in turn, be stored in pages containing up to 100 parts each. These stored parts can be called up from a screen display of the page that contains them and placed where desired in the currently active drawing. The user simply points to the desired symbol with a graphics tablet stylus and presses the Enter key. Once on the drawing screen, the symbol can be rotated, scaled, and moved. Until finally placed in the drawing, the symbol blinks as it is moved, making it easy to follow. The vendor makes about five thousand such symbols available. Others can be drawn and stored by users.

Dimension lines, with arrowheads, leader lines, and dimension text, are automatically created; the user specifies the units of measure.

A Repeat function allows objects to be duplicated along a line, around an object, or to fill an area. Whole drawing sections can be replicated using a Copy function. And portions of drawings as well as entire drawings can be deleted. The mirror imaging feature is a useful one. Text in mirrored objects will be displayed as if read in a mirror, but in plotter output the mirrored text can be correctly reproduced.

To aid in aligning drawing elements, an optional background grid can be displayed and the user can define the spacing of this grid. A snap function allows objects placed nearby to be snapped into precise grid locations.

Attributes such as line type, color, and plot-pen number are assigned to symbols and objects, rather than to drawings or layers of drawings. When these symbols and objects are moved, the attributes move with them.

An Inquiry feature allows the user to summon up stored and calculated information about specific objects in a drawing. This feature may include such items as lengths and areas, and the package is able to do some fairly complex area calculations.

An optional program allows the user to display sequences of drawings for presentation purposes. An optional 3D program includes a 3D box as a primitive element.

A bill-of-materials feature searches drawings for the objects they contain and creates a report. If told to search for something that does not appear in the drawings, the package will report finding no such object. Use of this feature, like similar features in other packages, is rather complicated and slow.

Speed of operation

In fact, these often excellent features tend to be offset by the fact that, like other personal computer-based drafting packages, VersaCAD runs rather slowly, even with the use of a special 8087 math coprocessor chip in the IBM/PC.

However, some built-in features are designed to speed up operation. Unused drawing layers can be switched off so that the package does not waste time scanning and redrawing material on the unused layers. Since recreating text after panning or zooming can be time consuming, the package allows the user to temporarily convert text lines into simple straight lines while doing a lot of moving around.

When problems arise

The HELP feature of the package provides advice on the use of selected program options. If material is accidentally deleted, it can be restored, not just the last item but all the objects deleted from the drawing during the current session. When the user enters a program option by mistake, pressing the Esc key will provide a quick means to back out.

If an equipment or system failure should occur, VersaCAD allows the user to completely recover the work files, so that valuable design effort will not be lost. Protection of work files involves frequent storage of current activity on disks, so the package is best used with a hard disk, rather than slower floppy disk storage.

VersaCAD is well maintained and is frequently updated to add new features. After the first 90 days of free support, an extended warranty service is available. Revisions and new software additions are sent to users as part of this service.

Chapter 14
Equation-solving Packages

In engineering analysis and design, a great variety of mathematical tools are used, and many computer-based solution techniques are available for dealing with polynomials, vectors and matrices, transforms, differential equations, and other types of mathematical entities that come up in engineering problems.

The ideal equation-solving package would be able to handle any mathematical problem thrown at it. It would allow the user to type the equations to be solved and define the necessary parameters. Then the package would choose an appropriate method, from among many it had available, and attempt a solution. All this would be done in a very simple manner as far as the user was concerned, yet there would be ample means for users to intervene and control the solution process if it were to go awry.

This ideal package may not yet be perfected, but the available packages discussed in this chapter provide at least some of these desirable features. In the near future, we can hope, more powerful yet useful packages will become available.

Using computer languages

Familiar computer languages such as BASIC and FORTRAN have very flexible mathematical capabilities. However, to use them for equation solving, the engineer has to become at least an amateur BASIC or FORTRAN programmer. Programming a solution means not only defining a solution method (algorithm) but also creating a means to input the needed information and to format and output the results.

What we can call the first generation of equation solvers assumes that users are willing to do more or less conventional programming. What these packages provide are conveniences to speed the programming process. For example, with the Calfex package, the user enters material in BASIC computer language format, using the BASIC mathematical operators and respecting all the rules of the BASIC language. The package offers input screens, output formatting, and some convenient storage and manipulation of parameters and previous work. Taking a somewhat different approach, The Scientific Desk provides a set of mathematical routines written in FORTRAN that can be tied together by programming-wise users; no standard input and output formats are provided.

BASIC and FORTRAN are not ideal tools for doing mathematics. If someone were to start out to design a special computer language for solving math problems, it would have commands and procedures not offered by these standard languages.

This is exactly what the creators of ASYST have done. ASYST is a computer language in the same sense that BASIC and FORTRAN are computer languages. It was designed with engineering and scientific problem solving in mind. For example, its commands include those for generating a polynomial from a list of numbers, performing polynomial addition, subtraction, multiplication, division, integration, and differentiation. ASYST has capabilities not just in mathematics but also in the areas of graphics and instrumentation. As with any other computer language, regardless of its intrinsic merits, the success or failure of ASYST will rest on the extent to which it is adopted and used by programming-minded engineers and scientists.

A similar approach using a somewhat less powerful mathematical language is provided by the Equate package. With it, the user has to set up the solution method as well as the expressions to be evaluated.

Applications modules

Recognizing that most engineers do not want to be programmers, the vendors of these packages have begun to make preprogrammed applications modules available. For example, The Scientific Desk has an optional linear algebraic systems-solver module called LINGEN that manipulates matrices and vectors. Applications modules that come with the Calfex package include those for location of maxima and minima of functions, numerical integration, and solution of up to ten simultaneous equations.

Equate comes with a potpourri of preprogrammed procedures, including production of calibration charts for thermistors, calculation of area and moments of inertia for geometric forms, vector operations, and structural column analysis.

Obviously, these available modules cover only a fraction of the math capabilities needed by engineers. The extent to which such packages can achieve wide acceptance among engineers will depend on the variety of the modules they offer.

Specialized equations only

Taking a more modest approach, some vendors provide packages with specialized math capabilities. Thus, Optisolve is designed specifically to deal with nonlinear equations. This package relies on one computation method, the steepest descent minimization technique. This avoids the complication of choosing among different techniques but imposes the weaknesses, as well as the strengths, of the chosen solution method. Even when a simple pair of equations with two solutions is to be solved, the package will find only one of these solutions. To find the other, it must be prompted or steered toward the second solution. Optisolve finds the nearest local minimum, which may not be the desired solution. User intervention is called upon to steer the computation toward regions where desired answers lie.

CALCU-PLOT gives the user a choice of specific equations that can be solved. The standard equations happen to be of limited utility to engineers, since they represent only a few classical curves, such as parabolas, circles, trigonometric functions, exponentials, and damped oscillations. Users also can solve and graph single-valued equations of their own choice with this package.

Typical of a number of available small, specialized solving packages is DIFFERENTIAL EQUATIONS, which solves only ordinary differential equations. The equations are limited to those that can be expressed using arithmetic operators EXP, ATAN, SIN, and COS, and the equation solutions must be unique.

User-oriented packages

In the real world of engineering design, problems often are solved by going through supplier catalog information and doing handbook calculations. Equation-solving packages such as TK!Solver and Varicalc bring the computer into this situation as a practical tool. These packages are not just for making the needed calculations but also for giving the engineer opportunity to rapidly try out alternative parameters and approaches. Using these packages, engineers can try out different design approaches and different components, discover problem areas, and modify their designs to find better solutions.

TK!Solver is by far the best example of a second-generation equation solver, oriented to meeting the practical needs of engineers. However, its mathematical capabilities are far from complete. TK!Solver can solve systems of linear and nonlinear algebraic equations, and it can deal with a variety of different functions, but the package has only limited ability for matrix algebra and numerical integration and has no capability to do nonnumeric calculus. The strength of TK!Solver lies in its ability to home in on bread-and-butter problem solving that can serve the bulk of engineering practice.

An example of the thoughtful features provided by TK!Solver is the top-of-screen available memory indicator, which keeps the user aware of how much computer main memory is being used by the current problem. TK!Solver error messages are aimed at the user, not at programmers. And there are many convenience features that speed and ease operation.

Many math packages ignore the use of units, leaving that to be solved by hand; not so with TK!Solver. Users can solve problems with one set of units and display the results in another set of units. Units can be freely changed, and they can be readily converted.

Equations, particularly where engineering work is involved, are not simply solved once and forgotten. Very often solutions involving many different values of the variables need to be found. The Varicalc package is uniquely equipped to provide this kind of capability. Varicalc assumes that these values can and will change and provides for connection of such changing inputs as those from game paddles, joysticks, keyboard arrow keys, and internally programmed loops. It can even accommodate external process variables if an instrumentation interface is used with the computer. Variables and calculated results will be updated continuously as these inputs vary, until the user orders a stop. In a similar but less flexible way, TK!Solver allows variable values to be input from prepared lists.

Equation-solver features

As personal computer packages continue to develop, users will come to expect an increasing number of useful features. The following are some of the features that have begun to emerge in the current crop of equation-solving packages.

Organized display screens. Instead of entering one line at a time, as a BASIC programmer would do, equation-solving packages can make use of screen displays. For example, there are eight different kinds of TK!Solver display screens, used in much the same way as sheets of paper would be in a manual calculation. The equations that define the problem are laid out on a Rule sheet; a Variable sheet contains variable names, values and units; a Units sheet defines the needed unit conversions.

Most of the other packages use similar approaches. In Calfex, names corresponding to the symbols used for variables are entered on a Labels screen, whereas input data are entered on a screen that displays all the independent variables. Working with this screen, the user can enter values and change them at will. Because many variables may be involved, this screen can be composed of several pages. Such screens can be stored and retrieved as needed.

Users key in the programs they design on an Equate worksheet display, but the worksheets are not limited in size to what the display screen can contain. With the computer's cursor and PgUp and PgDn keys, the display area can be moved around to expose all parts of the worksheet.

But not all equation-solving packages operate from screen displays. With the Optisolve package, input can be keyboarded line by line. Alternatively, prekeyboarded input prepared with the help of the user's own word processing or editing software can be used.

Handling complex numbers. Some packages are not equipped to handle complex numbers, but others, such as Optisolve, have been designed with this in mind. Normally, all variables in the entered equations are assumed to be real. An Optisolve command converts them all to complex variables so as to obtain complex solutions.

Calculator on the screen. While using an equation-solving package, it can be convenient to have standard hand-calculator functions available. Separate calculator packages, such as TexSolver, can be kept in the main memory of a computer and brought into play while other software is being used. However, several equation solvers have their own built-in calculator capabilities. For example, TK!Solver can operate as a calculator. If the user types an expression that contains only numbers and mathematical operators, the answer will be given as soon as the expression is entered. The Equate worksheet can also operate as a hand calculator. Users enter expressions such as 2 * 2 and get an immediate answer.

Clear error messages. Programmer-oriented packages will display cryptic messages, often undecipherable by engineering users. But as equation-solving packages become more user oriented, the quality of the messages they present when problems arise changes. For instance, the Equate package gives the user clear and useful error messages such as "THE FIRST CALC MARK MISSING," or "INPUT MUST BE A VALID NUMBER," and if these are not understood, the user can press a HELP function key and get further explanation. TK!Solver is another package with very clear and helpful error messages.

Constants from memory. An important computer asset, from the standpoint of user convenience and efficient operation, is the ability to store and recall items from the machine's memory. The Equate package makes particularly adept use of memory for handling constants. This package allows a pop-up constants window to be called by pressing a Function key. The package comes with about four hundred physical and conversion constants in this window, and users can add their own as they wish. Other packages, such as TK!Solver, have stored constants, along with functions and procedures that can be called into use, from storage, on command.

Worksheet storage and retrieval. The ability to store solution models and procedures for later use is essential for a well-designed package. Generally, users must give a file name to material to be stored. The system software conventions of the computer on which the equation-solving package is run are usually followed. The Equate package has a particularly handy approach to this kind of storage. Equate displays the file names of all existing worksheets, and it automatically creates a backup file when a revised worksheet is stored. Other packages have similar but often less convenient filing facilities.

Graphics. Some of the packages provide graphical display of the computed results. This is particularly useful in the Varicalc package, where input variables, and therefore the output results, can be changed so readily. As the inputs are varied, the display plot will take shape. The graph is automatically rescaled when the curves begin to go off screen. TK!Solver also provides graphical display of the computed functions.

Close to ideal

One package that aspires to the ideal of a universal equation solver is QuickSolver. Among the solution methods it can automatically bring to bear are substitution, Muller's iteration method, Gauss–Jordan for simultaneous linear solutions, the secant method for simultaneous nonlinear solutions, and specific intrinsic techniques such as linear programming. The lineup is potentially powerful, but it is implemented in a package that does not yet appear to have been perfected. For the user who is willing to endure incomplete documentation and somewhat puzzling operation, QuickSolver provides a window to the future of equation-solving packages.

ASYST: A LANGUAGE FOR EQUATION SOLVING

ASYST is a computer language in the same sense that BASIC and FORTRAN are computer languages. It was designed with engineering and scientific problem solving in mind and has a variety of capabilities, including those in the areas of graphics and instrumentation. Here we will discuss only the equation-solving abilities of this language.

ASYST runs on IBM/PC and compatible computers. To use its equation-solving capabilities, two modules are needed: the Systems module ($795), which also includes graphics and statistics, and the Analysis module ($495). The vendor is Macmillan Software Co., 866 Third Ave., New York, NY 10022. (212) 702-3241.

ASYST is not a package designed to solve equations conveniently. It provides the user with no menus of available functions, nor does it have predesigned formats to organize its use. Instead, like the BASIC language, it simply presents the user with an OK prompt and waits for commands to be keyboarded. As with any other language, these commands must reside primarily in the user's head, which means that the ASYST language must be learned before it can be used.

As with any other computer language, regardless of its intrinsic merits, the success or failure of ASYST will rest on the extent to which it is adopted and used by programming-minded engineers and scientists. If and when these users write programs that can, in turn, be put to use by the vast majority of engineers who have little or no interest in doing programming, then ASYST will indeed become a significant engineering tool.

Polynomial and matrix commands

The power of ASYST to perform equation-solving tasks rests in the specialized commands it makes available. For example, the POLY command generates a polynomial from a list of numbers, and the POLY[X] command evaluates a polynomial. Other commands perform polynomial addition, subtraction, multiplication, division, integration, and differentiation. The command ESTIMATE.ROOTS invokes a QR-algorithm to arrive at estimates, and the REFINE.ROOTS command uses the Laguerre algorithm.

Vectors and matrices are generated by the VECTOR and MATRIX commands from strings of supplied numbers. Among the manipulations that can be produced by related commands are vector dot product, tensor product, matrix product, matrix inversion, and determinant computation. The command SOLVE.SIM.EQS will generate a solution vector for a system of simultaneous equations, using the Gaussian elimination procedures. A number of matrix transformation commands also are available, including TRIDIAG, which transforms self-adjoint matrices into tridiagonal form, TRIDIAG.DIAG, which does the tridiagonal to diagonal transformation, and HESSEN, which produces upper Hessenberg matrices. The SPECTRAL.SLICE command computes the number of eigenvalues in a tridiagonal matrix.

CALCU-PLOT

CALCU-PLOT has a repertoire of 16 Cartesian and 9 polar coordinate functions that it will tabulate and graph. Single-valued functions entered by users can also be handled.

CALCU-PLOT runs on Apple II, and IIe computers. It sells for $150 from Human Systems Dynamics, 9010 Reseda Blvd. #222, Northridge, CA 91324.

CALCU-PLOT gives the user a choice of specific equations that can be solved. These represent the classical curves that are part of any descriptive geometry course, including parabolas, circles, trigonometric functions, exponential, and damped oscillation. By selecting menu-offered choices, the user can pick the desired equation type, then assign values to the equation constants. The package then asks for the ranges of x and y values that are desired. For the polar-coordinate equations, which include spirals, conics, and other curves, a similar procedure is followed.

With this input, the package produces tabulations of pairs of values for the defined function, and it will display a corresponding graph. If desired, two or three rectangular-coordinate functions can be tabulated and then displayed together on the screen.

Users also can enter formatted descriptions of single-valued equations of their own choice, and the package will handle these. Where second-order polynomials can be broken down into two single-valued functions, the package will solve and graph them as well.

The package is also capable of graphing arbitrary sets of data points that are provided to it. It will plot integrals and derivatives of equations and do log-log and semi-log plots.

CALFEX: BASIC PLUS ORGANIZATION

Calfex is a good package for those who practice BASIC programming. It provides a convenient way to organize the input and present the output, plus some convenient built-in features.

Calfex runs on Apple II, IIe, and III computers and on the IBM/PC. It sells for $175 from Interlaken Technology, 6535 Cecilia Circle, Minneapolis, MN 55435. (612) 944-2627.

Like most engineers, the staff at Interlaken Technology were frustrated with having to write a whole new program every time they wanted to find an equation solution using their personal computers. This package is a response to those frustrations.

When inputting equations, the user enters the material in BASIC computer language format, using the BASIC mathematical operators and respecting all the rules of the BASIC language. Therefore, to use this package, one must be familiar with BASIC.

Names corresponding to the symbols used for variables are entered on a separate Labels screen. As many as 120 independent and 120 dependent variables can be used, and these may be either simple or array variables.

Input data are entered on a screen that displays all the independent variables. Working with this screen, the user can enter values and change them at will. Because many variables may be involved, this screen can be composed of several pages. Such screens can be stored and retrieved as needed. Frequently used constants and BASIC input routines also can be saved and reused.

With input data complete, the package can be commanded, using one keystroke, to calculate. The results will appear on a display of dependent variables. They also can be readily organized into printouts formatted to the needs of the user, including comments along with the values of the independent and dependent variables.

Trading variables

A special feature of the package is the ability to trade places between any independent variable and a dependent variable. This feature is convenient in situations where the same design equations are being used to find answers for different design cases. It also offers an easy way to switch and experiment.

The vendor is developing specific applications of the package for mechanical and electrical engineering. A companion package also has been developed to make plotted results available.

EQUATE: GENERAL PURPOSE CALCULATION PACKAGE

Unlike some other equation-solving packages, Equate will not simply go ahead and find a solution to a problem like a set of simultaneous equations. The user must set up the the solution method as well as the expressions to be evaluated. The package comes with a few prepared procedures that can be used to directly solve the particular problems they address. When, and if, the vendor makes a more comprehensive set of such procedures available, Equate could be a very powerful engineering tool.

Equate runs on the IBM/PC and several compatible computers. It sells for $195 from Banyan Systems, 5632 East Third St., Tucson, AZ 85711. (602) 745-8086.

For those who, like most practicing engineers, do not wish to turn themselves into programmers and mathematicians, the only practical way to use this package is with already-prepared methods. The vendor does, in fact, provide some such solution methods in the form of stored worksheets that can be readily called up and used.

A look at an Equate worksheet disk that comes with the package reveals that a scattering of useful solution techniques are provided. However, the worksheets on this disk are all *protected,* meaning that the user will not see the actual solution method but only a prescribed procedure into which parameters can be plugged and that will produce whatever answers it is designed to turn out. Thus, the solution techniques that have been provided cannot easily be incorporated in procedures users may want to program themselves.

Engineering problem solutions provided by these prepared worksheets include:

THERMX, which produces calibration charts for thermistors using three known temperatures and resistances. It uses the Steinhart–Hart equation recommended in Yellow Spring's literature.

STATS, which calculates average, standard deviation, and variance as data are input.

MATRIX, which solves linear equations in 2, 3, or 4 unknowns using Cramer's rule.

MOMENTXY, which calculates the area and moments of inertia for rectangles, triangles, ellipses, circles, and concentric circles.

VECTOROP, which performs the vector operations of addition, cross product, dot product, and scalar product. It also converts between spherical and Cartesian coordinates and finds the angle between the two vectors.

COMBUCK, which finds properties of a slender compressive column, including critical bucking load, modulus of elasticity, minimum moment of inertia, and length of the column.

Making up worksheets

The package itself and the user's manual that comes with it are directed to those who wish to program their own solutions. Equate provides a programming language with 30 commands designed to invoke all the capabilities of the package. Users key in the programs they design on a worksheet display. Once set up, a worksheet can be used and then discarded, or it can be stored for future reuse. Worksheets are not limited in size to what the display screen can contain. Using the computer's cursor and PgUp and PgDn keys, the display area can be moved around to expose all parts of the worksheet.

What is entered on the worksheet can be edited, using convenient keyboard word processing commands. When the material has been examined and corrected, the entire worksheet is entered for use.

The worksheet can operate as a hand calculator. The user can enter calculations, such as 2 * 2, and get an immediate answer. But this calculator use is a rather trivial application of the package.

Mathematical operators and programming commands can be entered in a similar manner. All calculations can be carried out to 24 bits of accuracy (7 digits) if single precision is used or to 53 bits (15 digits) with double precision calculations. The user can choose whether to put the package into single or double precision mode. The user also has a choice of the number of digits to be displayed before and after the decimal point. Another user choice is to select either fixed-point or floating-point (digits plus an exponent) style of displaying the computed results.

Along with the mathematics and commands, users can enter notes and memos on the worksheets. These will not interfere with calculations but can be helpful reminders from users to themselves or to assistants.

Mathematical operations are denoted by the usual symbols (+, −, /, *, and ∧ for exponention). A variety of other operators, for trigonometric, exponential, and other calculations are provided. There also is a random number operator.

Handy constants

To assist in the proper construction of expressions, a constants window can be called up by pressing a Function key. The package comes with about four hundred physical and conversion constants in this window, and users can add their own as they wish. With many constants to look through, users may need searching help. Searching is eased by the fact that the constants are listed alphabetically. In addition, users can make a word-matching automatic search of the list. When first called up, the first page of the list is displayed. Users can view other portions of the list by paging up and down as needed. As the up and down arrows on the keyboard are pressed, one item after another on the list is highlighted. To enter a highlighted item into an equation on the worksheet, all the user has to do is press the Enter key.

Long calculations handled

Variables can be represented by alphabetic letters or by words interspersed with numerals, up to 60 characters per variable name. Each expression can be up to 200 characters long. Larger problems can be broken up into several expressions. Strings of expressions can be set up, and the package will use the results of each expression solution in the following expression. Calculations can be strung out at length since a worksheet can carry up to 799 lines of expressions and other information.

When a worksheet has been completed, the user presses a function key and the package does its work. If the worksheet has been prepared according to the rules, the results are calculated and displayed, and they can be printed or stored as well.

Clear error messages

Equate catches errors made in the format and content of the entries on worksheets. It gives the user clear and useful error messages such as "THE FIRST CALC MARK MISSING," or "INPUT MUST BE A VALID NUMBER." If these error messages are not understood, the user can press a HELP function key and get further explanation.

The package allows the user to control the number of lines printed on each page of output, and to print worksheets complete with equations or only the calculated results.

Create and store problems

Storing worksheets is a simple matter, performed with the help of a special Transfer window that can be called up onto the screen. Users must give each worksheet a file name, following the IBM/PC system software convention of limiting file names to eight characters. Using the same approach to files as that of popular word processing packages, Equate will display a directory listing the names of all existing worksheets. If the user gives a current worksheet the same name as one that is already filed, the package assumes that an updating of that file is intended. It changes the name of the old file, adding a .BAK ending to it. If a file by that name with a .BAK ending already exists, it is automatically deleted.

The similarity to word processing file procedures is not accidental. Equate worksheet files can be composed by popular word processing packages in standard (ASCII) format, then converted for use by Equate. The conversion process is a built-in and easy-to-operate feature of the package. Using a reverse conversion, worksheets can be put into ASCII format for editing by word processors.

Users can customize

Equate is a malleable package that allows the user to shape it to particular needs. It includes the capability to create prompting forms, allowing users to set up calculation procedures that then can be followed easily by assistants. Up to 256 such forms can be created and stored for future use. Custom table formats for output data also can be created, so that results can be easily included in reports. Automatic conversion of units and display of correct units can be incorporated as well.

OPTISOLVE: ONE SOLUTION AT A TIME

Optisolve solves nonlinear systems of equations and finds the maxima and minima of nonlinear functions with or without constraints. It is designed with little regard for the occasional user, and its use requires familiarity with the steepest descent solution techniques the package employs.

Optisolve runs on IBM/PC and compatible computers. Use of the 8087 math chip is recommended. It sells for $195 from Optisoft, 2920 Domingo Ave., #203, Berkeley, CA 94705. (415) 540-1224.

Equations are keyboarded, one to a line, using the usual mathematical function names and symbols. The only symbols that may seem at all unfamiliar, even to those who have never used a computer, are the * for multiplication and the ∧ for exponention. Variables are entered as alphabetic letters or names, constants as numbers. For lengthy constants a name can be defined (for example, PI = 3.141592653) and used as an abbreviation in the equations.

Along with the equations to be solved, the user enters some initialization information. Here, approximate solutions are indicated by assigning initial values to the variables.

The package is not interactive. Input can be entered from the keyboard or from prekeyboarded files prepared with the help of the user's own word processing or editing software. In any case, the input must be formatted to a set of rules prescribed by the manual.

Only one solution at a time

Optisolve produces only one solution at a time, and the solution it gives may not be a correct one.

The basic calculation technique used in Optisolve is the steepest descent minimization method. This finds the nearest local minimum, which may not be the desired solution. Most of the user intervention in the operation of this package is concerned with steering the computation away from regions where the solution diverges, or where undesired local minima exist, and toward regions where desired answers lie.

Even when a simple pair of equations with two solutions are to be solved, the package will find only one of these solutions. To find the other, it must be prompted or steered toward the second solution.

The basic Optisolve solution commands are MIN, MAX, and SOLVE. In addition, there is an ALSO command for adding constraints and a START command, which allows the user to rerun a calculation with revised initialization of the variables.

Solution times can be long

Perhaps the major limitation on the use of Optisolve is computation time. With sufficient prodding from the user, this package can find good answers, but that may take a very long time for some problems. For this reason, the vendor suggests using the 8087 math chip, which can speed up numerical solutions by a factor of about 30.

If a solution seems to be running forever, the user can press the Esc key. Optisolve will then halt and give the best solution approximation it has to that point. The search for a better solution can then be continued by giving the CONT command. Where several distinct solutions are expected, the OTHER command can be used after each solution to find the next. There also is a GLOBAL command that initiates global search strategy.

Normally, all variables in the entered equations are assumed to be real. The COMPLEX command converts them all to complex variables to obtain complex solutions.

In addition to the requested answers, Optisolve also provides the user with a "performance index" that indicates the extent to which the solution satisfies the input equations. However, even a perfect index of 100 does not mean that the answers given are necessarily adequate. For example, asked to find values of x between 2 and 3, Optisolve might return any value in that interval and give itself a performance index of 100.

The manual is poorly designed, with little regard for the convenience of the user. Operating instructions are intermixed with solution examples, and there is no separate, one-by-one detailed explanation of the use of the various commands. Under "Syntax Notes," buried in the middle of the manual, the user is given the needed formatting rules for entering information.

QUICKSOLVER: FOR A VARIETY OF PROBLEMS

Reviewer: Dr. Darrel G. Linton, University of Central Florida, Orlando, Florida.

QuickSolver has the potential to solve a large class of engineering-oriented problems using standard algebraic notation. Included are systems of linear or nonlinear equations, linear programming problems, multiple linear regression problems, and matrix algebra calculations.

However, incomplete documentation and program development appear to detract from its usefulness. This reviewer found that the package failed to solve a simple problem, and an example recorded on the package disk did not work. These experiences caused some loss of confidence in the package. Using a rating scale of 0 (low) to 100 (high), this reviewer's ratings of QuickSolver are: ease of use, 90; correct documentation, 70; freedom from error, 50; overall rating, 70.

QuickSolver runs on IBM/PC and compatible computers. Use of the 8087 math chip is optional. The package was designed by Conversational Computer Systems, 224 Summer St., Buffalo, NY 14222. (716) 886-0424. It sells for $125 from Technical Software, 3981 Lancaster Rd., Cleveland, OH 44121. (216) 486-8535.

Virtually no knowledge of computers or of special computer languages is needed to use this package. For the novice, some practice in saving disk files and using the full-screen editing features will be needed. Since the original QuickSolver disk is copy-protected, the user cannot make backup copies to protect against damage or loss of the original.

Users enter information in an algebralike format, one equation to a line. Special symbols are used to represent Greek letters, the constants pi and e, and the summation symbol. Keywords and other symbols are used to denote such items as *maxima* and *minima* of objective functions and the EST for the lefthand side of multiple linear regression model equations.

Almost any problem involving mathematical equations can be solved by this package, provided only that the equations conform to the QuickSolver format. For these problems, QuickSolver has the potential for not only solving them but also producing plotted output, saving the problem in disk storage, and allowing the problem to be edited and changed.

Included on the QuickSolver disk is a file containing supplementary documentation for using the package. This is an excellent idea for illustrating increased capability and correcting any faults in the printed documentation. However, it also carries the implication that QuickSolver is not yet a finished product and may have bugs as well as incomplete documentation. This reviewer observes that either or both of these implied characteristics may be inherent in the package.

Incomplete solutions

When two simple simultaneous algebraic equations (one containing a square root function) were entered, the package failed to obtain the solution for x. Instead, it produced a peculiar error message: "036 SINGULAR MATRIX." The message meant nothing to this reviewer since the package provides no documentation on error messages. The failure of QuickSolver to obtain a complete solution implies the possibility of an error in QuickSolver's algorithm for solving simultaneous equations.

Another simultaneous equation test problem produced the message "DIVERGED," followed by a display of answers that were correct to just two significant figures. The package apparently allows the user no way to specify convergence criteria.

Missing features and peculiarities

The linear-programming features of this package do not conform to standard notation. Only strict inequalities are allowed. Attempts to use greater (less)-than-or-equal-to notation resulted in error messages.

The documentation promised that a feature called ALL SLACK? could be used. Yet when this feature was tried with one of the example problems (QS.MAX) on the package disk, the response was another error message: "010 INVALID EQUATION."

The documentation implies (page 32) that there is no need to use any multiplication symbol between two (TABLE) variables when the summation operation is used. In fact, this reviewer found that the summation function would not work unless such symbols were inserted.

When the multiple linear regression capabilities were tested, this reviewer found it necessary to list the Y TABLE array last, after the independent variables, to get a correct result. There was no indication in the documentation that this order of things was necessary, but when it was not observed, the error message "129 MORE EQUATIONS OR VALUES ARE NEEDED" appeared and the value of beta-sub-zero was unknown.

Comment by Keith Beech, Conversational Computer Systems

Our acceptance of the reviewer's criticisms of the documentation for QuickSolver is reflected in extensions we have made to our supplementary documentation, based on his review. These extensions are included on disks being shipped from today onwards and illustrate the value of the thorough and informed reviews which appear in Engineering Microsoftware Review.

However, we feel that we should draw specific attention to a new section on non-linear systems which has been added to amplify the original information in the manual. We feel that if this had been available to the reviewer, he may not have felt a loss of confidence in the package's freedom from error and would not have felt compelled to give such a low rating on that score. The "simple" equations mentioned in the review (2X + 4Y = 12; X = SQRT Y) are nonlinear and were solved correctly when run according to the revised documentation. In addition, we feel the remarks about "answers correct to two decimals" are open to misinterpretation by those unfamiliar with numerical analysis and the meaning of divergence.

We would like to stress that we are not disputing the reviewer's conclusions based on the information made available to him, and that we accept that our incomplete documentation was the principal cause of the problems encountered.

THE SCIENTIFIC DESK: LOTS OF TOOLS FOR PROGRAMMINGWISE USERS

The Scientific Desk provides a set of mathematical routines that can be tied together by programmingwise users to solve a wide variety of engineering problems. For those who do not wish to program, there are some specialized problem-solving modules. The package does not seem oriented toward ease of use or learning.

This package runs on IBM/PC and compatible computers. It sells for $420; problem-solving modules are extra, $110 each except the Very Small Linear Algebra Algorithms module, which sells for $50. The package requires the use of FORTRAN and comes in four versions corresponding to four popular FORTRAN compilers available for the IBM/PC. The vendor is C. Abaci, 208 St. Mary's St., Raleigh, NC 27605. (919) 832-4847.

The Scientific Desk is essentially a collection of mathematical routines in areas useful to engineers, including linear algebra, optimization, nonlinear equations, and integral transforms, simulation, statistics, and probability. The most flexible way to use these routines is to incorporate them into programs written by the user. However, recognizing that most users will not want to do programming, the vendor has begun to make prepared applications modules available.

Matrix, array solver

LINGEN is an example of such a module. It is a linear algebraic systems solver that allows manipulation of and operations on matrices and vectors. LINGEN retains up to ten different arrays, keeping statistics on whether each array is a matrix or vector, on its dimensions, where it starts in the workspace, its length, and the remaining available space.

Factoring, solving, estimating matrix condition (and the number of correct significant digits in results), and inversion are allowed. In addition, vector and matrix arithmetic operations are provided for rectangular arrays.

Commands available to LINGEN users include those for performing array addition, subtraction, multiplication, and finding cross and dot products. A BUILD command constructs special arrays of uniform pseudorandom numbers, zeros, ones, or identity matrices, based on user prompts. Arrays also can be entered via the keyboard. Matrices can be manipulated in various ways. They can, for instance, be deleted, corrected, factored, inverted, and replicated. The matrix problem $Ax = B$ can be solved.

Other user-ready applications modules include one for Eigensystem analysis, which can find the roots of polynomials and estimate eigenvalues and eigenvectors. A Statistics module takes normal or pseudo-random samples and determines the basic statistics of the sample data. An Approximation and Quadrature module is designed to eliminate anomalies from curve data. Another module is for evaluating definite integrals in one variable. There also is a "Very Small" linear algebra package designed to take up little storage space.

An incomplete product

To a limited extent, this package and its modules are designed with user needs in mind. The commands for the solution modules are presented in menu form to make them more readily available to the occasional user. Included are commands for printing the results. However, the user is assumed to be willing to do a considerable amount of programming and experimentation with the tools provided by the package.

The documentation is whimsical, consisting of a melange of instruction, explanation, and listings of commands and other information. Apparently, no one has taken the trouble to spell it all out in a form that would minimize the time taken by a novice user to understand what is available and just how to go about solving problems. Nor is there any clear way for more experienced users to find quickly reference to command details and other needed information. Although the mathematical routines seem to have been carefully and knowledgeably constructed, the package does not fulfill what should be expected from a professionally crafted computer applications product for use by engineers.

In the basic mathematical routines of The Scientific Desk, there appears to be the potential for constructing a very useful package for solving a wide variety of engineering problems. The extent to which this potential will be fulfilled depends on how well the vendor designs and packages future problem-solving modules.

TEXSOLVER-10: HANDY CALCULATOR PACKAGE

TexSolver can be kept in the main memory of a computer and brought into play while other software is being used. Press the Alt and Ctrl keys together and a calculator screen appears. Press the Esc key and it goes away. TexSolver provides all the expected scientific calculator capabilities, with 100 storage registers and simple statistics as well.

This package runs on IBM/PCs. It sells for $44.95 from Microtex, PO Box 111054, Carrollton, TX 75011. (214) 980-9837.

TexSolver can either be run from the disk or stored in the computer main memory and called up at any time by pressing the Alt and Ctrl keys simultaneously.

The screen display shows all the calculator functions performed by the package, which are those offered by a competent scientific hand calculator. Operation is not self-evident, however, and the user must study the little manual to get the idea of how to properly enter the numbers and math functions in sequence.

The screen-displayed calculator also provides 100 data registers that can be accessed as needed, and arithmetic can be performed on the contents of any storage register.

A statistical capability calculates and stores statistics of paired sets of x, y data, depositing such items as summations of squares and products in specific storage locations, where they can be examined as needed. The mean and standard deviation of the data then can be calculated at will. A linear regression calculation will find the slope and intercept of the best fit line. A linear estimate and correlation coefficient also can be conveniently calculated.

A peculiar effect

A strange effect of this package is that while it resides in the computer main memory, the cursor used with a popular word processing package seems to be raised about a half line. This is disconcerting, but no other adverse effects were noted.

TK!SOLVER: IMPRESSIVE PROBLEM-SOLVING CAPABILITY

TK!Solver is well suited to meeting most of the practical equation-solving problems faced in everyday engineering work. User needs are kept in the forefront in the documentation and in the functional features of this package. Units are handled in as organized a manner as are the numerical parts of problem solutions.

TK!Solver is available in versions to run on most personal and microcomputers. It sells for $400 from Software Arts, 27 Mica Lane, Wellesley, MA 02181. (617) 237-4000.

The authors of this package have given some thought to the way engineers typically solve problems involving equations and have designed procedures that are computer-based equivalents of familiar manual methods.

There are eight different kinds of TK!Solver display sheets, used in much the same way as sheets of paper would be in a manual calculation. For example, the equations that define the problem are laid out on a Rule sheet; a Variable sheet contains variable names, values and units; a Units sheet defines the needed unit conversions.

At the user's convenience, two of these sheets, or just one, can be displayed at a time. Users type information onto these sheets, correct errors, and can store the sheets for further use. Each set of sheets is stored as a set in one file. The file names used are in the form prescribed by the operating system software of the computer on which the package is run, so users name the files in the same way they would name files for word processors and other packages used with their computer.

Three ways to solve problems

At its simplest, TK!Solver can operate as a calculator. If the user types an expression that contains only numbers and mathematical operators, the answer will be given as soon as the expression is entered.

Two methods of equation solution are available. With the Direct method, the user must supply values for all variables except one, in one or more equations. That one unknown variable must appear once and only once in any equation. Additional equations can be solved, even if they contain more than one unknown variable, provided that the number of unknown variables can be reduced to one by solving other given equations.

To deal with the many engineering problems that require, or can best be handled with, iterative solutions, TK!Solver also has an iterative-solution mode. The user starts with an initial guess and the package will then try successive values until it converges on a solution.

The Iterative solution method is able to handle sets of linear or nonlinear simultaneous equations in which none of the variables are known initially. This solution method also can solve equations in which unknown variables appear more than once. The price of a poor guess for an initial value is likely to be slow convergence of the calculation, and this may mean an excessively long computation time for the solution. If the answer is not a real number, the Iterative solution method will fail.

It helps if the user can do some algebra. Factoring equations before they are solved can often speed the solution. Manipulating the form of the equations, and possibly adding an extra one, can allow the Direct method to be used instead of the possibly slower and less dependable Iterative one.

Using a List sheet, the user can lay out a whole series of variable values for which a set of equations are to be solved. Portions of lists, as well as entire lists, can be processed. The answers can be displayed in tables or as plotted points on a graph.

Thoughtful features

Among the thoughtful features provided by TK!Solver is an indicator of cursor location at the top of each sheet; it is sometimes not easy to spot a cursor amidst a lot of text. More important is the available memory indicator, which keeps the user aware of how much computer main memory is being used by the current set of problem sheets. This is essential. Should available memory be inadvertently exhausted, it could take up to an hour of computer time to get the sheets stored.

Procedures are aimed at saving time for the user, and program features such as error messages are aimed at the user, not at programmers. For example, for each variable in the equations typed on the Rule sheet, TK!Solver will automatically put a variable name on the Variable sheet. If the user enters values for all variables in an equation save one, the package will find the remaining value and show it on the Variable sheet. However, if too few values have been assigned, the user will be notified that the problem is underdefined. If the values assigned are inconsistent with the equations, the package will notify the user of this fact. If input values are in conflict, the package will give an "OVERDEFINED" message.

Many engineering packages ignore the use of units, leaving that to be solved by hand. Not so with TK!Solver. Users can solve problems with one set of units, and display the results in another set of units. Units can be quite freely changed, and they can be converted, using the Unit sheet on which the user types unit names, conversion multipliers, and offset additions.

The manual is detailed and has an index for quick reference to needed information. On-screen HELP messages are also provided.

Keyboard commands

The ! in the name of this package refers to the fact that the user can initiate solution of an equation or model by entering the ! symbol. To conveniently manipulate information, users have to learn some other TK!Solver commands as well. For example, by typing a colon (:) users can command the cursor to go to specific locations on a sheet. To locate a specific item on a sheet, a " is typed, followed by the characters in the item. The package will then find those characters and move the cursor there. Items can be copied, replicated, as desired using the /C command.

For math operators TK!Solver uses the usual computer symbols (+, −, *, /, ∧). In addition to these operators, TK!Solver has many of the familiar scientific calculator functions built in. It also has a number of other useful built-in functions, including maximum, minimum, sum, and number of elements in a list of values. The min and max functions can be used to hold solution values within defined boundaries. Dot products of the entries in two lists also can be produced. A polynomial function allows coefficients to be defined by a list for convenient evaluation.

Like mathematical functions, stored constants and procedures come with the package and can be called into use on command, and users can define their own functions, constants, and procedures using the package's User Function sheet.

Custom solutions

TK!Solver is so flexible a tool that its commands can be used to create custom solutions to engineering problems. The vendor has already addressed two areas of engineering work with special preprogrammed solution methods. A package is available to solve some mechanical engineering problems, using TK!Solver, and another similar package addresses a group of building design problems. An extensive series of TK!Solver-based solution packages, under development, will cover many aspects of engineering design.

VARICALC: FORMULA-SOLVING WITH VARIABLE INPUTS

This package is set up to do iterative solutions of formulas with variable values incrementally changed. It includes a graphical display of computed values that changes as the variables are changed, and it can take changing variable values from related formulas.

VARICALC Runs on Apple II computers. It sells for $100 from Interactive Microware, PO Box 139, State College, PA 16804. (814) 238-8294.

The mathematical capabilities of this package are limited. Its basic function is to solve a formula with one independent variable, but it also can handle up to 18 dependent variables.

Users key the formulas they wish to solve onto the screen, using the usual notation for mathematical operators. Beyond +, −, *, /, and ∧ operators, the formulas also may contain Ln, Exp, Cos, Sin, Tan, Arc Tan, Arc Cos, and Arc Sin. There are some rather awkward restrictions on the way the formulas have to be written, but the rules seem to be of the sort that, once learned, will be easily remembered.

When a formula is entered on the screen, it can be solved for any variable, provided values are assigned to all the others. The unique feature of this package is in the way variable values are assigned. The package assumes that these values can and will change and provides for connection of such changing inputs as those from game paddles, joysticks, keyboard arrow keys, and internally programmed loops. It can even accommodate external process variables if an instrumentation interface is used with the computer. The calculated result will be updated continuously as these inputs vary, until the user orders a stop.

Built-in graphics

The package also provides graphical display of changes in the variables, with the calculated variable plotted along the y axis. The graphics display can be viewed alternatively with the VARICALC formula display. As the inputs are varied, the display plot will take shape. The graph is automatically rescaled when the curves begin to go off screen. Using the graphic display, along with the formula calculation process, the approximate minimum and maximum values of calculations can be obtained.

Once entered, formulas can be saved for future use. Up to 255 formulas can be stored on a floppy disk. In fact, the package comes with a number of frequently used engineering formulas already recorded on the disk. These formulas can be used or erased at will. Each formula is limited to two lines on the screen (79 characters). The user can enter a formula from the disk, edit and change it as needed from the keyboard, then get the solution right on the same screen display.

Injections from related formulas

The package can handle related formulas, along with the main formula that is being evaluated. When the related formulas contain some of the same variables as the main formula, the package will recognize this automatically and plug the appropriate values into all formulas.

A relatively powerful feature of the package is that values of a dependent variable in the main formula can be calculated from a related formula, then automatically "injected" into the main formula when the independent variable is calculated. This allows calculations to be chained. The example given in the package manual is calculation of the changes in mass of a rocket (as fuel is burned) by a related formula, with new values of mass injected into the main formula for the calculation of rocket thrust.

The package suffers from the limitation that it cannot deal with null quantities (zero amounts), and some fiddling with values and change intervals is sometimes required when negative values are to be handled. Outside of these anomalies, the operation seems quite straightforward.

Chapter 15
Linear Programming Packages

Linear programming has come into wider use as personal computers have become available to an increasing number of engineers. It is essentially a decision-aiding tool for use in situations where the choices are complex.

The problems it solves are often concerned with allocating available resources, such as raw materials, partly finished products, labor, capital, or time. The solution can describe how to make the allocation so that some measure of benefit, such as profit or efficiency, can be optimized.

With the computation speed offered by a personal computer, it is feasible to use linear programming to solve a number of problems that formerly were tackled with more approximate methods. For example, the technique is frequently used to find the most profitable mix of products to run through a production line.

Other optimization problems commonly solved with the help of linear programming include: best allocation of production at several centers, so that costs of transporting raw materials and finished goods will be minimized; setting up travel routes to make preassigned stops so that total distance traveled will be minimized; specifying optimum mixtures of materials from available stocks; selecting best alternatives from cutting and trimming schedules; even setting up investment strategies.

Linear programming concepts

To use this technique, the first task the user faces is to formulate the problem so it can be solved meaningfully by the computer. Not every decision situation can be handled by this technique but many can.

When applying linear programming, the decision situation has to be stated so that there is an overall goal or objective that the user wants to maximize or minimize. This will be expressed in terms of a mathematical relationship, a linear equation in which desired result is set equal to a function consisting of decision variables, each with its own coefficient. This objective function may represent a cost or any other result that is to be optimized.

Linear programming also provides for expressing the limitations or constraints under which the optimum result is to be obtained. These constraints are expressed as linear equations or inequalities in which the decision variables appear on the left with constraint coefficients, and a constant appears on the right-hand side (RHS).

There are a few additional linear programming concepts, referred to in this chapter, with which the reader should be familiar.

Linear programming solutions that satisfy all the constraints are called feasible solutions.

Whenever a less-than-or-equal-to constraint is used, there can be slack. That is, the right-hand side of the constraint can be decreased without having any effect on the solution. In a similar manner, a greater-than-or-equal-to constraint can have a surplus.

The computation method used in almost all the packages discussed in this chapter is called the Simplex method. One requirement for its use is that inequalities be converted into equations. For this purpose, an extra or slack variable is added to each inequality constraint relationship, allowing it to be treated as an equation.

The shadow price of a constraint is the amount by which the objective function will change if the right-hand side of the constraint is increased by one. A large shadow price is a danger signal, warning the user that a constraint is drastically affecting the solution.

Some problems have two or more optimal solutions, called alternate solutions. The packages usually signal this situation with an alternate solutions message, alerting the user to look for additional solutions.

The job of the linear programming package is to allow users to set up conveniently the necessary equations and inequalities, to solve them as quickly as possible, and to provide a simple way to change and rerun the solutions.

Available packages

Although all linear programming packages perform virtually the same task and use the same basic mathematical method, they offer quite different features.

One major difference is the size of the problems each package will handle. Linear programming problems can become quite complex, involving many variables and constraints. Many of the packages are able to handle only smaller problems. Thus the MPS-PC package handles 70 variables and

50 constraints. Others can find solutions to considerably larger problems. For example, the LPX88 package can handle 2,510 variables and 510 constraints.

Package prices are not, however, directly related to the size of problem they can handle. The MPS-PC package, which handles 70 variables and 50 constraints, sells for $150. Eastern Software, vendor of the LPX88, sells that package for $88 and charges the same price for each of its other packages. These include LP88, which solves problems with up to 255 constraints and 2,255 variables; BLP88, which solves problems with upper or lower bounded variables, handling 255 constraints and 1,255 variables; DLP88 for dual linear problems with up to 1,255 constraints; and SLP88/STSA88, which solves 45 constraints and 255 variables.

In fact, taking a variety of package characteristics into account, it is fair to say that at present, prices and capabilities of these packages are not necessarily related.

The only package discussed here that does not use some version of the Simplex algorithm to produce solutions is the Dynacomp Linear Programmer. This package solves simultaneous inequalities rather than the usual linear programming problem.

Data entry can be arduous

Once the linear programming problem is set up, the next task the user faces is to input enough data to the computer so that a solution can be attempted. For problems with more than a few variables and constraints, input tasks can be formidable and users need all the help they can get to make this part of the procedure as painless and accurate as possible.

From the user's standpoint, an ideal way to enter data would be to view a display of an objective function and a set of constraints, then move a cursor through this display, entering coefficients and other values as they were reached. However, none of the packages discussed here uses this particular method.

One that comes close is LP88, where entry is done on a row–column spreadsheet format. A labeled, largely empty spreadsheet is displayed. Using row–column locations to position the cursor on this display, the user can enter desired coefficients and other values into the spreadsheet cells. The cursor also can be moved about by the arrow keys. This display can show 15 constraints and 10 variables at a time, but users can move the display to any desired area of the overall spreadsheet.

Most of the other packages cannot be given high marks on their ability to provide user-oriented data entry. The LP1 package tries, by having the user assign a unique name to each variable. The package then displays these names, one by one, and prompts the user to enter objective function coefficient values for each.

MPS-PC data entry is on a column-by-column basis. This means that users will not be able to keep track of what they are doing by merely knowing what equations and inequalities they wish to enter. Instead, they have to make a careful handwritten layout of the entry data with all columns neatly and correctly lined up.

The Ramlp package tackles the input problem by offering users choices. They can enter or erase a single coefficient by first specifying its row and column location. Alternatively, they can enter an entire row, or column, or diagonal of data.

With the LP80 package, constraint coefficients are entered, one to a line, each accompanied by the keyboarded name of the constraint and variable to which it belongs. By the time this information has been repeated for each coefficient, a lot of typing has been done.

LP83 makes little use of the interactive capabilities of personal computers, requiring the user to come to the package with a complete set of input data, ready to be run. This allows the package to conform readily to input procedures traditionally used with large computers, where data and instructions are frequently punched onto cards, then fed to the equipment. LP83 will, in fact, accept input in the standard MPS format, used for linear programming based on large computers.

Even if they fail to take full advantage of the personal computer's ability to make things easier for the user, some of the packages do offer useful input aids. The Computer Heroes Linear Programming System package has a special sparse entry command for the situation where constraints to be entered have only a few non-zero coefficients. This command allows the user to skip around to those few positions in the constraint array where coefficients have to be entered.

LP1 has a similar method for rapid entry of scattered non-zero coefficients, and it also has a helpful data verification feature. At the completion of the objective function data, and as each set of constraint data is completed, the package prompts the user to verify that the entered data are acceptable. If not, the user can go through the appropriate set of prompts once again, skipping by those items that are Okay and stopping to change any incorrect items.

Most of the packages provide editing procedures for data and allow users to store data for later use. However, some of the simpler packages lack these features. For example, neither the Computer Heroes Linear Programming System nor the Dynacomp Linear Programmer have any provision for storing data.

File entry

An alternative to keyboard data entry is to create a data file using other software, outside the package, then read the file when the package is run. About half of the packages discussed in this chapter have this capability. In most cases the formats for such files are compact and formal.

The LP88 has this outside file capability. To the probable delight of users who are large computer programmers, this package can be run in a batch mode. To do this, a properly formatted command file, containing instructions on how the computation is to proceed, must be entered, along with a data file.

Calculation speed

Linear programming problems that involve many variables can take a while to run. The calculation time will depend not only on the number of variables and constraints but on the computational techniques used by the various packages as well as the equipment used to process them.

When the problems are very tiny, the run times typically will be only a few seconds, but even quite small problems can run for minutes. For example, reviewers of the LP88 package checked the run time with a 13-variable, 48-constraint problem. LP88 produced an optimized solution in 2 minutes, 15 seconds, whereas LPX88, which uses a revised simplex algorithm, produced an optimized solution in 54 seconds. With MPS-PC, a problem with 63 columns and 47 rows is said to have taken 68 iterations and 15 minutes to reach a solution.

Run times can be greatly reduced when packages allow the use of a math coprocessor chip, which seems to cut the run time for most problems by about a factor of ten. Type of personal computer also had an influence. The LP83 vendor found that, using an IBM/XT with math chip, a problem with 16 rows and 33 columns ran in 6 seconds, whereas another with 177 rows and 548 columns ran in 41 minutes, 33 seconds. These times were decreased to 3 seconds and to 29 and a half minutes, when the same problems were run on an IBM/PC/AT with math chip.

Users of large computers are fond of pointing out that these machines run much faster than personal computers. However, a more significant measure is often how long it takes from the moment inputting of the problem starts to the moment the user gets results. Elapsed times to get answers from the IBM/PC/AT and an IBM 3081 mainframe computer were compared by the LP83 vendor. For all but the very largest problems, the ready access the user had to the PC provided the answers more rapidly, although the large machine had an advantage of almost ten to one in raw computational speed.

Control over calculations

Users who are interested in how computer calculations proceed can get added satisfaction from several of these linear programming packages since they display ongoing partial computation data. Some also give users control over the computation process.

For example, the Ramlp package displays a log of partial computation results as iterations are made toward a feasible solution. These displays indicate the number of constraints satisfied at each iteration, along with a current number of infeasibilities (which gives the user some idea of how close a solution may be). If the solution process should fail, this record can then be used in deciding on possible fixes.

With LP88, users can control a number of computation parameters. They can set a limit on the number of iterations (pivots) that will be performed. They also can select the number of iterations performed before the inverse is recomputed and the tolerance values below which pivot elements, inverse elements, and basis variables will be rejected.

During its calculations this package shows a display that gives current information on the state of the solution. Solutions can be interrupted while in progress, allowing users to step through the pivots, reinvert, reset tolerances and controls, save the basis, and record the current solution and pivots.

Output options

Some of the packages provide only limited output information, but others give it in considerable detail. For example, the results for the MINIMAX package include displays of the optimal solution and of basic and nonbasic variables. (The nonbasic variables are those whose coefficient values are insignificant for a particular solution.) And the output includes an "entering value" to show at what point each coefficient value would have made that particular variable basic. A separate display shows the current values of the basic variables along with high- and low-range values, to show how much variation there can be before these become nonbasic. This output also displays slack variables, dual variables with shadow prices, and the ranges of feasibilty for the variables.

The added data are important because getting a single optimum answer to a linear programming problem is often insufficient for practical decision making. One most important consideration is: Supposing the input values do not accurately represent the actual situation, what then happens if those values are changed? The answer is given by a sensitivity analysis in which inputs are varied and different results calculated. Shadow prices, slack variables, and ranges of feasibility provide measures to guide this type of analysis.

A drawback of the MINI-MAX type of output is that the user has no control over the extent of the output information. In contrast, the LP88 package organizes its output into six tables. All or any of these can be called for by the user, printed, and stored in disk files.

Round-off and other errors

It would be nice if a user could simply run these packages and always be confident of getting a correct and usable answer. Unfortunately, there are many reasons why linear programming computations can fail to produce the desired results. Most often, such failures are due to mistakes the user makes in setting up the problem. However, there are also computational errors that can occur even when the input information is logically correct.

Round-off errors, for example, can accumulate over the numerous arithmetic computations in linear programming computations. Recognizing this situation, package designers have come up with various ameliorative approaches.

To help in identifying such problems, MINI-MAX displays a maximum error figure as part of the solution. To try to reduce this figure, users can experiment with changing or eliminating constraints, or they can adjust the rounding of data by the package. This can be set at anywhere from 6 to 12 places after the decimal point. However, users are advised that the results of such adjustments are unpredictable.

To handle round-off errors, LP/80 sets values less than an epsilon value of 0.0001 to zero. However, this does not satisfy all situations, and the package may incorrectly identify some problems as infeasible. To check whether rounding is the culprit, users can adjust the value of epsilon and try to get a better answer.

Inadequate user's manuals

Linear programming packages generally are poorly documented. Most of the manuals do not have such reference essentials as a section with detailed explanations of all command functions and menu choices, another section with explanations of all error messages, and a decent index to quickly find needed items.

In many cases, the manuals are skimpy and try to explain how to use the package only by example, without ever spelling out the options that are available to the user.

An exception is the manual for the LP83 package, which carefully explains each of the error and warning messages the user may see. The manual itself is also well organized, spelling out the information users need to know, and it has an index for reference once the package is in use.

LINEAR PROGRAMMING SYSTEM

Linear Programming System is for smallish problems, with up to 50 variables, including slack and artificial variables, and 35 constraints. There is no provision for storing data once a problem has been run, and all input is from the keyboard. The package seems simple to use and, within its limitations, a terrific bargain. It has not been reviewed here for software defects.

Linear Programming System runs on Commodore 64 computers. It sells for $9.99 on tape or $12.99 on disk from Computer Heroes, PO Box 79, Farmington, CT 06032.

The menu screen for this package shows abbreviated names of the 14 commands that can be used. These commands initiate sequences of screen-displayed queries to which the user responds, or, like the SOLVE command, they initiate actions.

Using these commands, which are explained in the manual, the user specifies the number of decision variables, calls either for maximization or minimization of the objective function, then enters the coefficients for each term in the objective function.

Constraint relationships may be set up as equalities or as < or > inequalities. The CON command invokes a sequence of queries that ask for entry of constraint coefficients and totals in an organized manner. What the user must keep in mind while entering these data is the form and notation the manual shows for the objective function and constraint relationships.

When constraints have only a few non-zero coefficients, a CONSP (sparse) entry command allows the user to skip around to those few positions in the constraint array where coefficients have to be entered. The procedure seems quite simple, as long as the user is able to identify the row and column location of items in the array of constraint relationships.

Display commands allow users to view all the data that have been entered or only selected types of data such as individual columns and rows.

A CH (change) command allows users to alter items that have been previously entered, including type (=, <, >) of constraint and type (max, min) of problem. The user also can specify "Big M," the number used to restrict the size of artificial variables. If not specified, M will be set to 5,000. Users are warned that if M is set too high, round-off error will be introduced.

When data entry is complete, the SOLVE command sends the package in search of a solution. If and when an optimal and feasible solution is found, a message will tell the user how many iterations have been made to reach it. Warning messages appear if there are any all-zero rows or columns, if no feasible solution is found, or if the system of relationships is unbounded. When there is a possibility of multiple solutions, the user is given the option of choosing to continue with further iterations.

The package offers displays of initial and final computed values. These are presented as short lists corresponding to the rows and columns of the array of constraint and objective function relationships. The interpretation of these results is laborious the first time but much simpler thereafter.

LINEAR PROGRAMMING SYSTEM

Linear Programming System has some inconveniences, but it does the essential linear programming job, and look at the bargain price. The vendor does not indicate how many variables and constraints can be handled. An example with just eight variables is given in the manual.

This package runs on Apple II+ computers. It sells for $29.95 from Microphase, PO Box 10461, Tallahassee, FL 32302. (904) 877-7794.

The main menu for this package gives the user the choice of creating a new data file, deleting an existing file, or EDIT/PRINT/SOLVE an existing file.

New files are created from the keyboard. The package prompts for each item of data. The user specifies the number of variables (including slack variables) to be used, the number of constraints, and whether the solution is to be maximized or minimized. The package then prompts for the coefficients of the objective function and for each constraint relationship.

Work for the user

Users have to come to this package with a clear picture in mind of the exact form of the objective-function and constraint relationships they wish to use. The prompts will call for a coefficient for every variable, including all the slack variables for every constraint and for the objective function. Any variable that should not appear has to be given a zero coefficient, so the procedure is methodical.

When this entry is complete, the user must still inform the package which (slack) variable in each constraint relationship is to be used as the basic variable in the computation.

At this point the package has enough information to make up a complete data file, and the user is given the option of storing the data.

Zero means trouble

To generate a solution, the user selects the EDIT/PRINT/SOLVE menu option and keys an S. The package asks for a data file name to run and, if it finds that file, goes on to a solution, diplaying a list of the variables and their computed values, along with a z value for the objective function.

If $z = 0$, something has gone wrong. Even if z is not zero, the user manual advises that "curious results" indicate that there is some difficulty, probably with the way the problem has been set up.

In such cases, the user will want to make changes in the data and can resort to the edit command, E, to do so. The appropriate file is then called for by name, and the first display will show the number of variables, constraints, and problem type. Unfortunately, nothing on this frustrating display can be changed, except by creating an entirely new file. What can be altered are the values of coefficients, constraint right-handside values, and basic variable designations.

LINEAR PROGRAMMER

LINEAR PROGRAMMER does not solve the conventional linear programming problem, but it does deal with simultaneous inequalities, and it is inexpensive.

LINEAR PROGRAMMER runs on IBM/PC, Apple II, TRS-80, and most other personal computers. It sells for $33.95 from DYNACOMP, 1064 Gravel Rd., Webster, NY 14580. (800) 828-6772, (800) 828-6773.

This package does not solve linear programming problems in their conventional form. Instead it solves for the linear variables in a set of $<$ or $>$ inequalities, with variables and coefficients on the left and a numerical value on the right.

LINEAR PROGRAMMER comes up with a "nearest" solution that satisfies as many of the inequalities as it can. The package also produces a partial solution that minimizes the largest difference between any variables expression and the right-hand side value. It uses a modified form of the Khachiyan technique.

Input data are entered on user-numbered lines. What is entered are: the number of inequalities and of variables; the coefficient values for the inequalities; and the right-hand side values.

To run the solution, the user types RUN and is prompted to scale the problem if the coefficients are large numbers since this can lead to an arithmetic overflow error.

The user also keys in the desired number of iterations. The manual advises that 100 iterations are usually enough, except for large problems. A sample problem with two variables and three inequalities ran through 100 iterations in about 70 seconds, according to the vendor.

LP1

LP1 can solve linear programming problems with as many as 100 variables and 100 constraints. It can accept input data either through menu selection, query-and-answer keyboard entry, or from files created by word processors.

LP1 runs on IBM/PC/XT computers. It sells for $99 from Micro Vision, 145 Wicks Rd., Commack, NY 11725. (516) 499-4010.

When this package is started, an input menu appears on the screen, offering the user the alternative of entering data through the keyboard or of reading data from a previously prepared file.

During the keyboard entry procedure, the user can assign names with up to seven characters for the variables. The package then will display the variable names, one by one, and prompt the user to enter objective-function coefficient values for each.

Users then can assign a name for each constraint relationship. The package will prompt for a constraint coefficient value for each variable and a right-hand side (RHS) value for each constraint as well as the nature of the constraint relationship (=, <, or >).

At the completion of the objective-function data, and as each set of constraint data is completed, the package prompts the user to verify that the entered data are acceptable. If not, the user can go through the appropriate set of prompts once again, skipping by those items that are okay and stopping to change any incorrect items.

The package also provides overall editing procedures for data. These can be used to further revise just-entered data or to alter data that were previously entered and stored. The provisions here for editing constraints are particularly helpful when only a few of the variables have non-zero coefficients. The user is allowed to move directly from one such coefficient to another, skipping over variables with zero coefficients. The edited data are automatically saved, in a user-named file, when this editing procedure is completed.

Can stop at each iteration

When a problem is run, the user has the option of making the package stop at the end of each iteration to display the current results.

When the computation is complete, the package will display whether the solution is optimal, infeasible, or unbounded or has a redundant constraint. In the last case, the user has the option of eliminating the offending constraint and rerunning the problem immediately. In all other cases, the package moves to an output menu after the computation. Through this menu, the user can display or print the results or save the solution in a disk file. The results are shown in a row-and-column format.

LP/80

LP/80 can solve small linear programming problems, for example, those with 70 decision variables and 40 constraints.

LP/80 runs on TRS-80 Model I or III computers. It sells for $70 from Decision Science Software, PO Box 7876, Austin, TX 78713. (512) 926-4899.

The first choice on the LP/80 main menu is to start a New Problem. A series of screen-displayed prompts follows this choice. Users reply to these by typing the desired information.

After the user supplies names for each of the variables in the problem, the package prompts for objective-function coefficients. As the name of each variable appears, the user keys in the corresponding coefficient value.

Constraint coefficients are then entered, one to a line, each accompanied by the name of the constraint and variable to which it belongs. By the time this information has been repeated for each coefficient, the user may be thoroughly exhausted but still will have to find enough energy to enter constraint types and right-hand side (RHS) values.

Solutions and sensitivity

The solution is started from the main menu. The display keeps the user informed of solution progress, and messages tell the user if the problem is infeasible or unbounded. Results are viewed by keying in a PRINT RESULTS command. These include the optimal value of the objective function, slack and surplus values for the constraint relationships, shadow prices for the constraints, and reduced costs for nonbasic variables.

The main menu choice to analyze sensitivity can then be made, producing current RHS values and bounds to maintain the current basis. Variables that will enter and leave the basis when these bounds are exceeded are also shown.

To handle rounding errors, LP/80 sets values less than an epsilon value of 0.0001 to zero. However, this does not satisfy all situations, and the package may incorrectly identify some problems as infeasible. To check whether rounding is the culprit, users can adjust the value of epsilon and try to get a better answer.

Another main menu option is to change the data, allowing objective function, constraint coefficients, and RHS values to be altered. The changes can be checked on a display that shows the objective function, then, one by one, the constraints. The data can be stored in disk files, and these files can be read to provide data for a new solution.

LP83

LP83 can handle large linear programming problems. For example, a problem with 513 variables and 340 equality constraints was run by one reviewer. It is not, however, an interactive package. Users prepare the input for each run in advance, using their own word processor or text editor.

LP83 runs on IBM/PC/XT/AT and compatible computers. Use of a math coprocessor chip is recommended. The package sells for $795 from Sunset Software, 1613 Chelsea Rd., #153, San Marino, CA 91108. (818) 284-4763.

LP83 makes little use of the interactive capabilities of personal computers. Instead it requires the user to come to the package with a complete set of input data, ready to be run. Users are expected to set up this input using their own word processing or text editor package, then use the input files they create to run LP83.

This allows the package to conform readily to input procedures traditionally used with large computers, where data and instructions are frequently punched onto cards, then fed to the equipment. Users familiar with large-computer procedures will find the way this package is set up, even the computer terminology used in the manual, quite familiar. LP83 will, in fact, accept input in the standard MPS format used for linear programming based on large computers.

Shaped for experienced users

Learning to use this package largely consists of mastering the input file formats that are required. It amounts to learning to use a simplified computer language. The result is that the package is inherently oriented toward experienced users who employ LP83 frequently enough so they remain familiar with the many details of its input formats.

Many of the features it provides are admirable. For example, users can exert control over about 36 different characteristics of this package by typing a string of commands on the input command line. These include determining the way the output reports will be laid out on the printed page and what reports and report information will appear. There also is a password arrangement to restrict access to use of the package.

LP83 has a number of convenient features such as use of shorthand symbolic names to represent long strings of repetitive input material and automatic conversion of numbers to percentages. Solution runs can be interrupted; the package will save the current solution so the iterative solution process can be restarted from that point.

Diagnostics

The package produces a large number of specific error and warning messages when things go wrong, and the manual provides an explanation for each of these.

The manual itself is well organized, spelling out the information users need to know, and it has an index for reference once the package is in use.

Run times

The run times for problems with this package depend on the available size of computer main memory and on the use of the 8087 math coprocessor chip, which seems to cut the run time for most problems by about a factor of ten. Using a larger main memory apparently affects the run time only for large problems. For example, the time for a problem with 340 rows and

513 columns was cut in half, but larger main memory had no effect on a problem with 60 rows and 93 columns.

The vendor found that, using an IBM/XT with math chip and 640K memory, a problem with 16 rows and 33 columns ran in 6 seconds, whereas one with 177 rows and 548 columns ran in 41 minutes, 33 seconds. These times were decreased to 3 seconds and 29 minutes, 30 seconds when the same problems were run on an IBM/PC/AT with math chip and 512K memory.

Elapsed times to get answers from the IBM/PC/AT and an IBM 3081 mainframe computer were also compared. For all but the very largest problems, the ready access the user had to the PC provided the answers more rapidly, although the large machine had an advantage of almost ten to one in raw computational speed.

LP88 AND RELATED LINEAR PROGRAMMING PACKAGES

Reviewers: George H. Brooks, and Deborah S. Fogel, University of Central Florida, Orlando, Florida.

The LP88 package solves linear problems with up to 255 constraints and 2,255 variables, using the revised simplex algorithm for quick and efficient solutions. The LPX88 package handles larger problems with up to 510 constraints and 2,510 variables. BLP88 solves problems with upper or lower bounded variables, handling 255 constraints and 1,255 variables. DLP88 is for dual linear problems with up to 1,255 constraints. There also is a student version, SLP88/STSA88, which solves simplex problems with 45 constraints and 255 variables and also handles transportation problems with 88 sources or sinks.

These packages provide users with a powerful set of linear programming tools. Operation is easy and efficient because the packages provide menu displays, function key controls, and a spreadsheet-style input editor.

All the packages run on IBM/PC/XT/AT computers with 192K main memory. Each package sells for $88. All but the student package are available in versions that make use of the 8087 math chip (these each cost just $11 extra). The packages are available from Eastern Software Products, PO Box 15328, Alexandria, VA 22309, (703) 360-6942.

The reviewers checked the speed of package operation with a 13-variable, 48-constraint problem that had 198 non-zero matrix coefficients. The computer used was an IBM/XT with 256K main memory. LP88 produced an optimized solution in two minutes, 15 seconds. LPX88, which uses the revised simplex algorithm with product form inverse, produced an optimized solution in 54 seconds. The increased speed probably was due to the sparse nature of the coefficient matrix. When the DLP88 package was used, the problem was first converted to its dual. That took just 5 seconds. Then the solution itself took 9 seconds.

An unbounded problem, involving 13 variables and 9 constraints, was also run on LP88. In 15 seconds, the result "NO SOLUTION" was produced.

A main menu initiates the principal functions of the package, and three more specialized menus control setup of problems and data entry, execution of the solution, and the output information. The package displays all four menus on a single screen, informing the user near the bottom of the screen as to which one is currently active.

Spreadsheet input

To input a new set of data, the user calls for a New Problem on the Setup menu. On the display that follows, the user answers questions to name the problem, state whether it is to have a maximum or minimum solution, and define the number of constraints and nonslack variables in the problem. LP88 names the constraints and variables (users can later edit these, assigning six-character names) and creates a slack variable for each constraint.

Data for the coefficient matrix as well as cost and requirement vectors can be entered by selecting the Display Editor from the Setup menu. The entry is done on a spreadsheet format. The user starts the data entry process by selecting an initial row and column. A labeled, largely empty spreadsheet is then displayed. Using row–column locations to position the cursor on this display, the user can enter desired constraint and objective-function coefficients. The cursor also can be moved about by the arrow keys. Up to six characters can be entered for right-hand side values and five characters can be entered for coefficients. The package automatically computes coefficients for the slack variables.

The Display Editor cannot show all rows and columns for large problems. It is, in fact, set up to display 15 constraints and 10 variables at a time. Users can move the display to the desired area of the spreadsheet by identifying the constraint name and variable name of the spreadsheet cell they wish to appear in the upper lefthand corner of the screen. If desired, an array similar to the spreadsheet display can be produced in printed form.

An alternative method of entry is to create a data file using other software, outside the package, then read the file when the package is run. The format for such a file is compact and formal, with the details explained in the manual. With this input method, ten characters can be used for coefficient values.

Control over the solution

The technique used for solving linear programs involves the explicit inverse variant of the revised simplex algorithm. The algorithm can be started in four different ways: generate basis; obtain a previously computed or saved basis and optimize; restart from the most recent point of execution; and solve for a given basis without maximizing or minimizing.

Users can set a limit on the number of iterations (pivots) that will be performed. They also can select the number of iterations performed before the inverse is recomputed, the tolerance values below which pivot elements, inverse elements, and basis variables will be rejected, and the scales for display of primal and dual variables and of the objective value. If the user does not make these selections, the package will use default values for these parameters.

During its calculations, the package displays current information on the state of the solution. Solutions can be interrupted while in progress, allowing users to step through the pivots, reinvert, reset tolerances and controls, save the basis, and record the current solution and pivots. User control is simplified by Function keys.

When operated by programming-minded users, the package can be run in a batch mode. Here a properly formatted command file, containing instructions on how the computation is to proceed, is entered, along with a data file. In the run that follows, users can exercise limited control from the keyboard.

Flexible output

Package outputs include the original problem data, primal and dual values, cost and right-hand side ranges, basis interval matrix, and inverse and nonbasis column products.

Six tables are used to display this information. All or any of these can be called for by the user, and printed and stored in disk files. The solution basis can also be stored.

To rapidly solve problems that are similar but not identical, a problem can be modified using the Display Editor; then it can be rerun starting with the previous solution. If a basis inverse has been computed, and there are no changes to any of the basis columns, the modified problem can be rerun even faster by starting from the previous inverse basis.

Some cautions

When using version 3.12 of LP88, DLP88, BLP88, and version 4.07 of LPX88, users should be aware of the possibility of program crashes. When errors occur, the package may suddenly terminate its operation, forcing the user to restart from scratch in order to get a solution. During the review procedures, the printer ran out of paper and such a crash occurred. The inconvenience of such incidents can be minimized if users store their information on disk immediately after it is produced by the package and before any printing is done. The package vendor has indicated that future versions will include error-trapping provisions to prevent such crashes.

The student version, SLP88/STSA88, already has such error trapping. This version offers most of the features of LP88. However, it is limited to six characters for coefficients, does not allow use of sequential files nor storage of results on disk, and does not support use of the 8087 math chip.

Before using the BPL88 bounded-variable version, users should take care to read the documentation covering solution output. The usual printout and display will show only those variables that appear in the basis. To obtain a complete statement of the solution, users must call for a special printout. This will mark nonbasis, nonslack variables with an asterisk if they have reached their bounded value. These must then be included in the determination of the optimized objective-function value.

Documentation is comprehensive

LP88 documentation is contained in a 42-page manual. It is comprehensive and includes a sample problem. The package disk includes three sample problems that allow users to experiment before running their own problems. Explicit instructions are included in the manual on how to use the package, the meanings of the Function keys, along with general information on the package. Documentation for BLP88, DLP88, and SLP88/STSA88 is recorded on the package disk and can be printed out, if the user wishes.

MINI-MAX

MINI-MAX can solve linear programming problems with up to 300 variables and up to 70 constraints, but not both at the same time. For example, when 132 variables are involved, only 30 constraints can be handled.

MINI-MAX runs on IBM/PC and compatible computers. It sells for $395 *from Agricultural Software Consultants, 1706 Santa Fe, Kingsville, TX 78363. (512) 595-1937.*

The package runs from a main menu and several series of screen-displayed queries to which the user keyboards responses.

Keyboarded input

One of the main menu selections is to set up a new linear programming problem. Replying to the queries that follow this selection, the user specifies how many variables and constraints will be used and whether the problem is to be run to find a maximum or minimum result.

Names are then keyboarded to describe the problem, each of the variables, and the disk file in which the input data will be stored. Then the coefficients of the objective function are entered. At the completion of this set of entries, the user is asked to verify that they are correct; if not, the set can be reentered. The next series of input queries is designed to allow the user to enter the coefficients for the constraint relationships. This is done for one constraint at a time, with a verification step when the coefficients of each constraint have been completed.

When this problem description has been completed, the user can call for a printout of the input data. The description is automatically stored in the named disk file, and it is made ready for running the solution to the problem.

Lots of output

To run the problem, the user presses the R key with the main menu displayed. The entire problem is run, and results are displayed, a screenful at a time. The user can rerun these displays or call for a printout of the results.

These results include displays of the optimal solution and basic and nonbasic variables. The nonbasic variables are those whose coefficient values are insignificant for the solution, and the output includes an "entering value" to show at what point each coefficient value would have made that particular variable basic. A separate display shows the current values of the basic variables along with high- and low-range values, to show how much variation there can be before these become nonbasic.

Sensitivity analysis

The output also displays slack and surplus variables, dual variables with shadow prices, and the ranges of feasibilty for the variables.

Whenever a less-than-or-equal-to constraint is used, there can be slack; that is, the amount the right-hand side of the constraint can be decreased without having any effect on the solution. In a similar manner, a greater-than-or-equal-to constraint can have a surplus variable.

The shadow price of a constraint is the amount by which the objective function will change if the right-hand side of the constraint is increased by one. A large shadow price is a danger signal, warning the user that a constraint is drastically affecting the solution.

Some problems have two or more optimal solutions, called alternate solutions. MINI-MAX will signal this situation with an "ALTERNATE SOLUTIONS" message. It is then up to the user to locate additional solutions by making small changes in the coefficients of the objective function.

When solutions fail

If the problem has no solution, the package will display a "NO SOLUTION" message and show the shadow prices. This may have happened because one or more of the constraints cannot be satisfied. In this situation, it may be enlightening to examine the shadow prices, since those constraints with zero shadow prices will not be the ones preventing the solution. The user can then try altering or eliminating the remaining constraints, one at a time, until a satisfactory solution is obtained.

Another error message that may appear is "UNBOUNDED SOLUTION." This usually indicates that the problem has been set up incorrectly, but it also may be due to a computational arithmetic error. To help in identifying such problems, MINI-MAX displays a maximum error figure as part of the solution. One recommended remedy is for the user to experiment with changing or eliminating constraints. Another is to adjust the rounding of data by the package. This can be set at anywhere from 6 to 12 places after the decimal point. However, users are advised that the results of such adjustments are unpredictable.

Changes and reruns

To make changes to an existing problem description, the user has to call for the file by name. If in doubt, a directory of stored files can be consulted.

A listing of the variables and coefficients in the objective function is then displayed, with variable names, and the user can choose to make changes to any individual variable or coefficient by answering a short sequence of displayed queries. In a similar manner, changes can be made to displays of any or all of the constraint relationships.

MPS-PC

MPS-PC solves linear programming problems with up to 70 variables and 50 constraints. It has procedures that could make it very easy to learn and use but somehow falls short of achieving all it could in this area.

MPS-PC runs on IBM/PC and compatible computers. It sells for $150 from Research Corporation, 6840 E. Broadway Blvd., Tucson, AZ 85710. (602) 296-6400.

MPS-PC is a personal computer version of the MPS-X program used on large computers. It operates from three menus and various displayed prompts that follow from menu choices.

Column-by-column data entry

The data entry process is an organized one, but the organization does not seem to give high priority to user convenience. In several ways, users are asked to compensate for programming peculiarities and oversights by special procedures.

Data are entered from the keyboard or read from previously prepared files. When entering data, users have to be aware of the particular rules this package uses. Thus, the number of rows the user calls for does not include the row for the objective function, and the number of columns does not include the one for the right-hand side (RHS) data. The package does not recognize <, >, and =, so users must enter L, G, and E for these relationships.

Actual data entry is on a column-by-column basis, which means that users will not be able to keep track of what they are doing by merely knowing what equations and inequalities they wish to enter. Instead, they must make a careful handwritten layout of the entry data with all columns neatly and correctly lined up.

MPS-PC will not tolerate negative RHS values, so users must multiply all the terms in such a constraint by minus one to enter it.

Users must be sure not to inadvertently press the F1 function key followed by Enter. That will erase whatever data entry has been done. This feature can be helpful, but it lacks a verification step in which users are asked if they really want to take that step.

Data storage and editing

Sets of input data are stored in disk files with names that can be eight characters long, plus a three-character suffix. With enough use of the package, these files can accumulate. As with most personal computer storage arrangements, it is up to the user to keep the disks from filling up by using the PC system software to erase those files that are no longer needed.

Correcting an error is easy if it is caught during data entry before a column is complete. All the user has to do is enter the row name and then the correct coefficient value. Once the column is entered, the user must do corrections with editing procedures that can be used after the data entry process.

To examine what has been input, the user can call for a printout of the data in row–column format. There also is an alternative printed format that first lists the row names, then shows the non-zero coefficient values for each column, with each value identified by its row name. This printout also lists RHS values, each identified with a row name. Only the alternative form can be shown on the display screen. The vendor excuses this fact by stating that this is due to "limited screen space."

Data editing is done from an editing menu where the user calls for specific types of items, such as objective-function values, RHS values, and constraint names. The package will not tolerate duplicate names for columns or rows. To transpose two names that have been wrongly placed, a third temporary name must be used.

How solutions run

When the data for a problem are complete, the user can start the solution portion of the package. During the solution calculation, the user is given a running account of each iteration on the display screen. Added information will be printed at each iteration.

It is not uncommon, we are told, for this package to run through 100 iterations to solve a problem with 50 rows. Users are warned by the manual that they may be "somewhat disconcerted" by how long it takes to solve problems. A problem with 63 columns and 47 rows is said to have taken 68 iterations and 15 minutes to reach a solution. However, all the constraints in this problem were equations, and $<$, $>$ constraints are said to take longer.

Some tricks can speed things up a bit. For example, problems with many greater-than-zero constraints can be speeded up by changing these to less-than-zero. The only effect on the solution will be changes of sign in the coefficients of the affected rows. Could not the package have included a built-in feature to do all this automatically?

MPS-PC notifies the user when it fails to solve problems because of infeasibility and identifies the row where the logical problem arises. It also notifies the user of unbounded conditions that prevent a solution. In this case, it is entirely up to the user to find the trouble.

Occasionally, solutions with this package may not diverge but instead may fail to converge to a solution. The designers of the package decided to tolerate this situation because the procedures that could avoid it would slow down the entire solution process. The manual advises on various measures users can take if faced with this kind of solution failure. To stop the package while it is cycling through one of these infinite iteration sequences, it is necessary to terminate package operation and exit to the computer system software.

Options with cautions

The user then has the option of performing calculations that will produce an upper and lower bound for each objective-function coefficient and for the RHS values of the constraints. This provides a measure of how much these quantities can change, for one variable at a time, without changing the optimal solution.

Another optional calculation allows the user to make changes to objective-function coefficients and RHS values, then rerun the solution. The procedures for this seem annoyingly complicated and indirect. Instead of making specified changes to selected items, the user has to manipulate a change vector, with operations that could better have been done internally by the package. Even more constricting is the requirement that the user make up these change vectors before running the problem, whereas what the user needs is to make the changes on the basis of the calculated results.

However, when a solution is completed, the package automatically stores the solution information on a disk file, so it is not too difficult to make a rerun with an appropriate change vector.

An existing solution file can be used as the starting basis for a solution to another problem, but here the package offers little control over the situation. An automatic check assures that the number of columns and rows in the problem file and the solution file are the same. However, the package will not check to see that the nature of the constraints and the values of the constraint coefficients match up, and, unless they do, the calculated answers will be wrong.

OPERATIONS RESEARCH PACKAGE

Reviewer: Darrell G. Linton, University of Central Florida, Orlando, Florida.

The four programs in this package are not intended for solving very large and complex problems. Instead, they permit the user to obtain quick and inexpensive answers to problems, provided that the underlying assumptions approximate those of the model. The programs are written with user convenience in mind. The examples and documentation are easy to follow. This reviewer's grade rating for this package is 95 on a scale of 100.

Programs included are for linear programming, maximum flow analysis, inventory models, and queueing models. The package as reviewed runs on an IBM/PC. Versions for TRS-80 I, II, III, and 4, and for Apple II, II+, and IIe are available, all from The Institute of Industrial Engineers, 25 Technology Park/Atlanta, Norcross, GA 30092, (404) 449-0460. Price of the package is $175, or $140 to IIE members.

The linear programming portion of the package optimizes a linear objective function subject to linear constraints. Problems with up to 30 constraints and 40 decision variables can be run. Typical objective functions involve total cost or profit, whereas the constraints impose limits on such items as raw materials and personnel.

A typical maximum flow problem would be one where a commodity is to be shipped from one point (the source) to another (the sink). There also may be several intermediate locations, called transshipment points. The problem is to maximize the flow of the commodity from source to sink, subject to the paths connecting the various points. The total flow into a transshipment point must always equal the total flow out. Although this problem can be solved by linear programming, the maximum flow algorithm is usually more efficient. The total number of points that can be handled is limited by the available amount of computer memory.

Input to the inventory models program may include the total annual demand for a product, the per-unit price, procurement cost per order, annual holding cost, shortage cost, maximum storage space, and number of price breaks. Depending on user input, the program selects from among eight possible models, including a basic economic-order-quantity model, a price-break model, a shortage-cost model, and a storage-limitation model. The program outputs include the optimum order quantity, total inventory costs, number of order cycles per year, cost per unit, maximum inventory, and number of units back ordered per cycle.

The queueing models program analyzes single- and multiserver waiting line problems. It assumes a first-in first-out serving procedure and steady-state operation of the queue. Problems with exponential interarrival and service time distributions are handled. Users can choose from four models: a single-server, infinite waiting room model; a multiple-server, infinite waiting room model; a parallel-server, finite waiting room model; and machine-servicing model.

Typical outputs for the queueing models include the steady-state probability of N units in the system, the average number of units in the system and in the queue, and the average waiting time in the system and the queue. The limitations on the program are the built-in assumptions of the models, such as Poisson arrivals, exponential service, and steady-state results.

Clear documentation

Instructions and examples in the 83-page loose-leaf manual are clear. There is a separate section for each of the four programs, and each section concludes with an IBM/PC source listing for programming-minded users. The reviewer found a few minor errors in the text of the manual. These will undoubtedly be corrected in the next edition.

As for recovery from errors, the reviewer found that the programs are designed to recover from almost every conceivable error and allow easy editing of input data.

RAMLP

Ramlp runs fairly large problems, with the size limited by the main memory of the computer. For instance, with a 128K main memory, Ramlp can handle up to about 100 constraints and 100 variables.

Ramlp runs on IBM/PC and Colombia computers. It sells for $149 from Miniware, 205 Winchester Rd., Annapolis, MD 21401. (301) 757-5626.

For problems involving a lot of input data, the main task the user faces is putting the data into the computer and verifying that it is correct. Ramlp offers some alternative ways to do data entry, but the basic input format is to enter rows and columns of data.

To enter data from the keyboard, the user has to be aware in advance of how these rows and columns will look and specify how many of each will be needed.

In the rows and columns, everything must be in its place, and each has a name. The first row contains the coefficients for the objective function. Subsequent rows each contain data for a constraint function. The first column contains constraint levels and is called the right-hand side (RHS) column. Subsequent columns, each assigned to one decision variable, contain constraint coefficients. Users can assign convenient names to all but the first row and column.

Shown on this data array is the user's choice of either a maximum- or minimum-value solution for the objective function. Whether a constraint function is an inequality or an equality, the appropriate symbol is shown in the RHS column. At the bottom of each of the user-named columns, lower- and upper-bound values for each decision variable can be entered.

To set up the rows and columns for their problem, users call for an Edit menu. The choices on this menu allow changing row or column names, bounds, and =, <, > designations. They also provide for editing coefficient values once they are entered.

Data entry aids

A coefficient menu provides some aids for the main data entry task, keyboarding the coefficients for the objective-function and constraint expressions. Choices here allow users to enter or erase a single coefficient by first specifying its row and column location. Other choices allow entering an entire row, column, or diagonal of data. Then, as each coefficient is entered, the package will display location numbers for the next.

However, keyboard entry is not the only means for getting data in. When the package is first started up, a displayed menu allows users to choose the type of data input they want. In addition to input from the keyboard, data can be taken from previously stored disk files.

When a problem is run, the data that went into it are discarded. Therefore, if users want to reuse those data, they have to deliberately store the information. This choice is one of several that can be made from the package main menu, from which users can also order printed listings of rows, columns, and bounds or of the entire array of input data.

Getting a solution

The main menu also offers a Solve choice to be selected after a complete array of input data is in place and ready to be processed. With Solve underway, the computer display shows a log of the partial results as iterations are made toward a feasible solution. Remarks shown on these displays indicate the number of constraints satisfied at each iteration, along with a current number of infeasibilities (which gives the user some idea of how close a solution may be). If the solution process should fail, this record can be used in deciding on possible fixes.

Based on computation results, the user may decide to change the input description of the problem by deleting or adding data array rows or columns, which can be done through the main menu.

If and when a solution is found, the package prints out a final report showing the computed value of the objective function and of each of the rows and columns in the input array.

Ramlp runs problems entirely in computer main memory, thus avoiding the need to store and retrieve data from disk storage during the computations. This eliminates slow disk accesses, giving a fast computation. However, it limits the size of problems that can be run, since main memory space is limited. In a 128K main memory, the package can run problems of a size equivalent to about 100 constraints and 100 real variables. The package automatically checks to see if the problem size fits the allowed limits before it is run.

Chapter 16
Comprehensive Statistical Packages

Chapters 16 and 17 discuss statistical packages. In this chapter we will take a look at comprehensive statistical packages, those that offer the user choice of a wide variety of statistical tools. Such packages seem to come in two basic varieties. In the first, the procedures are invoked by keyboarded commands. The second type of package operates from displayed menus.

Command-based packages are for serious users who are willing to invest the time to learn to use a new language. For those who will find frequent, consistent use for such a package, this will be a worthwhile investment. With the use of commands, there is no need to wait for menus to be shown nor to move from one menu to another to reach a desired function.

Other users, who may turn only occasionally to personal computer-based statistical tools, will probably feel that learning to use a command-based package is an overly budensome task. With menus always there to remind the user of the available choices, there is no need to memorize commands or the syntax rules of use that inevitably surround them.

The ideal package would make menus available to those who need them, while it offered direct commands to more practiced users. Some word processing packages have such combined facilities, but there do not seem to be any statistical packages that do.

Some command-based packages do have features designed to ease the process of learning and use. For example, the SYSTAT package uses about one hundred commands, but the package is divided up into modules, one for handling data and others for various types of statistical procedures. Each module has a set of commands peculiar to its operations, so the SYSTAT language can be learned piecemeal, starting with those procedures the user needs first. At the end of each user session, SYSTAT prints a log of all the commands that were used, providing a convenient reference for performing similar procedures in the future.

The language of statistics

For those readers who may not be familiar with the terms used to describe statistical procedures, it seems worthwhile to put some of the terms used in the reviews into context.

Statistical analysis starts with a set of data, and the first step is to weed out errors in the data. Packages may include procedures for scanning the data to see that no inappropriate symbols are included with the numeric data. Histogram displays of individual variables can be used to spot "outlier" values. Statistics such as the Mahalanobis distance can reveal unusual variation in a row of data, compared to other rows. Since not all data sets are complete, there also should be some way to handle missing data.

Descriptive statistics can include about a dozen measures such as the mean, median, maximum and minimum, variance, standard deviation, skewness (left- or right-handed tendency), and kurtosis (peakedness) of a set or "sample" of numeric data.

Often the user wants to compare one data sample to another or to a larger data "population." The so-called null hypothesis is that the sample is representative of the population, and a number of statistical tests are used to test this hypothesis.

The choice of tests depends, in the first place, on whether the population and sample are normally distributed along the bell-shaped curve that is typical of many populations. Statistical tests such as the Kolmogorov–Smirnov and Wilk–Shapiro are used to test for normality.

The null hypothesis for normal distributions ordinarily is tested by "parametric" procedures such as the various types of t-tests and the F-test. However, there also are nonparametric tests appropriate for use with data sets that are not normally distributed.

In some cases, it is desirable to transform such data sets by taking the logarithm, or square root, or performing some other mathematical operation on each of their members, such that the resulting distribution is normal. The parametric tests can then be used, and these are considered more sensitive than the nonparametric in detecting departures from the null hypothesis.

Two types of errors occur in these tests: (1) the null hypothesis may be rejected when in actuality it is true; and (2) the hypothesis may be accepted when it is false. The probability of an error of type 1 is called the significance level of the test; values on the order of .05 to .01 are considered small.

Correlation tests measure the extent to which two data samples are linearly related to one another, so that knowledge of data points in one can be used to predict the data points in the other.

Analysis of variance (ANOVA) is used to determine to what extent different factors contribute to the total variance of a data sample. Thus, an engineer might be interested in the extent to which maximum temperature or processing time contributes to the variance of the dimensions of the items in a production sample. One-way ANOVA tests the effects of a single factor on the sample. Two-way ANOVA can test the simultaneous effect of two factors.

Analysis often involves separating the data into common-characteristic groups. The software can provide for coding and segregation of such groups. One-way or two-way analysis of variance and other procedures can be used to check for differences and similarities between groups. Cross tabulations can be used to summarize the ways that data separate out into groups.

When two or more variables are involved, graphical displays can be used to reveal whether or not relationships exist. Regression analysis, linear and nonlinear, can shed further light on these relationships.

There can be situations where it is difficult to identify groups or subpopulations. Clustering algorithms provide a way to identify possible groupings of variables and data. Where there are complex functional relationships between variables, factor analysis can be used to study the correlations between them.

One of the key assumptions underlying ANOVA tests is that the variables involved are normally-distributed and independent of one another. The chi-square test is used to draw inferences about their dependence or independence.

For data that clearly violate normality, there are nonparametric tests, such as the Kruskal–Wallis one-way analysis of variance test and the Friedman two-way analysis of variance, that can be used.

Specialized procedures are available for dealing with time-series data, and such analyses may use either frequency or time-domain approaches. Survival analysis techniques can be used to study outcomes over various time intervals.

A variety of statistical procedures

At a minimum, every one of these comprehensive statistical packages offers some basic, standard procedures. There are always some descriptive statistics, a suite of tests for normally distributed variables, nonparametric tests, correlation, regression, and analysis of variance (ANOVA) procedures. Graphics such as histograms and scatter graphs also are standard. Some of the packages also include procedures such as cross tabulation, factor and cluster analysis, time-series analysis, and survival estimation.

The predominant approach is simply to present a kit of statistical tools, a kind of statistician's smorgasbord, from which the user can choose whatever seems to meet the current need. Two command-operated packages that exemplify this approach are BDMPC and SPSS/PC. Both are personal computer versions of mainframe computer packages. Both offer such a large number of procedures that it takes user manuals running over six hundred pages to describe what is available, and both offer powerful procedures. Thus, although most packages provide one-way, perhaps two-way, ANOVA, the SPSS/PC ANOVA command will handle up to five dependent variables, ten factors, and ten covariates.

This approach is not restricted to command-based packages. For example, the NCSS menu-based package also aims to be a wholesale supplier of everything in statistics that a user might need. The NCSS manual is thinner, mainly because its descriptions are more succinct.

Working with a more restricted set of procedures, the DAISY package (command-based) is organized to meet the needs of users who want to analyze their data by first plotting to get a preliminary impression of whether there is a relationship between the variables. Product-moment, partial, and time-shifted correlations, along with four nonparametric measures, are provided as means of further measuring relationships between variables. Linear and stepwise regression procedures are provided to establish trends in the data and to set up predictions. The other statistics offered by the package are selected specifically to support these basic capabilities by testing the results.

The SYSTAT package proudly offers two unique procedures. An MLGH (Multivariate General Linear Model) procedure can estimate and test any univariate or multivariate general linear model, allowing linear contrasts on vectors of dependent as well as independent variables. There also is a multidimensional scaling procedure for up to five dimensions.

Convenient data entry

Using a package for statistical manipulation is a relatively easy matter. Once having decided which procedure to use, a few keys are pressed and the computer churns away until results are produced. Entering and checking the data that are to be analyzed are much more arduous.

Some of these packages seem to recognize that personal computer users will often be faced with the task of keyboarding large amounts of data, then correcting and checking what has been entered. For example, the DAISY package has spreadsheet-like data entry and editing capabilities. A table for data entry is displayed, and users can make entries one after another, moving from one location to the next by pressing the appropriate arrow key. Other keyboard actions make the cursor jump to the start of the next row or column. Editing consists of typing a new number over the existing one at the cursor position. Users can even split the screen and compare nonadjacent columns side by side.

Most of the packages provide for entering data in a much more piecemeal manner, usually a row at a time. Thus, the ELF package displays an entry form for the first row that shows the variable names, each followed by a space for ten characters. As each data item is filled in, pressing the space bar will move the cursor to the next item-space. When the data for that row are complete, the user can call up the display for the next row by pressing the Return key, or, if there are any typing errors, the process of entering that row can be repeated. MICROSTAT users follow a similar procedure, straightforward, but not as convenient as working with a full-screen display of a data table.

Less desirable procedures in other packages prompt the user for one item at a time, never giving an overall view of even one complete row of data.

StatPac users can devise their own input screens. This capability permits either row-by-row or column-by-column input procedures. The idea appears to be that operators with little or no knowledge of statistics will do data entry and that these screens can be designed for their use.

Emulating punched cards

When punched cards are used to input data into large computers, the processes of verifying and correcting the data take place on the punching equipment, not in the computer. The authors of the SPSS/PC package apparently have assumed that the data input to the personal computer will be as pure and correct as that from punched cards. There seem to be no convenient built-in procedures in this package that would allow users to correct errors in the data they type into the keyboard.

BDMPC users will find that this package takes a similar approach. It seems, incredibly, to be set up to take data as if the information were coming from punched cards. The user first has to write a short program that reserves storage space and specifies the format in which the information is being presented as well as describing the information itself (title, number of variables, maximum and minimum values, and so forth). The procedure, appropriate for mainframe processing, is inappropriate for personal computer use, but there seems to be no alternative.

Manipulating files

Given the choice, any user would rather work with existing files of data than have to enter new data through the keyboard. An important part of most of these packages is their ability to store data files on disk and to retrieve and manipulate those files to present just the desired data for analysis by statistical procedures.

These data manipulation facilities can be quite flexible. For example, with the NCSS package two tables may be joined, one above the other or side to side. Individual columns may be copied from one table to another. Portions of tables may be selected out. Tables can even be rotated so that the rows become columns and vice versa.

Another important capability is to be able to take data from other computer packages and adjust the formats so these data can be treated by the statistical package. Most of the packages discussed here can accept files in standard ASCII format. In addition, the DIF format used by some personal computer spreadsheets can be accepted and used by packages such as MICROSTAT, ELF, and DAISY. Some of the packages also will accept files in the popular dBase II database format. ABSTAT can accept files produced by computer languages such as BASIC and reformat its own files so they can be used in BASIC programs.

Transforming the data

Often statistical analysis involves transforming the data before actually carrying out statistical procedures. All of these packages provide some facilities for making such transformations. Usually this involves operating on a table of data recalled from storage.

One of the most extensive facilities is offered by the NCSS package where transformations of the data may be done by adding, subtracting, multiplying, or dividing the items in one column by those in a second column, or by a constant, to form a transformed column. Single-column transformations include taking the absolute value, cumulative totaling, forming truncated or rounded-off integers from decimal numbers, calculating modulo remainders or signature values, taking Tukey's folded log or root, natural logs, exponents, square roots, and trigonometric functions. Columns of uniform-random or sequential numbers can be generated. There also are smoothing and lagging transformations, and values within a column can be sorted or ranked.

The ABSTAT package uses a single, flexible procedure that allows users to construct transformations using algebraic functions that can include expressions in parentheses; arithmetic, trigonometric, exponential and log functions; modular arithmetic, absolute values, truncation; and a variety of constants, including the mean and standard deviation of the variable. Transformations also are provided to convert times and dates.

The ELF package takes this idea one step further, requiring users to write short programs in a special computer language in order to make data transformations. Allowed operations include arithmetic, selections based on $<$, $>$, $=$, and odd conditions, trigonometric, exponential, logarithmic, modulo 10, rounding, truncation, and absolute value.

Statistical Graphics

Graphics provided with these packages generally are crude, with points represented by printed characters. The SYSTAT package makes the most of such plotting capabilities by offering variety. Scatter plots, function plots, and histograms can be produced easily. With the help of some BASIC programming, crude contour plots can be made. There also is a stem-leaf plot, a histogram-like display made up of numerals, that can give users insight into the repeated appearance of certain discrete values in the data. Box plots are used to identify quickly median, hinges, and outside values in the data. SYSTAT also will produce probability plots, which show a variable plotted against corresponding percentage points of a standard normal variable.

The DAISY package is unusual, providing high-resolution plotting for those users whose computers will accommodate it. There are histograms, useful for examining the data items in a single variable, and sequence plots that can help to discover trends, repeating cycles, and unusual data points in a variable. To compare one variable to another, high-resolution scatter plots are available in linear, semi-log or logarithmic form. If requested, the package will draw a regression line of best fit for the scatter plot.

ABSTAT

ABSTAT provides most of the standard statistical procedures, and it also does correlation, regression, and cross tabulations. Operation is based on stored files of data and about 64 keyboarded commands are used. There are graphics, but they are primitive. ABSTAT will handle up to 64 variables and can make use of a math coprocessor chip.

ABSTAT runs on IBM/PC/XT/AT and compatible computers. It sells for $395 from AndersonBell, PO Box 191, Canon City, CO 81212. (303) 275-1661.

"Since the number of commands is too large for an effective single menu" (a rather lame excuse), the authors of this package have elected to require users to remember keyboard commands, rather than work from menu choices. ABSTAT is therefore a specialized computer language, oriented to statistical procedures. As with any computer language, the user must examine carefully the functions of the commands and remember as many as possible as a prerequisite to serious use of this package.

Once a user knows the command names, there is a HELP facility that can call up and display explanations of any named command.

To ease the learning process, the commands are divided up into five categories: data manipulation, statistical analysis, graphic functions, report writing, and miscellanous. In the first two categories there are about 50 commands, and only about 14 more in the other categories. Most of the user manual is taken up with explanations of what these commands mean and how to use them. A run-through of how to use the package to solve a typical problem is given in a short appendix. Another appendix gives the calculation formulas used by the package. A third lists the errors the package may trap and gives a brief explanation for each of them, a handy feature. Appropriate to a manual that is primarily a reference source, there is an index.

Serial data entry

Data entry is accomplished with the EDIT command, after the user has set up a file using the CREATE command. Entry is made using a serial query and reply procedure in which the user keys in each of the variable values for case one, then enters case two, and so forth. During the entry process, the user can back up to previous queries to correct typing errors. Once the data have been entered, there is a command for making changes in selected cases. A LIST command causes the whole data table or selected variables to be displayed or printed out for detailed examination and verification. There also is a command that allows data entry one variable at a time. The package does not seem to provide a means for entering data onto a displayed tabular form.

A flexible transformation procedure allows users to construct transformations using algebraic functions that can include expressions in parentheses; arithmetic, trigonometric, exponential, and log functions; modular arithmetic, absolute values, truncation; and a variety of constants, including the mean and standard deviation of the variable. Transformations also are provided to convert times and dates.

ABSTAT can accept files produced by the BASIC computer language and reformat its own files so they can be used in BASIC programs. It also can exchange files with the dBase II and III data base packages. However, one reviewer found that this capability had some undocumented defects.

Statistical procedures

Most of the statistical commands involve a specification by the user of the data file and the variables within that file that are to be operated upon. The DESC command will produce a standard set of descriptive statistics. There are t-tests for comparing matched and independent variables and a t-test for comparing a sample mean to a population mean. The package also has a group of probability calculations based on user-keyboarded information. For example, the user can enter chi-square and degrees-of-freedom values and get a probability for the combination. Similar calculations are provided for F distribution Student's t distribution, binomial and Poisson's distributions and Z-scores.

Other commands perform tests such as chi-square goodness of fit, the Kolmogorov–Smirnov two-sample test, the Kruskal–Wallis one-way analysis of variance for ranked data, the Mann–Whitney U test, a one-sample runs test, the Wald–Wolfowitz two-sample runs test, and a sign test.

The SRANK command will rank data and produce a Spearman rank correlation matrix. The CORR command will produce a correlation matrix for selected variables.

There is an ANOVA command that comes up with a summary table for up to a four-way ANOVA. The capacity of this procedure is limited. When the numbers of values for each of the factors are multiplied together, their product must be no greater than 400.

The REGR command performs multiple linear regressions on 1 dependent, and up to 19 independent variables. The XTAB command produces cross tabulations of selected variable pairs, calculates chi-square, and performs Fisher's exact test where appropriate. There is no provision for stepwise regression.

Rough graphs

The package can produce histograms and scatter plots for selected variables. These are made up of printed characters: Xs for the histograms and numerals that indicate point locations on the scatter plots.

If printed output or storage of the results in a disk file are desired, the user must specify these options before a statistical procedure is run.

BDMPC

BDMPC contains just about any and all statistical procedures that one might want to use. The problem is that the user is assumed to be willing to learn an extensive special computer language and to be familiar with mainframe-style procedures. BDMPC runs on IBM/PCs and compatible computers and is designed as if the personal computer were never invented. The package requires use of 640K main memory, hard disk storage, and a math co-processor chip.

There are 21 software modules available. The price for all of them is $1,500. They also can be purchased in groups or individually from BDMP Statistical Software, 1964 Westwood Blvd., #202, Los Angeles, CA 90025. (213) 475-5700.

BDMPC is a personal computer version of the BDMP statistical package, long a standard for mainframe users. The BDMPC modules are essentially the same programs used with the mainframe version, and they require the full capabilities of the IBM/PC, including a 640K main memory, hard disk storage, and a math coprocessor chip. Still, there are certain limits to the statistical solutions that can be run with BDMPC modules. For example, the limit on number of data values is 16,000.

The 1985 *BDMP Statistical Software Manual* is a 734-page compendium, a lot of paper for $17.50, but it is not a guide to using the BDMPC package itself. There also is a small *Users Guide to BDMP on the IBM PC.* This explains how to load the package into the computer and gives some other information as well, but it is not a user manual in the usual sense.

Many commands

BDMP is a statistical computer language, and a large number of commands must be mastered by those who want to use it. Mainframe users of BDMP customarily enter data and commands by first producing punched cards, then feed the cards to the computer equipment. The BDMPC version makes a concession to personal computer use by offering an interactive mode of input.

To carry out this interactive entry, the user has to learn 30 or so line-editing commands and variations and get acquainted with the procedures for invoking these commands. With this knowledge, it is possible to begin to use the actual BDMP commands.

These are outlined in a 177-page *User's Digest.* To use any particular BDMP module, one has to look it up in the table of contents of this digest and turn to the appropriate pages. There, a brief description of the procedure will be found, followed by setup instructions for keying in the general commands to get the procedure going, and then by descriptions of the particular commands that invoke the various features of that procedure.

For more details and for examples of how that procedure has been applied, the user turns to the full-size manual.

Getting the data into the computer

The first question a BDMPC user may ask is how to get the data into the computer. The answer is that there is no convenient way to do this from the keyboard. If the data are already stored in a computer file, there are commands to get and use that file. However, BDMPC seems, incredibly, to be set up to take new data only as if they were coming from punched cards.

The user has to write a short program that reserves storage space and specifies the format in which the information is being presented as well as describing the information itself (title, number of variables, maximum and minimum values, and so forth). The procedure, appropriate for mainframe processing, is wildly overcomplicated for personal computer use, but there seems to be no appropriate alternative.

Statistical procedures

Despite their inaccessibility, a brief rundown of the statistical procedures seems in order. Assuming that the user does get the data to be analyzed into a readable file, the list of those procedures is formidable.

A simple set of descriptive data can be produced for each variable, or the user can call for detailed data description, including frequencies and estimates of location. There are one- and two-group t-tests, single-column frequency counts, and flexible histograms, probability, frequency distribution, and scatter plots. For data groups, there are similar capabilities, and four methods are provided for doing correlations when data are missing.

Frequency tables available to users include two-way and multiway cross sections. Two methods are provided for estimating survival, and there is a method for analyzing survival data when explanatory variables are present.

Cluster analysis of variables includes a choice of four measures of similarity and three criteria for linking, and there are similar cluster analyses for observations and for blocks of observations. Factor analysis provides four methods of initial extraction and several rotation methods.

There is a canonical correlation analysis for two sets of variables, a stepwise discriminant analysis between two or more groups, a Boolean factor analysis, scoring based on preference pairs, a procedure that describes the pattern of missing values for multivariate data, and another that partitions a set of observations into clusters.

Regression procedures include multiple linear regression, forward and backward stepwise regression, nonlinear regression with six functions built in, regression on principal components, polynomial regression, partial correlation and multivariate regression, all possible subsets regression, derivative free nonlinear regression, and stepwise logistic regression.

There is a Box–Jenkins time-series analysis procedure that can be used to build a parametric model and also a univariate and bivariate spectral analysis for time series. A multipass transformation procedure provides data editing and univariate statistics, along with transformations.

Nonparametric procedures include the sign test, Wilcoxon signed ranks test, Mann–Whitney rank sum test, Kruskal–Wallis one-way ANOVA, Friedman one-way ANOVA, Kendall's coefficient of concordance, and the Kendall and Spearman rank correlation coefficients.

ANOVA procedures include one for one-way analysis of variance and covariance, another that handles repeated measures models, a general mixed model analysis of variance that uses maximum and restricted maximum likelihood approaches, procedures for univariate and multivariate analysis of variance, and a mixed model for any complete design with equal cell sizes.

DAISY PROFESSIONAL

DAISY Professional uses manageable four-letter commands to invoke its procedures. Data entry is onto a convenient full-screen display. Statistical procedures are organized around a sensible concept of user needs.

DAISY Professional runs on Apple II, IIe, and Apple III computers. It sells for $203.45 from Rainbow Computing, 8811 Amigo Ave., Northridge, CA 91324. (818) 349-0300.

Like most statistics packages, DAISY handles data in tables with the variables in columns and with a row for each observation. The number of rows that can be handled depends on the available computer main memory space. For an Apple IIe with 128K main memory, about 830 rows can be used. As many as 20 columns can be handled.

Four-letter commands

Users give instructions to this package by typing four-letter commands. Thus the ENTE command invokes a sequence of displayed queries that ask for the data, one row at a time, and for the names of the columns. The QUIK command gives the user a look at a column of data, identified by name or number. The FILE command calls up a menu of file-manipulation choices.

Plotting the data

The DAISY manual wisely suggests that a good first step in many computer-based statistical investigations is to plot the data to get an immediate impression of whether there is a relationship between the variables.

The package offers the user histograms, useful for examining the data items in a single variable. It also allows the user to set up sequence plots that can help to discover trends, repeating cycles, and unusual data points in a variable. For those whose computers allow high-resolution plotting, either the sequence points or their logarithms can be graphed.

To compare one variable to another, high-resolution scatter plots are available in linear, semi-log, or logarithmic form. On these plots the axes will go through the means of the two variables, and the tick marks for the plot will denote units of standard deviation. If requested, the package will draw a regression line of best fit for the scatter plot.

Correlation capabilities

Another way of measuring the relationship between two variables is to calculate a correlation. The CORR command produces a table of product-moment correlation coefficients for any columns the user selects. Partial correlations also can be computed when it is desirable to look at the relationship between two variables after the effects of other variables have been considered. The COVA command calculates covariances between columns. Finally, the AUTO command does a correlation between a variable and a time-shifted version of that same variable. The user selects the desired time shift (number of rows). For interpretation, standard errors and Box–Pierce scores are shown with the calculated autocorrelation figures.

In addition to the parametric correlation measures just mentioned, DAISY offers four nonparametric measures: the Spearman rank correlation and the Kendall rank correlation, partial rank correlation, and coefficient of concordance. These are all based on the ranking of data items, rather than on the data themselves. Therefore, to use them it is necessary to first transform the data using the RANK command. The package will display an error message if a user tries to use these correlation measures with unranked data.

Regression models

Often, users want to use their data to predict the future course of events. Regression procedures allow trend lines to be fitted to existing data. These give an indication of data trends that may extend into the future. The REGR command calls up a sequence of choices and queries that allows users to do simple linear regression on two variables and multiple regression. There also is a forward stepwise regression procedure in which independent variables are added, one by one, to the regression calculation, with results calculated at each step. A backward stepwise regression procedure reverses the order, starting with all the variables and eliminating one independent variable at each step. Statistics designed to test the regression results are in the reports generated by these procedures. These include the multiple correlation R, R-squared, and the standard error of estimate.

If the user wishes, an ANOVA table will be displayed for the regression showing the sums of squared deviations accounted for by the regression and the residuals that are left unexplained, along with the F-ratio.

To check on whether or not the residuals are autocorrelated, Durbin-Watson and Von Neumann statistics can be generated. For those who are interested, the beta weights for the variables are also calculated.

The package sets up special data columns for both the fitted and residual values of the regressions. These can be subjected to runs tests or autocorrelations, and the columns can be plotted against one another.

Users are warned that regression calculations may strain the capabilities of the package and are easily misinterpreted. Several internal checks are run by the package to turn up possibly serious rounding errors. In addition, the package warns the user if very high correlations turn up as these may indicate multicollinearity.

Once a regression solution has been obtained, the package uses the resulting equation to make predictions. These procedures are organized under the PRED command. This command also can be used to carry through cross-validation procedures, where the data are divided into two subsets. One set is used to develop a regression equation; the other data subset then is used to check the predicted values produced by that equation.

Working with data

DAISY has spreadsheet-like data entry and editing capabilities. The DATA command causes the first 16 rows in the first two columns of a data table to be displayed. The cursor can be moved about to any position in these columns.

Users can make data entries one after another, moving from one location to the next by hitting the appropriate arrow key. Other keyboard actions make the cursor jump to the start of the next row or column. Editing consists of typing a new number over the existing one at the cursor position. Users can even split the screen and compare nonadjacent columns side by side.

The manual warns that this package may take considerable time to do its calculations and that only the first three to five digits of the results can be considered "dead accurate."

To use data from other packages, such as VisiCalc, Lotus 1·2·3, and Multiplan, the vendor of DAISY offers a package called File Freedom. It runs on Apple II, IIe, and IIc computers and sells for $99.95.

ELF

ELF, a menu-operated package, offers standard statistical procedures along with stepwise regression and discriminant analysis.

Versions of ELF run on Apple and CP/M computers. These sell for $250 each. An IBM/PC version sells for $350. The Winchedon Group, 3907 Lakota Rd., PO Box 10339, Alexandria, VA 22310. (703) 960-2587.

Up to 750 variables can be entered and stored in an ELF data file using an IBM/PC, and there can be up to 10,000 data items in a file.

One row at a time

Once the user names the data file and the variables that are to be in it, ELF displays an entry form for the first row that shows the variable names, each followed by a space for ten characters. As each data item is filled in, pressing the space bar will move the cursor to the next item space.

When the data for that row are complete, the user can call up the display for the next row by pressing the Return key. If there are any typing errors, the process of entering that row can be repeated.

After all data for a file have been entered, a review and editing procedure allows the user to leaf through displays of row data or jump to any row designated by its number. Here, the user can move the cursor, using the arrow keys, to any item and type over it with a corrected value. New rows can be added as desired.

New variables also can be added, using the review and editing procedure. However, there seems to be no way to enter a new column of data without reviewing and editing complete rows of data, one at a time.

There seem to be no built-in procedures for merging files or extracting portions of files, but complete files can be translated to and from the DIF format used by packages such as VisiCalc.

Transformations

ELF requires users to write short programs in a special computer language in order to make data transformations. Allowed operations include arithmetic, selections based on $<$, $>$, $=$, and odd conditions, trigonometric, exponential, logarithmic, modulo 10, rounding, truncation, and absolute value.

Statistical procedures

Users select desired statistical procedures from a menu. In each procedure there is a series of screen-displayed queries. The user replies by typing appropriate answers, entering the files and variables to be processed, and selecting procedure options, if any.

The Statistics procedure produces a set of descriptive measures, including the mean, variance, standard deviation and error, minimum, maximum, and range, sum, kurtosis, and skewness. Significance and level of a normal distribution, t, F and chi-square statistics are calculated by a Probabilities procedure. There also is a separate procedure for t-tests on means for variables and groups of variables.

A Correlation procedure produces a matrix of correlation coefficients for normally distributed variables. There do not appear to be any rank correlation procedures in this package.

There are procedures for one-way and two-way analyses of variance. Both are simple procedures that produce tables showing sum-of-squares and mean-square values and F statistics.

A cross-tabulation procedure produces tables for up to three variables.

There is a discriminant analysis procedure for separating rows into categories, using one or more of the variables to perform the classification. A factor analysis procedure performs principal-components or principal-factors analyses. There is a choice of rotations.

A multivariate linear regression procedure uses a stepwise technique to enter and remove variables at the user's choice. A crude graph of actual and predicted values (standardized) can be produced. For time series, data there is a procedure that corrects a regression for autocorrelated error terms. Up to 24 independent variables can be handled.

MICROSTAT

The authors of the MICROSTAT package have chosen to make all of its operations available through menu displays. Its data files have relatively large capacity. A good variety of statistical procedures are available. Data entry is on a query-and-answer basis.

MICROSTAT runs on IBM/PC and compatible computers and on CP/M computers. It sells for $375 from Ecosoft, 6413 N. College Ave., Indianapolis, IN 46220. (317) 255-6476.

MICROSTAT data are stored in files, then recalled from storage when statistical procedures are to be applied. Data storage is controlled from a data management menu display. This display provides options for a number of data transformations, including arithmetic, logarithms and exponents, conversion to Z-values, a linear transformation, rounding, truncation, random numbers, grouping, scaling, deletion, rank-ordering, sorting, and lag transformations.

The package also allows factorials, permutations, and combinations to be calculated and will generate data for a number of different probability distributions, based on user-selected parameters. Available distributions include binomial, hypergeometric, Poisson, exponential, normal, F distribution, Student's t, and chi-square.

There is a file directory display, and files can be renamed and erased by MICROSTAT menu choices. Files from some other packages and computers can be used by MICROSTAT. It will accept certain ASCII files, DIF-format files, and dBase II files.

As for the capacity of MICROSTAT files, an IBM/PC with 128K main memory can handle about 120 variables for most of the procedures. For multiple regression, about 60 predictor variables can be handled with such a computer.

Data entry

To start a new file, the user enters a file name and descriptive label, specifies the number of variables to be used, and gives a name for each of these.

Data are then entered on a row-by-row basis; the package will prompt with the row number and variable (column) numbers and names. Adding rows to the end of an existing file follows the same procedure. There also is a procedure for inserting new rows into an existing file. Editing is done on a row-at-a-time basis; the user calls for the desired row, gets a display of the current entries in that row, then responds to queries asking for new values. All this is straightforward, but not as flexible and convenient as full-screen entry and editing.

The package allows display and printing of any subset of rows or columns in a file. The user can elect to change the number of places to be displayed to the right of the decimal point in each entry, to improve the appearance of the display. There is a batch mode of operation for those so inclined.

Statistical procedures

When users ask for descriptive statistics, there are two main choices. They can call for an abbreviated output that includes only the mean, standard deviation, and minimum and maximum values for all of the variables in the chosen file, or they can choose from a wider range of descriptive statistics and specify exactly those variables for which the statistics will be supplied. The additional choices include the mean, variance and coefficient of variance, standard error, sum and sum of squares and sum of deviations around the mean, moments about the mean and skewness and kurtosis, and a normal curve goodness of fit.

MICROSTAT will produce frequency distributions showing grouped categories of data values or counts of coded values.

Tests for means include checking the mean against a hypothesized value, the difference between means for paired observations, and the difference between two group means with known variance or with a polled estimate of variance.

There is a one-way and a two-way ANOVA procedure, and an ANOVA for randomized cases.

MICROSTAT will produce a correlation matrix for all variables in a file or for selected variables. It also will display the raw and adjusted sums of squares and cross products and the covariances.

The regression analysis features perform simple, multiple, and stepwise-multiple regression on selected variables and subsets of variables. This involves a two-step computation procedure that can be lengthy. Results include means and standard deviations. The package allows the user to conveniently change the dependent variable or use another set of predictor variables and then rerun the calculation.

The regression output contains the regression coefficient, standard error, F value, a partial r2, the constant term, standard error of estimate, R-squared or r-squared, Multiple R or r, and an analysis of variance table. Display of residuals and Durbin–Watson test results are optional. For stepwise regression, variables entered or removed are indicated, each with a tolerance and F to enter.

Time-series analysis procedures calculate a moving average, or center moving average with seasonal indexes or will perform exponential smoothing for selected variables.

Choices of nonparametric tests include the Wald–Wolfowitz runs test, the Wilcoxin signed ranks test and rank-sum test for two groups, the Kruskal–Wallis one-way ANOVA by ranks, the Kolmogorov–Smirnov goodness of fit and two group tests, an absolute normal scores test, the Friedman test and the Kendall coefficient of concordance test, a sign test, the Fisher exact test, and the Spearman rank-order correlation.

A Crosstabs procedure, limited to 20 rows and 5 columns, generates a two-way contingency table by counting selected values from two variables. It also will calculate the chi-square statistic for the table and perform a Kolmogorov–Smirnov goodness of fit test.

Finally, the package will perform hypothesis tests for two proportions from independent groups, or for a sample proportions versus a hypothesized value, or for two proportions from one group with overlapping or with mutually exclusive categories.

Crude graphics

MICROSTAT provides a crude histogram display in connection with its frequency count. There also are scatter plots, with the points represented by asterisks, numerals, and # signs.

With the results of the multiple regression calculation, a similar graph displays standardized residual values.

MINITAB

Minitab is a statistical computer language with about 150 commands. The Minitab Fundamental version does descriptive statistics, simple and multiple regression, analysis of variance, nonparametric statistics, and cross tabulations. It produces scatter plots and histograms, generates random data, and has macro and looping capabilities. The Standard version does all that and also has time-series analysis, stepwise regressions, and does exploratory data analysis and matrix operations.

Minitab runs on IBM/PC/XT/AT and compatibles, on the Victor 9000, *the Fortune 32:16, and the DEC Professional 350. The Fundamental version sells for $250, the Standard version for $500 from Minitab, 215 Pond Laboratory, University Park, PA 16802. (814) 865-1595.*

Data are input, a row at a time, with columns defined by the user. There is, however, no display of this data entry worksheet. What the user sees are the lines of data, with spaces or commas between items. There also is a column-by-column data entry mode. For convenience, single commands can be used to generate a column with all row entries equal to a prescribed constant or filled with selected sequences of numbers.

These data entry procedures are each initiated by specific commands. Apparently, if any data entry errors are made, they are corrected by inserting a new, correct row of data and deleting the incorrect row. There seems to be no way to go directly to an incorrect data item and just correct that item. FORTRAN-formatted data files also can be used.

Simple histograms for individual variables can be displayed, and there are special histograms with Gaussian distributions fitted on them.

Available plots include scatter plots of two and three variables. On the three-variable plots, positions in the Z-dimension are indicated by the symbols used to mark the points. This feature is ingenious but of dubious value.

Multiple two-variable plots can be displayed on the same set of axes. There also are time-series plotting formats for equally spaced data and for periodic data.

Data transformations include arithmetic, exponential, logarithmic, and trigonometric functions performed on columns. Other transformations include absolute value, sign, normal scores, and scaling.

The exploratory data analysis procedures include a stem and leaf display of selected columns as well as box plots.

Statistical procedures

Minitab has commands for calculating descriptive statistics for rows or columns: counts, sums, means, standard deviations, medians, maximums and minimums, and sums of squares. The package also will calculate t-confidence intervals and do two-sided t-tests and two-sided runs tests on column data.

Wilcoxon signed-ranks tests, rank estimates, and confidence intervals can be calculated for sets of columns.

A chi-square test can be performed on data in specified columns. The CONTINGENCY command constructs a contingency table from specified data and performs a chi-square test.

The REGRESSION command fits regression polynomials to selected columns of data. There also are stepwise regression procedures.

The CORRELATION command calculates the Pearson product-moment correlation between two columns, and the package will perform a Kruskal–Wallis nonparametric test.

Time-series operations include computation of autocorrelations and partial autocorrelations as well as cross correlations between two time series. There also is an ARIMA procedure that uses the Box–Jenkins methodology. Smoothing can be performed by a choice of methods using running medians and Hanning averages.

Other procedures include generation of tables of binomial and Poisson probabilities, a two-sample t-test, a Mann–Whitney two-sample rank test, and one-way and two-way analyses of variance.

NCSS: NUMBER CRUNCHER STATISTICAL SYSTEM

NCSS pulls together a large number of statistical procedures, makes them relatively easy to apply, and supplements them with helpful graphical displays.

This package runs on IBM/PC, CP/M, TRS-80, and Macintosh computers. It sells for $202 from NCSS, Dr. Jerry L. Hintze, 865 E. 400 N, Kaysville, UT 84037. (801) 546-0445.

NCSS is menu-based. Starting with a main menu, the user moves to other menus that show the statistical and data procedures the package offers.

When it come to data handling, NCSS works from stored tables of data in which the columns represent the variables and the rows are individual observations. There may be up to 250 columns in a table, and the user must choose how many are to be used. The storage space the package allots to the table will depend on that choice and on the choice of the number of rows in the table. There can be up to 32,000 rows in a table.

Flexible data tables

The tables are set up with user-selected titles and the data items filled in by making selections from displayed menus, then responding to a series of screen-displayed queries with keyboard entries. This procedure could have been improved by presenting the user with a full-screen display of the table and having entry take place in a tabular format.

Transformations of the data in the tables may be done by adding, subtracting, multiplying, or dividing the items in one column by those in a second column or by a constant to form a transformed column. Single-column transformations include taking the absolute value, cumulative totaling, forming truncated or rounded-off integers from decimal numbers, calculating modulo remainders or signature values, taking Tukey's folded log or root, natural logs, exponents, square roots and trigonometric functions. Columns of uniform-random or sequential numbers can be generated. There also are smoothing and lagging transformations, and values within a column can be sorted or ranked.

The package can generate a number of different probability distributions including the normal, inverse normal, chi-square, Student's t, F, Weibull, binomial, negative binomial, and Poisson distributions.

Users are given considerable power to manipulate the stored tables. Standard ASCII files can be easily transformed into NCSS tabular files. Two tables may be joined, one above the other or side to side. Individual columns may be copied from one table to another. Portions of tables may be selected out. Tables can even be rotated so that the rows become columns and vice versa.

Descriptive statistics

Any column can be chosen for summary by descriptive statistics. Included in the unusually long list of measures generated will be the mean, standard deviation, variance, coefficient of variation, standard error of the mean, number of observations and of missing values as well as the sum of weights, and the adjusted sum of squares, of cubes, and of quartics. Other measures include coefficients of kurtosis and skewness, the range and inner quartile range, a percentile, and a t-value for testing whether the mean is or is not equal to zero.

In addition to these single-column statistics, the package will produce a smaller number of descriptive statistics for rows selected from a column, where the selection is based on corresponding values in a second column.

Correlation and regression

In this package correlation coefficients and information on least-squares regression are generated by the same procedure. The idea is that both represent measures of the linear relationship between two variables (columns of data). In the same vein, this procedure also generates the t-statistic for testing the significance of the slope.

A multiple regression procedure is included in the package, along with a procedure that provides for selecting and changing the variables to be included in a multiple regression model. Since outlier points can have a profound influence on regression analysis, NCSS provides a procedure, called Robust Regression, to locate such points and minimize their influence on the results. This procedure can be used as a check on regular multiple regression analysis.

NCSS provides a discriminant analysis procedure for cases where the dependent variable in a regression analysis is discrete rather than continuous.

When a set of data is analyzed, more variables involved may be than can be readily plotted, and several variables may represent the same general factor. NCSS provides a Principal Components Analysis procedure designed to determine the underlying factors of the data and to reduce the number of variables to be handled.

ANOVA procedures and t-Tests

NCSS includes both paired and unpaired t-tests. In connection with a one-way analysis-of-variance procedure, Fisher's LSD test can be used to determine which of the means are significantly different from each other. After the one-way ANOVA has been completed, Duncan's New Multiple Range Test and the Newman Keuls Multiple Range Test can be used to check the differences between two means.

The package also provides a two-way ANOVA procedure and an *n*-way ANOVA that will handle models with three or four factors. A Repeated Measures ANOVA is provided for designs where one treatment is an error term and another is a repeated measure of an experimental unit.

Tests of independence

To check for the independence of rows and columns in a data table, NCSS has a Chi Square procedure that produces such measures as chi-square, Phi, Cramer's U, Pearson's contingency coefficient and product-moment correlation, Tschupro's T, Lambda A and B, Symmetric Lambda, Kendall's tau-B and C, Gamma and Spearman's Rho.

An alternative way to check for normality of the data is to do a Z-score or inverse-normal quartile transformation. When the resulting scores are plotted against the original data, normally distributed data will produce a straight-line plot, and the degree of departure from a straight line is a measure of the departure from normality.

Nonparametric tests

Among the nonparametric tests included in the NCSS package are the Sign or Binomial test for determining that the mean of a variable has a particular value or that the median of the difference between two variables is zero.

The Wald–Wolfowitz test is used to check for the randomness of a sequence of numbers made up of only two distinct values. The Wilcoxon Matched Pairs test checks on the difference between two sets of paired observations. The Mann–Whitney test checks on the difference between the means of two unpaired groups. The Kruskal–Wallis test does a similar check for groups with unequal sample sizes. The Friedman Block/Treatment test checks whether the means of two or more treatment groups are equal.

Nonparametric correlations can be performed using Spearman's rank correlation coefficient and Kendall's tau coefficient.

Graphic tools

NCSS allows the user to plot the values of the rows in a column in histogram form, so that the data can be easily scanned for questionable values. With a graphics board plugged into the computer, the package will produce scatter plots of the row values from a pair of columns to visually reveal relationships and outlier points.

Box plots also are available. These show the location and spread of groups of data in relation to the 75th and 25th percentiles, and extreme points are displayed as circles. These plots generally are used with treatment groups of ANOVA or t-test data.

As for the user's manual, it provides brief and often repetitious material on the various procedures, relying on many references to published works on statistics to fill in the gaps. The overall message to the users seems to be that if they do not understand what the package can do, they should go to the library and read another book.

PC STATISTICIAN AND STATS PLUS

A variety of different statistical procedures are offered by these two packages, along with regression and cross-tabulation features. Package procedures are activated by choices from menu displays, with data entered either from the keyboard or from stored files.

STATS PLUS runs on Apple II, IIe, and IIc computers; it sells for $200. PC STATISTICIAN is for IBM/PCs and compatible computers; it sells for $300. Both packages are from Human Systems Dynamics, 9010 Reseda Blvd., #222, Northridge, CA 91324. (818) 993-8536.

Both STATS PLUS and PC STATISTICIAN perform about the same statistical procedures. PC STATISTICIAN, the more recently written package, seems a bit easier to access and operate, and the data base functions seem more thoroughly integrated with the statistical computation procedures. Several features have been added to PC STATISTICIAN, including the ability to use files from popular spreadsheet packages such as Lotus 1·2·3 and VisiCalc.

Statistical features

Descriptive statistics can be computed by both packages for up to eight data variables at a time, with output including sample size, sum mean, standard deviation, and minimum and maximum values. PC STATISTICIAN gives added details on the means of dependent variables.

The packages will produce a frequency distribution from a set of data values, but the values must all be whole numbers. The output can include basic descriptive statistics, frequencies or proportions, and a percentiles report, with intervals chosen by the user. The packages also will display a frequency bar or polygon graph.

Any of three types of t-tests can be performed, including those for a single sample mean against a population mean, the difference between two correlated or dependent sample means, or between two independent means.

Nonparametric test procedures include: the Mann–Whitney U test, the Signed Ranks test, Kruskal–Wallis test, Friedman ANOVA by ranks, Spearman Rho, and Kendall Tau. PC STATISTICIAN adds the Wald–Wolfowitz Runs test. The input data for these tests are ranked automatically, and the sorting procedures involved can be lengthy.

Correlations can be calculated for up to five sets of data at a time. If the data sets are not all of the same size, the program will truncate them to produce equal size. Results are shown on the upper right portion of a four-row, four-column table of correlation coefficients. There are some display peculiarities. Correlations of +1 and −1 will appear as 999 and −999. The coefficients will appear as three digit numbers with no decimal point, and correlations smaller than .001 will appear as 000. PC STATISTICIAN also provides for multiple correlations involving three to seven variables and has a partial correlation procedure.

An analysis-of-variance procedure produces a full summary table with F-ratios, p-values, and descriptive statistics. Either one or two factors with equal or unequal sample sizes may be used.

A contingency tables procedure allows the user to perform the Fisher Exact test (STATS PLUS only) and one- or two-variable chi-square tests. These procedures accept keyboarded data in the form of frequency counts.

Other package features

A STATS PLUS linear regression procedure performs a least-squares calculation for one independent and one dependent variable. Multiple regression can be performed for two independent and one dependent variable. The output includes correlations and first-order partial correlations as well as an analysis-of-variance table, regression coefficients, and significance tests. PC STATISTICIAN has a unified least-squares regression procedure that will handle one to five predictor variables. That package also has a curve-fitting option that uses power polynomials. In addition, PC STATISTICIAN has an *x, y* plotting capability that produces either scatter or line plots.

Finally, a cross-tabulation procedure analyzes stored data files, identifying each record in a file (based on the contents of its fields) as belonging to one of a set of categories, then generating reports on category occurrence frequency or percentage, along with basic descriptive statistics.

Using the database files

The example of STATS PLUS use given in the manual is for a set of ten measurements of the strength of a material before and after a stressing procedure. The before-and-after data are stored in two fields of a database file. They are entered as data pairs. After the last pair is entered, the user simply keys in an asterisk in response to the query for the next pair, and the package creates the desired file. This stored file can then be called upon to calculate t-test results or to be run through any other appropriate procedure that the package offers.

Keyboard data entry for the statistical procedures offered by the STATS PLUS package is through serial query and answer. When long columns of data have to be entered, this can be clumsy, time consuming, and unforgiving of keyboard errors. However, database procedures include editing as well as entry. Apparently, for all but the simplest data sets, users are expected to do their data entry into the database portion of the package. In fact, with the PC STATISTICIAN package, all data entry is done through the database route.

As might be expected from a package that gave its author a chance to update an existing product, PC STATISTICIAN improves upon and extends the database capabilities of STATS PLUS. For example, it will display a listing of available data files when the user presses a Function key, and it offers additional data transformations, such as percent and Z-scores, beyond those in the older package.

SPSS/PC

SPSS/PC is a statistical language with over 175 commands and subcommands and a 600+ page manual. It provides an extensive list of statistical procedures but poor facilities for interactive data entry and editing. Flexible report generation is provided.

SPSS/PC for IBM/PC/XT computers sells for $795. A version for DEC Professional computers sells for $595. From SPSS Inc.,444 N. Michigan Ave., Chicago, IL 60611. (312) 329-2400.

SPSS/PC is a personal computer version of a statistics package long used with large mainframe computers. To operate the package, users give the computer commands and SPSS/PC responds. In other words, this is a specialized computer language. Users must be familiar with a total of about 175 commands and subcommands to exercise the full capabilities of the package, and many details of syntax and usage must be mastered as well.

Probably because of its large-computer origins, the authors of this package could not resist providing a batch-processing mode of operation in which the user can enter all commands and data, then let the computer run them without further user interaction.

To document the use of this language, the package comes with a manual that is well over 600 pages. It describes how each statistical procedure works and shows many of the equations on which the computations are based. A good part of the manual is devoted to spelling out the commands and related procedures.

SPSS/PC is a package for serious users who are willing to invest the time to learn to use a new language. For those who will find frequent, consistent use for the package, it will be a worthwhile investment. Other users, who may turn only occasionally to the statistical tools offered by SPSS/PC, will probably feel that learning to use this package is an overly burdensome task.

Errorless data

The DATA LIST command is used for interactive data entry. Up to 200 variables can be defined. For fixed-format input, the user specifies the exact row locations in which each variable will appear; this is done as if the data were being entered on a punched card, so each numeral is assigned to a specific card-column position. There also is a freefield format in which exact row locations of the data need not be spelled out. Freefield data items are entered with a space or comma between each item. They can be strung out to any length, on as many lines of the input display as desired. The only requirement is that they be in row-by-row order.

When punched cards are used to input data into large computers, the processes of verifying and correcting the data take place on the punching equipment, not in the computer. The authors of the SPSS/PC package seem to have assumed that the data input to the personal computer also will be pure and correct. There seem to be no convenient built-in procedures in the package that would allow users to correct errors easily in the data they type on the keyboard.

Data entered with the DATA LIST command can be displayed, using the LIST command. They can be stored and retrieved, and files of data from mainframe versions of SPSS or files produced by other packages can be used by SPSS/PC, provided they are in compatible formats.

Data transformations available with SPSS/PC include arithmetic, exponents, logarithms, trigonometric functions, absolute value, truncation, rounding, modulo 10, lag, and conversion of dates into day intervals. Conditional transformations can be controlled by Boolean logic operators. Pseudo-random numbers can be generated. There also is a transformation that counts occurrences of the same value across a list of variables.

For those who want to turn out neat summary reports on their statistical findings, SPSS/PC provides a report generation capability that allows result and input data to be combined, with added headings, breakouts, and comments, on printed pages formatted by the user. As with the other features of this package, a report generation program, using SPSS/PC commands, has to be written.

Statistical procedures

A Frequencies procedure produces frequency tables and descriptive statistics, including the mean, standard error of the mean, standard deviation, median, mode, variance, skewness and standard error of skewness, kurtosis and standard error of kurtosis, range, minimum and maximum, sum, and percentiles, for selected variables. It also produces bar charts for discrete variables and histograms for continuous variables.

Means, standard deviations, and group counts are generated by the MEANS command for a dependent variable with groups defined by one or more independent variables.

A t-test procedure calculates Student's t and displays the two-tailed probability of the differences between the means. Statistics can be produced for independent or paired samples.

Nonparametric tests available in SPSS/PC include one-sample binomial, chi-square, Kolmogorov–Smirnov, and runs tests. For two related samples there are Mcnemar, sign and Wilcoxon tests. For k-related samples there are Cochran, Friedman, and Kendall tests. For two independent samples, there are Mann–Whitney, Kolmogorov–Smirnov, Wald–Wolfowitz, and Moses tests. For k-independent samples, there are Kruskal–Wallis and median tests.

A Correlation procedure does Pearson product-moment correlations with one-, or two-tailed probabilities. It also will produce univariate statistics, covariances, and cross-product deviations.

Multiple regression equations, associated statistics, and plots can be produced by the Regression procedure. Predicted values, residuals, and related statistics are calculated. The automatic variables are available for analysis by various plots.

The ONEWAY command does a one-way analysis of variance for an interlevel dependent variable, with groups defined by one numeric independent variable. The ANOVA command handles up to five dependent variables, ten factors, and ten covariates.

A Cluster procedure produces heirarchical clusters of items using any of six measures of similarity or distance to one or many variables. The Crosstabs procedure does *n*-way tabulations with up to 250 rows or columns for each variable.

The Factor procedure does factor analyses using a choice ofseven extraction methods. A Hiloglinear procedure fits heirarchical loglinear models to multidimensional contingency tables using iterative proportional-fitting algorithms. This procedure is for data from populations that are not normally distributed with constant variance. Up to ten factors can be handled.

Plots and graphs

SPSS/PC can do a variety of printed plots, with plot symbols chosen by the user including simple bivariate scatterplots, scatterplots with a control variable, contour plots, and overlay plots.

STATPAC

StatPac is very well organized, sometimes tediously so. It allows users to set up their own input procedures. A resonable selection of statistical procedures is provided and the documentation is excellent.

StatPac runs on IBM/PC and compatible computers and on CP/M com-

puters. It sells for $495 from Walonick Associates, 6500 Nicollet Ave So., Minneapolis, MN 55423. (612) 866-9022.

Each of the procedures of this package is accessed through a menu display. Thus, on the Data Management menu, the user can access Enter New Data and be presented with four ways to enter data.

User-controlled input procedures

In the row-by-row keyboard entry procedure, variable names are displayed for the first row, and the user types in the data items after each name. Then the same names are displayed for entry of the values for the second row, and so on. There also is a Keypunch Emulation procedure in which the user types data on the screen in a rigidly prescribed format.

Data editing, after a data file has been stored, is carried out with a keypunch like procedure. Three rows of data are displayed, and the user can scroll or page up or down to the other rows, or the package will search for a row designated by number. The cursor can be moved or tabbed along the current row and, at the cursor position, a character can be deleted or added.

Users also can devise their own input screens. This capability permits either row-by-row or column-by-column input procedures. The order of inputting the data and the labels used for columns can be varied by the user. Data inputs can be restricted to assigned ranges of numbers or other symbols. The screens can include whatever instruction or reminder text the user wishes to add. The idea appears to be that operators with little or no knowledge of statistics will do data entry, and that these screens can be designed for their use.

A separate editing procedure for these screens involves locating the desired screen, positioning the cursor, then deleting or adding characters.

Managing files and procedures

To keep stored data in order, users are required to create an overall index (called a codebook) to all the files and variables.There also is an overall management routine for the statistical procedures that controls the sequencing of these procedures and also is used for any needed selection and recoding of data.

What this amounts to is that if users want to run just one statistical procedure, they have to give it a title, select the procedure and the desired options for that procedure, identify by number the variables to be processed, and specify any needed data recoding or selections. All this has to be done before any actual statistical computation can begin.

It is a very formalized routine that may tend to discourage the exploratory approach to statistical analysis permitted by personal computers.

Statistical procedures

StatPac has a Frequency Analysis procedure that will sort selected variables by value or by frequency of response, calculate cumulative percentages, and produce tables and bar-chart or pie-chart frequency displays.

The Descriptive Statistics procedure calculates all the usual measures and also gives the 95 percent and 99 percent confidence intervals around the mean. A chart of number of cases versus scores can be displayed.

Cross tabulation can be done for one variable against a list of variables. A table of up to 50 rows by 50 columns can be produced, along with appropriate statistical measures.

The Correlation procedure produces a coefficient and other statistics for two normally distributed variables. Spearman's rank-order coefficient also will be calculated. This procedure also can be used to display a scattergram showing the relationship between the variables. A regression line will be calculated and displayed.

The t-test procedure includes tests for matched pairs or for two independent groups.

Stepwise multiple regression can be performed for a dependent variable and a list of independent variables. This procedure will produce descriptive statistics, regression statistics and coefficients, simple and partial correlation matrices, and an analysis of the residuals.

One-way and two-way analysis-of-variance procedures operate from grouped data. These produce ANOVA tables, t-tests between groups, critical F and T probabilities, and descriptive statistics for the two-way ANOVA.

Good documentation

The user's manual spells out very clearly what the user has to do. Formulas used in the computations are given and explained, and there is a list of the textbook sources for these formulas. The manual also lists the error messages that the package may produce, along with a brief explanation of each one. Finally, there is a well-organized index to the manual so it can readily be used as a reference.

SYSTAT

SYSTAT offers the user a statistical computing language with about 100 commands, plus a version of BASIC. A full-screen entry and editing procedure eases data input. Up to 200 variables can be handled.

SYSTAT runs on IBM/PC and compatible computers and on CP/M computers. It sells for $495 from Systat, Inc., 603 Main St., Evanston, IL 60202. (312) 864-5670.

SYSTAT is a statistical computer language with almost 100 commands used to exercise its full capabilities. In fact, the package includes a version of the BASIC computer language that can be used to put together routines for manipulating data, sampling, and performing complex transformations. Most calculations are done in double precision with 15-digit accuracy.

The package is divided up into modules, one for handling data and others for various types of statistical procedures. Each module has a set of commands peculiar to its operations, so the SYSTAT language can be learned piecemeal, starting with those procedures the user needs first.

At the end of a procedure, SYSTAT prints a log of all the commands used during the session, which is convenient for users who expect to perform similar procedures.

Full-screen data entry

SYSTAT takes advantage of personal computer capabilities in that it uses full-screen data entry and editing. Columns for variables can be set up and filled with data by moving the display cursor to the appropriate positions on the display. When data are methodically filled in, a row at a time, the cursor will automatically jump to the next input position as each item is entered. If a change needs to be made, the cursor can be moved, using arrow keys, to the desired item, and the correct data can be typed over the existing data. Variable names can be changed in the same way.

The data display can contain as many rows as can be stored in the disk used with the package. It can accept numbers with up to 15 digits. Once entered, the data are stored and the files can be read, merged, sorted, and transposed for use with the statistical procedures of the package.

There is an alternative data entry procedure in which the user creates data files without the full-screen display. SYSTAT also has provisions for receiving similar files from packages such as Lotus 1·2·3, dBase II, and

WordStar and converting them to SYSTAT format. In some cases, these files will have to be pre-edited with the user's own word processor or editing package.

Transformations are handled as part of data editing procedures, using the LET command together with the name of the variable to be transformed and the mathematical operation to be used. Available operations include arithmetic, exponents, logarithms, and trigonometric functions. Ranking and standardizing are handled by separate commands.

Statistical procedures

For basic descriptive statistics, the STATISTICS command is used. This displays counts, maxima and minima, means and standard deviations. The mean, skewness, kurtosis, range, variance, and standard error of the mean are available on request.

A correlation module calculates symmetric triangular matrices of correlation and similarity for data sets. Types of matrices that can be calculated include sum of squares and cross products, covariance, Pearson correlation, gamma coefficients, Guttman mu2 coefficients, Spearman rank correlations, Kendall tau-b coefficients, and normalized Euclidean distances.

Nonparametric statistics for groups of rows or pairs of columns include the sign test, Wilcoxon signed ranks test, the Friedman and Kruskal–Wallis one-way analyses of variance, and the Kolmogorov–Smirnov one- and two-sample tests.

A tables module produces multiway tables and can fit them with a log-linear model. Fitted models are tested with Pearson and likelihood ratio chi-square statistics.

An MLGH procedure can estimate and test any univariate or multivariate general linear model, allowing linear contrasts on vectors of dependent as well as independent variables. This procedure can perform multivariate profile analysis of repeated measures, ANOVA, multivariate ANOVA, discriminant analysis, simple regression, multiple and multivariate regression, stepwise regression, analysis of covariance, principal components analysis, and canonical correlation.

A Factor Analysis procedure produces a principal-components analysis with optional rotations and factor scores. An MDS procedure provides nonmetric multidimensional scaling of similarity or dissimilarity matrices in one to five dimensions. This procedure is said to usually be able to fit an appropriate model in fewer dimensions than can be done using principal-components or factor analysis.

There also is a Cluster Analysis procedure for detecting natural groupings in data. The JOIN command produces heirarchical clustering with a tree diagram showing the relationships between columns or rows. There also is a command for k-means clustering.

To aid in time-series analysis, a SMOOTH command can calculate moving averages and running medians. It also does linear filtering with Hanning weights and scatterplot smoothing. There also is Fourier transform smoothing, and an ARIMA (AutoRegressive Integrated Moving Average) procedure.

Stem and leaf plots

SYSTAT can produce a variety of crude plots made up of printed characters. Scatter plots and function plots can be produced by the PLOT command. With the help of some BASIC programming, crude contour plots also can be done. The HISTOGRAM command produces plots with bars made up of asterisks.

The STEMLEAF command produces a histogramlike display made up of numerals, which can give users insight into the repeated appearance of certain discrete values in the data. The BOX command produces box plots, used to quickly identify median, hinges, and outside values in the data.

Probability plots, which show a variable plotted against corresponding percentage points of a standard normal variable, are produced by the PPLOT command. The QPLOT command compares a sample to its quartiles or the quartles of two samples to one another.

Chapter 17
Statistics II

The specialized statistical packages discussed in this chapter include some that concentrate on particular statistical procedures or on only a few procedures. There also are packages capable of handling only limited amounts of data, though they may do that job very conveniently.

Several of the packages are what can be called statistical add-ons, that is, they are packages that perform nonstatistical functions but that have been equipped with at least some statistical capabilities.

Languages with statistics

RS/1 is such a package. It is, in fact, a specialized engineering-scientific computer language that has interactive procedures for data manipulation, graphs, spreadsheets, and system modeling. Our interest in RS/1 here is that it also includes descriptive statistics, statistical tests, correlation, and analysis of variance procedures.

RS/1 handles statistical data with a dialogue sequence that leads the user to name a table and its rows and columns, and to fill in data one column at a time. The package allows for intermixing of such dialog sequences with regular keyboard commands plus single-key actions of various sorts. All this gives the user very flexible control over the package functions, but also makes that control a complex process and one that can be confusing, particularly to the novice or occasional user.

For those who will use a package like this one on a day-to-day basis, learning its intricacies will be a worthwhile task. Less frequent users will have to consider whether packages that are less flexible but simpler to learn and use might better meet their needs.

ASYST is another engineering–scientific language package with a repertoire of statistical procedures that is not as large as that of RS/1.

The data management function, involving the building, storing, and manipulating of data files, is an essential part of most personal computer-based statistical packages. This fact has encouraged vendors of packages primarily designed to handle database operations to enhance their products by adding statistical procedures.

Like some other database packages, PC/FOCUS aims at providing the user with a capability that can handle many different computation tasks. Toward this end, the package offers many different commands and functions more as a computer language than as a simple storage facility. With this kind of general database package, learning how to set up a file and enter data is no simple task. If prospective users need a flexible database, and decide that PC/FOCUS can meet that need, then the statistical features of the package can be a valuable add-on. However, for those whose only interest is statistical analysis, a package such as PC/FOCUS is probably not what they want.

STATA is another data manipulation language that can be used to perform statistical procedures. It also offers spreadsheet capabilities. Stata is less flexible than PC/FOCUS, and it seems to have less powerful statistical capabilities, but it does have the advantage of being simpler to learn.

Small packages

The INSTAT-SC statistical calculator is a neatly concieved, compact, and handy statistical tool that makes intelligent use of personal computer capabilities. It works with only limited amounts of data, nevertheless it produces dramatic results. When the package is started up, it displays a four-column framework of lines. As the user keys data into a column, statistics are displayed to summarize the data characteristics, and they change with each added entry the user keyboards. Furthermore, the package gives the user as wide a choice of different statistical calculations.

EPISTAT is a more conventional statistical package but is designed for users who have relatively small amounts of statistical data to work with. It will handle less than 28 variables and under 2,000 total data entries.

Specialized statistics

Of the packages that concentrate on particular statistical procedures, several are devoted to regression calculations. REGRESS II, for example, performs multiple correlation and regression. The package includes procedures for simultanous regression, forward selection, backward elimination, stepwise regression, and a regression procedure using power polynomials. The Dynacomp group of regression packages include one for fitting one-dimensional data to polynomials, another for fitting one-dimensional data to other functions, a third for handling several independent variables, and a fourth for performing stepwise regression. The BIOSTATISTICS 3PC package also concentrates on regression analysis, although it can perform one- and two-way analysis of variance and several statistical tests as well.

Analysis of variance is the main function of PC ANOVA, which handles problems involving one to five variables. This package solves simple and multivariable randomized or repeated-measure problems, between, within, or split-plot problems, randomized complete block problems, and latin square problems. It also performs analysis of covariance and nonparametric analysis of variance.

Nonparametric tests are the exclusive concern of the BASIC Statistical Subroutines package, which includes 29 of these tests.

The Statistical Analysis package from IIE is oriented to some of the needs of industrial engineers, but may be useful to others with similar interests. It includes procedures for generating descriptive statistics, performing a Mann–Whitney test, doing a distribution-fitting procedure, and for polynomial curve fitting.

From pull-down to do-it-yourself

Some of these packages include features designed to make things easier for the user. Others are bare-bones packages that assume only programmers are interested in using personal computers for statistical calculations.

The StatView package uses some of the latest personal computer techniques that were developed to help business people apply personal computers. While the user is viewing a table of data, overlay menus listing statistical procedures can be displayed, the desired procedures can be selected, and results will be displayed.

StatView data entry is from the keyboard, a row at a time, onto a tabular display. The table can be larger than the screen since additional rows and columns can be exposed by scrolling. Each cell of the table can be individually edited by simply typing over what is displayed. There are cut, copy, paste and undo functions to simplify the editing process. Selection of cell locations is simplified by the use of a mouse pointing device.

STATPLAN also does data entry onto a full-screen display. The initial display shows 100 numbered data locations for the variable, arranged in five groups of 20 locations each. A second screenful of 100 data locations can be called into view by pressing the PgDn key. The user can move the cursor to any desired point on the screen, adding, removing, and changing data items until the display contains the correct data.

The other packages have not incorporated the convenience of full-screen entry. They use procedures where the user deals with one or just a few items at a time. In the Statistical Analysis package from IIE, for example, the user enters each item for the first sample, one to a line, then repeats the process for the second sample. When completed, the input can be viewed and corrections made by entering item numbers and updated values.

The procedure for the Regression I package follows a simple running dialog between the user and the computer. The package prompts the user for the *x,* and *y* values for each point. When completed the entered data are displayed in *x* and *y* columns with each point numbered. The data are edited by answering a query for the desired point number, then changing the *x,* or *y,* or both values, again in response to screen-displayed queries.

In the BIOSTATISTICS package each statistical procedure includes its own data entry dialog. This can be used to enter data from the keyboard, or a peviously stored data file can be used. However, if data prepared for one procedure are to be used for another, it may be necessary to go through one of the conversion procedures displayed on the Data File Manager menu.

EPISTAT is a do-it-yourself package. Users enter data and run statistical procedures keeping fairly close to BASIC computer language procedures. Data are entered, a row at a time, by pressing the enter key twice after each item is keyboarded. If the user wants to name the records, letters instead of numbers are entered in the first column.

Plotting capability

For the most part, the graphics offered by these packages are crude and restricted to scatter plots and histogram displays. For those willing to learn their special languages, the RS/1 and ASYST packages provide more sophistocated graphics and output to plotting equipment.

Only one of the other packages has plotting capability. The BIOSTATISTICS package can send output to Houston Instruments DMP-29, DMP-40, and PC series plotters.

Manual on a disk

Most of the packages offer conventional printed user manuals. An exception is INSTAT-SC. The author of this package, probably in an attempt to limit the expense of producing copies, provides a user manual only in the form of files on a disk. Users with less than perfect memories will probably want to print out the manual, and the package provides a simple way to do this (but makes no provision for single-sheet printing). Those who prefer to consult a written manual while making decisions about how to carry through package procedures will also find a printout necessary.

Speed and convenience

No comparison of the times needed to run various statistical procedures by different packages was made, but it is evident that some of the more complex procedures, such as regression, correlation, and analysis of variance, can take long minutes to calculate.

Speed is an admirable quality, but it does not necessarily increase the user's work efficiency. Frankly, this reviewer finds it more convenient to cope with a 200-second wait than with a 10-second wait. During the shorter wait about all one can do is sit at the keyboard and watch helplessly while the computation grinds on. There is no time to turn to other tasks. However, during a wait of two to five minutes, there is time to read letters or notes, to do some of those small tasks that are always begging to be taken care of.

For the personal computer user, speed is really appreciated when it reduces the waiting time for a calculation down below five seconds. At such speeds, the interactive process between user and computer retains its continuity. Once that is lost, the user will probably want to turn to other tasks while waiting for the computer to finish its processing.

ASYST: SYSTEM, GRAPHICS, STATISTICS MODULE

ASYST is a computer language in the same sense that BASIC or FORTRAN are computer languages. It was designed with engineering and scientific problem solving in mind and has a variety of capabilities, including some limited statistical calculations.

The ASYST System, Graphics, Statistics module runs on IBM/PC/XT/AT computers. It sells for $795 from Macmillan Software Co., 866 Third Ave., New York, NY 10022. (212) 702-3428.

ASYST provides the user with no menus of available functions, nor does it have predesigned formats to organize its use. Instead, like the BASIC language, it simply presents the user with an OK prompt and waits for commands to be keyboarded. As with any other language, these commands must reside primarily in the users head, which means that the ASYST language must be learned before it can be used.

As with any other computer language, regardless of its intrinsic merits, the success or failure of ASYST will rest on the extent to which it is adopted and used by programming-minded engineers and scientists. If and when these users write programs that can, in turn, be put to use by the vast majority of engineers who have little or no interest in doing programming, then ASYST will indeed become a significant engineering tool.

ASYST appears to be a unique language, capable of more rapidly implementing its procedures than more standard languages such as BASIC, but it has its own shortcomings. It does not seem oriented toward programmers, the very people it must please if it is ever to amount to anything. For example, programs written in ASYST seem to work efficiently, but they are very difficult to understand and therefore will be hard to debug and modify. Since a large portion of most programming projects goes into correcting and changing what has been written, it seems unlikely that ASYST will gain much popularity among those who do this work.

Data entry and statistical calculations

The ASYST ENTER command uses a number list to guide the user in entry of data items into a table. The entry format is a continuous sequence of data items with commas separating the items. These are stored, filling a blank table one row at a time. Like all other functions of this package, data checking and editing must be done with the commands provided by the ASYST language; there are no convenient special procedures for this purpose.

There are ASYST commands to calculate the mean, variance, mode, minimum, maximum, and sum of a column of data. Histograms can be displayed. There also is a command to compute fractiles of a data array.

The package also can compute cumulative distributions and fractiles of normal, chi-squared, and Student's t distributions.

BASIC STATISTICAL SUBROUTINES

The statistical procedures in this package consist entirely of nonparametric tests. There are 29 such tests, a goodly number. The problem will be to make them useful together with the data and other statistical tools a user may have.

This package is available in versions that run on most personal computers. It sells for $99.95 from Dynacomp, 1064 Gravel Rd., Webster, NY 14580. (716) 671-6160.

The statistical procedures in this package are in the form of BASIC computer language subroutines. They are for use by those who are generally familiar with the BASIC language, and the procedures can be incorporated into the user's own BASIC programs.

The package includes a routine that gives access to the statistical procedures through a menu display. There also is a procedure on the menu for creation and editing of data files which then can be used by the statistical procedures.

A package of nonparametric tests

The procedures are all nonparametric tests. For the one sample case, there is a binomial test, chi-square one-sample, Kolmogorv–Smirnov one-sample, and a one-sample runs test.

For two related samples, there is a Mcnemar test, a sign test, a Wilcoxon matched-pairs signed ranks test, a Walsh test, and a randomization test for matched pairs.

For two independent samples, there is a Fisher exact probability test, a Fisher–Tacher test, a chi test, a Yates 2 × 2 chi, a median test, a Mann–Whitney U test, a Kolmogorv–Smirnov two-sample test, a Wald–Wofowitz runs test, a Moses test of extreme reactions, and a randomization test.

For *k*-related samples, there is the Cochran Q test and the Friedman two-way ANOVA by ranks. For *k*-independent samples, there is a chi test, an extended median test, and the Kruskal-Wallis ANOVA.

Finally, there are some measures of correlation with their tests of significance. Included here is the contingency coefficient *C*, the Spearman–Rank correlation coefficient, the Kendall rank and partial rank correlation coefficients, and the Kendall coefficient of concordance.

BIOSTATISTICS 3PC

BIOSTATISTICS 3PC concentrates on regression analysis. It also performs one- and two-way analysis of variance and several statistical tests and can display output on plotting equipment.

BIOSTATISTICS 3PC runs on IBM/PC/XT computers. It sells for $125 from A2 Devices, PO Box 2226, Alameda, CA 94501. (415) 527-7380.

The package operates from a main menu with statistical procedures, data handling, and some housekeeping operations as options. Each statistical procedure includes its own data entry dialog. This can be used to enter data from the keyboard, or a previously stored data file can be used. However, if data prepared for one procedure are to be used for another, it may be necessary to go through one of the conversion procedures displayed on the Data File Manager menu.

The files also can be converted for use with other computer packages, such as spreadsheets, that use the DIF format, and BIOSTATISTICS 3PC can accept DIF files.

An unusual feature of this package is its ability to send output to a plotting device. Houston Instruments DMP-29, DMP-40, and PC series plotters can be used.

Statistical procedures

A Standard Deviation procedure produces a few descriptive statistics: mean, sum of data, and square sum, in addition to the standard deviation. Test procedures include paired and unpaired Student's t tests, Wilcoxon paired test, and the Mann–Whitney U test.

Regression procedures include linear, curvilinear, polynomial, and multiple regression. The one- and two-way ANOVA procedures will each handle up to 500 variables.

To reduce computation time and save on storage space, the package requires the user to use some intermediate results to look up answers on tables provided in the manual. Included are tables for the t-distribution, Wilcoxon's T, and the q distribution.

EPISTAT

EPISTAT is a package for those who have relatively small amounts of statistical data to work with. It will handle less than 28 variables and under 2,000 total data entries.

EPISTAT runs on IBM/PC and compatible computers. It sells for $25 from Epistat Services, 1705 Gattis School Rd., Round Rock, TX 78664.

EPISTAT is a do-it-yourself package. Users enter data and run statistical procedures keeping fairly close to BASIC computer language procedures. Data are entered, a row at a time, by pressing the Enter key twice after each item is keyboarded. If the user wants to name the records, letters instead of numbers are entered in the first column.

When entry is complete, the package displays the mean, median, and standard deviation of the data, and that seems to be the extent of the descriptive statistics.

A normal distribution routine compares sample and population means and does other calculations. Several chi-square tests are provided. In addition to tests based on Yates correction and Cornfield's method, there are Mantel–Haenszel tests for the relationship between two variables while a third is controlled and also for multiple controls. There also is a paired chi-square test. For situations where the expected value of a cell is greater than five, Fisher's exact test is provided.

The package performs t-tests. Nonparametric tests of significance include a rank sum and a signed rank test, and there is a Bayes theorum calculation.

Pearson's correlation coefficient and Spearman's rank correlation calculations are provided. There is a linear regression routine for paired samples. One-way and two-way ANOVA procedures also are available. The package has calculations for the binomial and Poisson distributions.

Sampling is facilitated by a random sample generator, a routine for adjusting sample rates, and a sample size calculation. A crosstabs routine generates one-, two-, or threeway reports. There are scattergram and histogram routines for graphical display.

INSTAT-SC: STATISTICAL CALCULATOR

INSTAT-SC is a neatly concieved, compact, and handy statistical tool that takes advantage of personal computer capabilities. It can handle only limited amounts of data, up to 200 values in each of four samples, but it does what it can quite well.

The package runs on IBM/PC/XT/AT and compatible computers. It sells for just $29.95 from Statistical Computing Services, 517 East Lodge Drive, Tempe, AZ 85283. (602) 838-7784.

INSTAT-SC is a kind of statistical spreadsheet. When the package is started up it displays a four-column framework of lines. As the user keyboards data into a column, statistics are displayed to summarize the data characteristics, and they change with each added entry to the column.

Simple operation

There are a set of about ten one-key commands the user has to learn, such as D to delete a number and C to clear a column. With just a little use of the package these can be easily mastered.

Once the data are entered, the package allows column data to be stored in a disk file, and such files can be called up to fill data into columns.

By selecting the Transformations procedure, users can perform certain computations on the data in the columns. A display shows the transformations available, listed by number. The user selects which type is desired, then replies to prompts, indicating which columns of data are to be involved in the transformation. There is a caution here. Results of the transformations will be stored in a specified column and, since there are only four columns, this may already contain data. Any preexisting data in the results column will be erased by the transformation operation.

The user then has a choice of a number of statistical procedures that can be run. The names of these procedures are displayed, one at a time, at the press of a key. All the user has to do is stop at the right one, then follow a few displayed prompts to bring it into action. Results are immediately displayed on a blank portion of the table.

Available procedures

A one-sample t-test and confidence-interval procedure is used to test the null hypothesis. The procedure asks that the user give the number of the column containing the data and the null value for the mean. It will then report the test statistic, its degrees of freedom, and three P-values corresponding to various alternative hypotheses. A confidence interval is derived from the data so that with some stated probability it will cover the true value of the mean.

There also is a two-sample t-test and confidence-interval procedure. In it the degrees of freedom of the t-distribution are determined by a formula due to Aspin and Welch, and the confidence intervals for the difference between the two means are derived from the Aspin–Welch statistic.

A paired sample t-test is carried out with the help of the Transformations feature of the package. This is used to compute the difference between the paired values in two columns of sample data, and the results are stored in a third column. Then the user selects either the one-sample t-test or confidence interval for a mean and uses it to specify the third column.

When dealing with a population of objects that can be classified as being either *successes* or *failures,* the user is interested in the proportion of the population consisting of successes. The procedure for dealing with this problem has the user enter the data into a column so that the first entry is the number of successes observed and the second is the number of failures. Results can be computed for either one or two populations.

The package also performs tests and finds confidence intervals for correlation coefficients. The confidence intervals are produced by performing Fisher's z-transform, determining the upper and lower endpoints of the confidence intervals for the transformed parameter value and then untransforming these endpoints.

For regression calculations, the package asks which stack contains the *y* values and which the *x* values. Standard deviations are reported. The two tests that usually are carried out in the regression situation are that alpha = zero and beta = zero. The P-values reported by the package are two-sided P-values for these tests. To make tests of other hypotheses about alpha and beta, the user transforms the *y* variable, using the package's Transformations procedure. Included in the regression results are the correlation coefficient, its square, and the residual standard deviation.

The package has an analysis-of-variance procedure for two, three, or four data samples. It is designed to test the null hypothesis that the means in the populations, from which the samples were drawn, are all equal. The output display shows the two sums of squares SSB and SSE and their associated degrees of freedom. It also displays a P-value for the hypothesis that the population means are all the same. The mean squares, F-statistic, and contrast values (P-values) for each column also are displayed.

A chi-square analysis is provided for testing whether two factors are independent of each other. The package reports the chi-square statistic, its degrees of freedom, and the P-value.

Population parameters that can be tested by Hypothesis Tests are the mean, a difference between two means, a proportion, a difference between two proportions, a correlation coefficient, the slope and intercept of a regression line, and up to four group means. The package asks for the null value of the parameter. It then reports three P-values.

Manual on the disk

The user manual for this package is comprehensive and clearly written. However, it is made available only as information recorded on the package disk. This undoubtedly saves the vendor the cost of printing a manual for each user, a necessary economy for a package with a price as low as this one has.

Users with less than perfect memories will probably want to print out the manual, and the package provides a simple way to do this (but makes no provision for single-sheet printing). Those who prefer to consult a written manual while making decisions about how to carry through package procedures will also find a printout necessary.

PC ANOVA

Using menu-operated procedures, PC ANOVA provides for data entry, file manipulation, and analysis-of-variance procedures that include a variety of designs for one to five variables.

PC ANOVA runs on IBM/PC and compatible computers. It sells for $200. A similar package for Apple II computers sells for $150. Human Systems Dynamics, 9010 Reseda Blvd., #222, Northridge, CA 91324. (818) 993-8536.

PC ANOVA handles problems involving one to five variables, solving simple and multivariable randomized or repeated measure problems, between, within, or split-plot problems, randomized complete block problems, latin square problems, and performing analysis of covariance, and nonparametric analysis of variance.

The package produces ANOVA summary tables, descriptive statistics, marginal means for all effects, and also provides graphical display of treatment means.

Data manipulation menu

The package operates from a main menu on which the choices are mostly statisical procedures, but there also is a Data File Procedures choice. This calls up a data menu that allows users to enter data sets from the keyboard, review and edit them, store them in named files, perform transformations on the data, and manipulate the files to select data for processing by a number of statistical procedures. There also are provisions for converting files from ASCII and DIF formats for use by this package.

Data are entered from the keyboard one row at a time; the user is prompted by the package with the variable name for each item to be entered. There is an initial editing step, in which the user can correct the row that has just been entered, by keying in the column number of the item and a new value, and there is a similar procedure for editing a selected row of data after all the data have been entered. Rows and columns can be deleted or added; new data sets can be created using subsets of existing row and column data. The package will search the data to find items that meet selection criteria the user defines.

There is a Data Transformations menu that can be used to transform specified columns of data. The operations include arithmetic (using a constant), exponents, logarithms, arcsin, ranks, Z-scores, and percent. There also can be arithmetic between two or more columns of data.

ANOVA procedures

When one of the statistical procedures on the main menu is called for, the user is presented with a group of fill-in items that identify the data file and the variable or variables, rows that are to be processed, and so forth. Where variables are to be divided up into groups, blocks, treatments, or levels, the user indicates a classification variable to use for that purpose. The user is then presented with a fill-in display of available output options, such as descriptive statistics, post hoc tests, graphics, or printed copy for the results, and the desired items are selected.

The package then reads the data, reporting on the number of records and usable records it finds, and the computation is run, producing an ANOVA summary table. If more information is needed to define a graphical output, the user is prompted for that, and the results for the procedure are completed.

PC/FOCUS

PC/FOCUS is a data base package, designed primarily for those who wish to design and implement new computer procedures and programs. It includes a competent selection of statistical procedures.

PC/FOCUS runs on IBM/PC computers. It sells for $1,500 from Information Builders, 1250 Broadway, New York, NY 10001. (212) 736-4433.

Data management, building, storing, and manipulating data files are important in most personal computer-based statistical packages. This has not gone unnoticed by vendors of packages designed primarily to handle database operations.

PC/FOCUS is a database package that incorporates some statistical procedures. Like some other database packages, PC/FOCUS aims at providing the user with a capability that can handle many different computation tasks. Toward this end, the package offers many different commands and functions more as a computer language than as a simple storage facility.

Setting up a file and entering data is no simple matter in a general database package such as PC/FOCUS. The user manual for this package is formidably thick, and its bulk signals the kind of learning effort needed to master its database operations. In large part, the procedures are complicated by the need to provide prospective users with the ability to tailor database operations to any of a myriad of possible uses.

If the prospective user has a need for a flexible database, and decides that what PC/FOCUS can offer meets that need, then the statistical features of the package can be a valuable add-on. For users whose only interest in a database is as a vehicle to handle data for statistical analysis, PC/FOCUS is probably not the package they want.

In this review, we are concerned mainly the ability of the package to perform statistical analysis procedures. For this purpose PC/FOCUS can handle up to 64 variables (columns). The number of cases (rows) that can be handled depends on the available main memory space in the particular computer on which the package is being run.

Statistical procedures

PC/FOCUS can perform statistical analyses on data already filed by the package, or it can use data from external files or combinations of the two kinds of files. In any case it is necessary for users to master the commands of the PC/FOCUS language in order to enter the data into the system and bring those data into a position where they can be analyzed statistically.

Once the desired files exist, the package provides a convenient command, STATSET, that allows users who are PC/FOCUS novices to follow an interactive procedure to select the right file, set the width of printed output for the reports, and do some other housekeeping tasks.

The actual statistical analysis is started by the ANALYZE command, which makes a variety of procedures available.

Descriptive statistics produced by the Stats procedure can include the mean, median, range, standard deviation, variance, square root of the variance, the mode, maximum, minimum, standard error, skewness, and kurtosis. Tables of quartile or decile values also can be produced.

The Corre procedure produces a symmetric correlation matrix for all variables in the analyzed file. Means and standard deviations also are displayed.

A multiple linear regression can be performed on any choice of variables from a file, using the Multr procedure. The output report gives the mean, standard deviation, correlation (versus the dependent variable) regression coefficient and standard error, and computed t-value for each independent variable. Also listed are the intercept, multiple correlation coefficient, and standard error of estimate. An analysis of variance for the regression lists the sources of variation, degrees of freedom, sum of squares, mean squares, and F values. The regession equation itself is displayed and a table of residuals or Durban–Watson statistics can be produced.

The package also can perform a stepwise multiple regression, processing up to 64 variables. A polynomial regression procedure calculates polynomials of increasing degree until there is no further reduction in the residual sum of squares or until a polynomial of specified maximum degree has been reached.

A factor analysis, that performs a principal component solution and a varimax rotation of the factor matrix, is available.

The Mdisc operation performs a discriminant analysis on files in which groupings of rows are defined. Outputs include the mean of each variable by group, the pooled dispersion matrix, the common means for each variable, the generalized Mahalanobis D-square, numbered discriminant functions, and an evaluation of classification functions for each row.

An ANOVA procedure performs an analysis of variance for an equal-cell factorial design. The Xtabs procedure generates contingency tables and displays summary and cell statistics.

The Timeser procedure includes forecasting, curve fitting, and smoothing for time-series data. Among the transformations provided by this procedure are lead, lag, moving average, center moving average, moving total, center moving total, linear interpolation, exponential interpolation, lead minus current value, and lag minus current value, and there also is a triple exponential smoothing procedure.

REGRESSION PACKAGES

Included in this group of four regression packages are two for one-dimensional data and two for multidimensional data. The procedures are relatively simple and limited in scope.

These packages are available in versions that run on almost any personal computer. The prices are: Regression I and II, $23.95 each; Multilinear Regression, $28.95; Stepwise Multilinear Regression, $29.95. All from Dynacomp Inc., 1064 Gravel Rd., Webster, NY 14580. (716) 671-6160.

Regression I is a curve-fitting package for analysis of linear and nonlinear one-dimensional data. For computers with minimal main memory, only about 10 to 15 data points can be handled in each full calculation.

The procedure itself follows a simple running dialog between the user and the computer. The package asks how many data points are to be entered, then prompts the user for the *x* and *y* values for each point. When completed the entered data are displayed in *x* and *y* columns with each point numbered. The data are edited by answering a query for the desired point number, then changing the *x*, or *y*, or both values, again in response to screen-displayed queries. The data can be stored on disk under a user-assigned file name, and a plot of the data points can be displayed.

The user then chooses, from a menu, the function to which the data is to be fitted. Choices include ten different functions, among them are linear, exponential, logarithmic, and trigonometric functions.

Regression I calculates the regression coefficients and will display a plot of the fitted curve. The user also can ask for standard error of the coefficients; this calculation assumes that the true functional form is of the second order. If it is not, the answers will be inappropriate.

Least-squares curve fitting

The Regression II package is for fitting one dimensional data to curves other than those available in Regression I. Here, the package starts by listing a short BASIC language routine in which the user enters the function to which the data are to be fitted. The example given in the manual is a Weibull function. As in Regression I, a query and answer procedure is followed.

Regression II uses a least-squares fit with a modified steepest descent algorithm. Although the computation is said to be generally stable, the user is warned that some data may yield only approximate answers and that it is possible for the calculation to "blow up." The user controls the calculation by providing a convergence criterion, maximum number of iterations, and starting values for the parameters.

Multiple variables

The Multiple Linear Regression (MLR) package analyzes the relationship between several independent variables and one dependent variable. The fitting equation is linear but may include functions of the independent variables.

The procedure follows a query and answer sequence. Data can be entered row by row, in response to prompts, or loaded from a previously prepared file. Data editing also is on a row-by-row basis.

There is a choice of ten functional forms. A different one can be fitted to each independent variable to set up the regression equation. The package then calculates the regression coefficients and, on request, estimates of the error associated with those coefficients.

Additional calculations for multiple variables are offered by the Stepwise Multiple Linear Regression (SMLR) package. This allows the user to add or delete independent variables easily and to run a calculation where the independent variables are added one by one. SMLR also allows extrapolation and interpolation with Student's t-value confidence limits, and it produces a correlation matrix.

Unlike the other regression packages described here, SLMR operates from a menu of procedural choices. The procedure for adding a variable from the keyboard is somewhat tedious. The user is queried for the first point, then asked if the entered value is correct before the query for the next point is displayed. This is in lieu of an editing procedure. If there are any errors in the completed variable, the only available editing procedure is the one in the MLR package.

There is no capability in SLMR to calculate the standard error of the estimates for the regression coefficients.

REGRESS II

REGRESS II will handle up to 20 variables, performing five different types of regression procedures and producing a variety of output statistics. File manipulation and data transformation capabilities are included.

REGRESS II runs on Apple II, IIe, and IIc computers. It sells for $150 from Human Systems Dynamics, 9010 Research Blvd., #222, Northridge, CA 91324. (213) 993-8536.

REGRESS II performs multiple correlation and regression. There are five regression procedures in the package: simultanous, forward selection, backward elimination, stepwise, and a procedure using power polynomials.

The package produces correlation and covariance matrices, descriptive statistics, and a number of plots, in addition to the regression statistics.

A row at a time

From the main menu of this package, the user can go to a data preparation menu to set up new files and review, edit, and manipulate existing ones. Data are entered a row at a time. The package prompts for each entry with a variable name. Editing follows a similar procedure, but rows to be changed can be called for by number. There can be up to 20 variables in a file. The number of rows must be such that the total number of characters in the file does not exceed 32,000.

File manipulation includes creation of subfiles, merging of files, and automatic search for rows that meet user-selected criteria.

Available data transformations include arithmetic, exponents, logarithms, moving average, and ranks. There also is exponential smoothing.

Follow the prompts

When a regression procedure is selected, the user is presented with a series of screen-displayed questions to answer. Data for the procedure can be entered from the keyboard, but the user is then limited to 7 predictor variables and 50 rows of data. If the data for the procedure is taken from a previously stored file, the full capabilities of the package can be used. Correlation data can be taken from a file created with the help of a built-in correlation matrix calculation.

Once the data are in place, the various regression procedures call for additional information from the user such as F-to-enter values and orders of polynomials.

The outputs from the regression procedures will include: an ANOVA summary table, squared multiple correlation, and adjusted squared multiple correlation; a list of regression coefficients, standard errors, t-ratios and p-values; actual, predicted and residual scores; the serial correlation of residuals and the Durbin–Watson statistic; and a file of predicted and residual scores.

RS/1 MULTIPURPOSE SOFTWARE

RS/1 combines a specialized engineering–scientific computer language with simple interactive procedures to provide data manipulation that includes display in tabular and graphical formats, statistical manipulations and tests, linear and nonlinear regression analysis, spreadsheet, and system modeling.

RS/1 runs on IBM/PC/XT/AT computers. It sells for $2,000 from BBN Software Products, One Alewife Place, Cambridge, MA 02140. (617) 491-8488.

Of RS/1's many capabilities, only those concerned with statistics will be considered here. Use of the package, like that of any other computer language, involves keyboarding the proper commands, and these must be learned in order to get the package to perform statistical operations.

Statistical data, like all other data handled by this package, is stored in tabular format. The command MAKE TABLE starts a dialogue sequence that leads the user to name a table and its rows and columns and to fill in data one column at a time.

RS/1 allows for intermixing of such dialog sequences with regular keyboard commands plus single-key actions of various sorts. All this gives the user very flexible control over the package functions but makes that control a complex process and one that can be confusing, particularly to the novice or occassional user.

The implicit assumption behind RS/1 is that users will frequently turn to this package, using it on a day-to-day basis, as a regular part of their work. For those users who need to work with tabular data in this way, the time needed to become familiar with RS/1, perhaps a week or two, will be an insignificant investment, compared to the rewards of constant use.

However, those whose use will be less frequent will have to consider whether packages that are less flexible but simpler to learn and use might better meet their needs.

Statistical functions

Once data have been stored in tabular form, RS/1 allows the user to perform a variety of statistical procedures on those data. Some of the commands involved are very simple and powerful. For example the command MEASURE, followed by the name of a table, will produce a set of 12 descriptive statistics for the numerical data in that table. These include a count of the items and their sum, mean, standard error of the mean, median, variance, standard deviation, maximum, minimum, range, skewness, and kurtosis measures.

It is up to the user to see to it that the table, or portion of a table, on which the MEASURE command operates contains just those data items that are to be analyzed. More specific commands cause descriptive statistics to be calculated for individual rows or columns of a table.

The COMPARE command invokes a screen-display dialog that runs the user through a sequence of choices designed to test the null hypothesis. If the user is uncertain about the normality of the data distribution, that will be tested and the results displayed. Then the user can choose either parametric or nonparametric testing and can test either paired or unpaired data.

With this COMPARE command, if one table is being tested, the tests used will be either a t-test or the Wilcoxin signed ranks test. If two tables are being compared, and the user calls for parametric tests, the package will check for pairing with the F-test for equality of variance; then either a pooled or unpooled variance t-test will be applied. If the user chooses nonparametric tests for two tables, the package will use the Ansari–Bradley test for equality of dispersions, then apply either the Mann–Whitney test for pooled samples with equal dispersions or the Wilcoxin signed rank test for paired samples. Using more specific commands, users can invoke these various tests directly.

The package also has a CORRELATE command that calculates the correlation coefficient and its square, the coefficient of determination, for any two specified table portions. For ranked or ordered data, the command RANKCORRELATE will cause Spearman's Rho, the rank correlation coefficient, to be calculated.

Analysis of variance is invoked with ANOVA ONEWAY or ANOVA TWOWAY commands. There is an ANACOVA command that combines one-way analysis of variance with linear regression on a single independent variable, and there is a CHISQUARE command as well.

Although it runs on personal computers, the RS/1 package is very much oriented toward traditional mainframe and time-sharing computation. The package therefore contains provisions for transferring data to such IBM mainframe computer statistical analysis programs as BDMP and SAS. The RPL (Research Programming Language) can be used to create custom additions to RS/1 functions.

STATA

STATA is a statistics and data manipulation language that can be used to set up and store data tables and perform some statistical procedures using these data. It will handle 99 variables and about 1,300 rows of data using single precision calculations but only about half as much data using double precision.

STATA runs on IBM/PC/XT computers. It sells for $395 from the Computing Resource Center, 10801 National Blvd., Los Angeles, CA 90064. (213) 470-4341.

A set of about 50 commands are used, along with arithmetic, relational, and logic operators. The command words are easy to use and remember, nevertheless this is a computer language with a syntax and many rules that the user has to learn before it can be effectively applied to data and statistical manipulation.

Keyboard entry of data into STATA is done a row at a time. Following the INPUT command, the column names are entered. The package then prompts with a row number and data items are entered, separated by one or more spaces, and ending the row with a return. The next row prompt appears and the process continues.

A good part of the STATA language is devoted to formatting, sorting, and performing other data manipulation, display, and printing functions. However, in this review we are concerned mainly with its statistical capabilities.

Statistical features

Once data are stored under a user-assigned file name, the SUMMARIZE command will produce a few descriptive statistics (mean, standard deviation, maximum, minimum). A DETAIL option of this command adds information on skewness, kurtosis, and the median. The CORRELATION command produces a correlation matrix.

Scatter plots that use asterisks to denote point positions can be generated with the PLOT command.

The REGRESS command performs linear multivariate regression, producing summary statistics, a table of regression coefficients with standard errors, an ANOVA table, t-statistics, F-statisic, R-square, and adjusted R-square.

The TEST command performs F-tests of linear restrictions applied to regressions.

The TABULATE command produces one-way and two-way tables of frequency counts and performs a chi-square test for independence of the variables.

Speed is not necessarily convenience

One of the main virtues highlighted by the vendors of STATA is its speed compared to that of other packages that perform similar functions. According to figures provided by the vendor, STATA will run a 10×10 correlation matrix in about 37 seconds, whereas some other packages take from 105 to 240 seconds to perform a similar calculation.

Speed is an admirable quality, but it does not necessarily increase the user's work efficiency. Frankly, this reviewer finds it more convenient to cope with a 200-second wait than with a 10-second wait. During the shorter wait about all one can do is sit at the keyboard and watch helplessly while the computation grinds on. There is no time to turn to other tasks. However, during a wait of two to five minutes, there is time to read letters or notes, to do some of those small tasks that are always begging to be taken care of.

For the personal computer user, speed is really appreciated when it reduces the waiting time for a calculation down below five seconds. At such speeds, the interactive process between user and computer retains its continuity. Once that is lost, the user will want to turn to other tasks while waiting for the computer to finish its processing.

STATISTICAL ANALYSIS

Statistical Analysis includes four separate analysis procedures selected for their relevance to the interests of industrial engineers. The procedures are straightforward and simple to use. Included are descriptive statistics, a Mann–Whitney test, a distribution-fitting procedure, and polynomial curve fitting.

The package runs on IBM/PCs and sells for $175 from the Institute of Industrial Engineers, 25 Technology Park/Atlanta, Norcross GA. (404) 449-0460.

This package runs under the BASIC computer language, and must be started by entering BASIC, then calling for the particular procedure that is desired. This is a slight inconvenience for those unfamiliar with BASIC. The package instructions make it clear how to enter BASIC but not how to return to the IBM/PC operating system when finished.

Descriptive and other statistics

The Statistical Data Analysis procedure calculates descriptive statistics, including the mean, median, variance, and standard deviation for a set of data. It elaborates by calculating five confidence intervals for the value of the true mean and computing the skewness and kurtosis of the data distribution. Quartile breakdowns of the data set are given and a histogram, made up of printed characters, shows percentage of observations per interval and cumulative distribution.

The Mann–Whitney test determines if two sets of sample data are significantly different or whether the samples seem to come from the same population. It is useful in checking to see if a production process has gone out of control and may also be used to verify manufacturer's quality claims.

Up to 200 items for each sample will be accepted as input.The user enters each item for the first sample, one to a line, then repeats the process for the second sample. When completed, the input can be viewed and corrections made by entering item numbers and updated values. This data entry method is used in all the procedures in this package.

With the data entered, the package then does its computation. For the Mann–Whitney test, results give the value of the Z statistic and tell whether or not to reject the hypothesis of coming from the same population, at significance levels of 1, 5, 10, 15 and 20 percent.

There also is a distribution-fitting procedure designed to help decide whether a set of data can be considered to come from a particular probability distribution and thus be used as a basis for making predictions. Distributions may be normal, exponential, or uniform.

Up to 200 observed items of data are entered, viewed, and corrected. The user then chooses which type of distribution will be tested. The package calculates the maximum absolute difference between the the observed data and that type of distribution. For alpha significance levels of 1, 5, 10, 15, and 20 percent, the output indicates whether or not the observed data are from the chosen distribution.

Curve fitting

To use the Polynomial Curve Fit procedure, the user enters a set of up to 150 x and y values to represent the curve to be analyzed. These are keyboarded in pairs, with a comma to separate the x and y values in each pair. The package then fits a polynomial and displays a table showing the input x and y values, along with corresponding polynomial values and the difference between the y and polynomial values. This is followed by a display of the polynomial coefficients. If the user does not like the fit, the procedure can be rerun with a higher-degree polynomial.

STATPLAN

STATPLAN uses menu choices and full-screen data entry to provide a number of statistical procedures, including descriptive statistics, correlation, and regression. The number of variables and data items that can be handled depends on available main memory.

STATPLAN runs on IBM/PC and compatible computers. It sells for $99 from the Futures Group, 76 Eastern Blvd., Glastonbury, CT 06033. (203) 633-3507.

STATPLAN works from displayed menus. Making the choice "ENTER new data from the keyboard" starts a procedure where the user defines the first variable, then enters all the data for that variable. The procedure is then repeated for each of the other variables to be entered.

The data are entered on a full-screen display over which the user can move the cursor to any desired point, adding, removing, and changing data items until the display contains the correct data. The initial display shows 100 numbered data locations for the variable, arranged in five groups of 20 locations each. Eight numerals or other characters can be entered at each data location. A second screen of 100 data locations can be called into view by pressing the PgDn key.

When data entry is complete, the package returns the user to the main menu, where the "SAVE data now in memory" choice can be used to store the data under a file name selected by the user. Another menu choice allows stored data files to be retrieved.

A data editing procedure allows changes in existing data, using essentially the same display and operations as for data entry. Data items also can be inserted into a variable by item number, and entire variables can be deleted.

STATPLAN can translate its files into ASCII or DIF formats for use in other packages and also can translate files in these formats for use with STATPLAN procedures.

Data transformations that can be made to a selected variable include adding or multiplying by a constant, lagging, leading, taking first differences and percent first differences, logarthms, inverse ln, Z-scores, moving averages, LOGIT (for linearizing). Arithmetic operations can be performed between two variables. Variables can be sorted and the sort on one variable can be extended to another. Variables also can be categorized and subgroups created.

Variables can be set up with values following sine or cosine distributions or with evenly or normally distributed random values.

Selected statistical analysis

Selecting the main menu "ANALYZE data" option brings up a menu of statistical analysis procedures. The first of these produces a set of descriptive statistics that includes a count of the data points in the selected variable and its maximum, minimum, range, mean, standard deviation, coefficient of variation, median, skewness, and kurtosis.

A correlation procedure produces a matrix of correlation coefficients for selected variables. Therc also is a one-way ANOVA procedure.

A multiple regression analysis procedure uses an ordinary least-squares approach. The user selects a dependent variable and a set of independent variables, and an ANOVA and set of statistics are calculated and displayed. If the variables are too highly correlated to yield an accurate result, the package will inform the user of this fact and ask the user to try again with a different set of independent variables. The user can optionally examine a table of residuals, look at graphs of actual versus calculated values, and use the regression equation for forecasting values of the dependent variable.

An exponential smoothing procedure is provided for time-series variables. There also is an Autocorrelation and cycle analysis procedure to check data for repetitive cyclic values.

Curve-fitting procedures allow data to be fitted to any one of eight curve shapes or to a straight line. Actual and calculated values, residuals, and an R-squared statistic aredisplayed, and a graph of actual versus calculated values can be shown.

A cross-tabulation procedure is limited to showing the joint frequency distribution of data items from a selected variable, according to two classificatory variables, along with appropriate statistics.

STATPLAN will display a scatter plot of any variable against any other variable. It can display a histogram that shows frequency of occurance for values in 20 intervals. There also is a time-series plot capability.

STATVIEW

StatView features convenient data entry and editing. It provides a limited selection of standard statistical capabilities.

StatView runs on Macintosh computers. It sells for $199.95 from Brainpower, Inc., 24009 Ventura Blvd., Calabasas, CA 91302. (818) 884-6911.

StatView uses pull-down menus. While the user is viewing a table of data, overlay menus listing statistical procedures can be displayed, the desired procedures can be selected, and results will be displayed.

The StatView user enters data from the keyboard into a tabular display. Data entry is done a row at a time. Additional rows and columns can be created and exposed by scrolling. The maximum capacity of a data table depends on the main memory size of the computer. For example, with a 128K memory there can be 50 columns and 150 rows containing integer numbers or 35 columns and 50 rows of real numbers.

Each cell of the table can be edited individually by simply typing over what is displayed. There are cut, copy, paste, and undo functions to simplify the editing process. Selection of cell locations is simplified by the use of a mouse pointing device.

Available data transformations include roots, exponents, logarithms, trigonometrics, ranks, standard scores, and recoding.

Statistical capabilities

Descriptive statistics generated by this package include the mean, standard deviation, standard error, variance, coefficient of variance, count, minimum, maximum, range, sum, sum squared, and number of missing values. The median, mode, geometric mean, harmonic mean, skewness, kurtosis, and frequency distributions all are available.

The package will perform t-test for one sample or two sample data. Chi-square testing produces a contingency table with up to eight rows and four columns.

A correlation procedure produces the covariance, correlation coefficient, and R-square values. Nonparametric tests include the Mann–Whitney U, Wilcoxin signed rank, Spearman and Kendall rank correlation coefficients, Kolmogorov–Smirnov, and Wald Wolfowitz statistics.

StatView performs simple, multiple, or polynomial regressions. Up to 18 independent variables can be handled. One-way or two-way ANOVA with up to 50 groups can be performed. StatView graphics include scattergrams and a bar chart display that shows Z-scores.

Chapter 18
Instrumentation Packages

Instrumentation software packages are designed to collect and interpret information from equipment outside the computer and also to exert control over that equipment. More often than not the incoming information is in the form of continuously varying analog signals.

To convert these signals into digital form acceptable to the computer and to properly interconnect the computer to outside instrumentation, interface circuits are needed. Generally, the interface takes the form of circuit boards that plug into the computer.

Many different brands and types of such circuit boards are available, some designed to accept direct connections from individual instruments, others built to feed into an instrument "bus" (wires or cable to which many instruments can be connected).

Specialized instrumentation languages

The most flexible way to exercise computer control over intrumentation is to use a set of general commands that can be sequenced to perform whatever task is at hand, in other words, to use a computer instrumentation language.

The AMPRIS package is designed for just that purpose. It provides a set of special commands that can be used as an extension of the BASIC computer language. Thus, AMPRIS analog input commands include those for taking single and serial readings and for averaging inputs from specific channels. Multiple channels, sequentially or in arbitrary sequence, also can be scanned on command.

AMPRIS also includes analog output commands that send prescribed single signals or series of signals to specific channels, and there is a command to send such signals to a set of sucessive channels. For digital devices, those with on–off output, there are commands to take single or multiple input readings from specific channels and to generate digital outputs on specific channels.

There are actually several versions of AMPRIS, one for each specific type (analog or digital, with 2–16 channels) of interface circuit supplied by the vendor.

PC-488 is designed for use with the IEEE-standard 488 instrumentation bus. Like AMPRIS, this package is for use by engineers who are adept at programming, but there are some striking differences. Everything PC-488 has to offer is contained on a circuit board that plugs into the computer. The software (called, in this form, firmware) is packaged along with the interface circuitry on this board.

PC-488 commands provide an instrumentation extension to BASIC, Pascal, C, and other computer languages. These commands implement the functions provided by the circuit board. On an IEEE-488 bus all communications are centrally controlled and PC-488 equips the computer to function as such a central controller. For example, sequences of commands and data are sent using the TRANSMIT command. LISTEN and TALK designate the instrumentation device that will do the sending and those that will be receiving. SEND and ENTER establish the computer as a talker or a listener. There are SPOLL and PPOLL commands for serial and parallel polling of devices attached to the bus. In all there are about 15 of these PC-488 commands.

A more universal language

Because instrumentation software has been so closely tied to interface circuits, the software packages are usually designed to work only with one particular type of plug-in circuit. An exception to this rule is the ASYST package.

Like the packages already mentioned, ASYST provides a set of programming commands for reading and controlling instrumentation. These are actually part of a larger set of ASYST commands designed to perform data analysis, statistics, and a variety of mathematical operations useful in engineering and scientific applications.

Unlike these other packages, though, ASYST includes the ability to work with a number of different interfaces. At the time this report was written, ASYST included customized routines for using the Data Translation DT2801 board and the Techmaster Lab Master, Lab Tender, and DADIO boards. The Keithley DAS Series 500 instrumentation system also can be handled. ASYST is not a perfect universal instrumentation language, for there are some ifs, ands, and buts about using specific interfaces, but the basic idea is a powerful one: once a user learns to program the ASYST commands, they can be used with many different interfaces.

However, like any intrumentation package that simply provides programming commands, ASYST is not a package designed to conveniently control instruments. It offers the user no menus of available functions, nor does it have predesigned formats to organize its use. Instead, like the BASIC language, it simply presents the user with an OK prompt and waits for commands to be keyboarded.

User-oriented packages

Instrumentation software packages designed for use by engineers who are not programmers have begun to appear. Thus far, most of these packages are directed to specific types of instruments. Thus the CHROMATOCHART package collects data from up to four chromatographs, massages the data, displays stripcharts, and stores the data. It also produces printed reports on the characteristics of integrated peaks and compares unknown to standard data.

CHROMATOCHART operates from a series of screen-displayed menus. The user makes selections by keying in the number of the desired item on the menu. When information is to be input, the user is presented with a screenful of information and choices, makes them or changes them until satisfied, then can look the screen over and enter it into active use. The keyboard arrow keys are used to maneuver around the screen while editing.

With the help of a plug-in interface board, the RTFSA package allows a personal computer to be used as a spectrum analyzer with 10-KHz bandwidth. The package accepts filtered or unfiltered data, performs fast Fourier transform analysis and plots the magnitude of frequency spectra. Another package called SPECSYSTEM works with external acoustical spectrum analyzer equipment to implement applications such as analysis of noise, engine and machinery diagnostics, music, and sound and voice analysis.

More general instrumentation capabilities are provided by the Analog Connection package. This package logs and displays the maximum, minimum, and average values of analog signals and signal differences. Digital, on–off, inputs and outputs also are handled. The package works with plug-in interface boards, each of which connects to up to 16 different inputs.

The Analog Connection package operates through well-designed and easy-to-use menus and display screens. There are two main menus, one for setting up the operation of the package, the other for actually using the package to monitor or manipulate instrumentation readings.

Particular attention is paid to the laborious details of setting up the individual parameters for each input. These include items such as the range that will be used, the units that will be displayed, types of thermocouples that will be used, scale and offset factors, and alarm limits. By following the menu procedures, all these parameters can be set up in an organized manner. There is virtually nothing for the user to memorize, so that neophytes and occasional users of this package should have an easy time following these procedures whenever needed.

Although use of the package is quite simple, it does take experience to most effectively use some aspects of the Analog Connection package. For example, the user can specify the resolution of input readings, the rate at which readings are taken, and the A/D conversion filtering time, but all these, in turn, affect the time required to cycle through a set of readings, so some knowledgeable juggling may be needed.

It is important to contrast the capabilities of packages intended for engineering instrumentation applications with a similar class of packages designed strictly for educational use in student laboratories. Consider, for example, the simple temperature-reading package TEMPERATURE PLOTTER. It is very low cost, only a fraction of the price of the other packages discussed here, and very easy to use. However, the package has limitations that make it unusable for most engineering applications. Thus, each reading takes a minimum of one second, only four probes can be monitored, and a total of only 512 readings can be stored. There also are limitations, which may be unacceptable, on the accuracy of the results and of the calibration procedures.

Timing considerations

As in any instrumentation situation, accuracy is a key consideration when using a computer software package, and one way the accuracy issue arises is in terms of timing.

Two main sources of timing generally are used in connection with these packages. Timing may be derived either from the computer and the software or from separate timing circuits that usually are mounted on the instrumentation interface circuit board.

Because of the inherently unpredictable nature of computer software timing, the AMPRIS vendor recommends that all such times should be considered to have a 5 percent worst-case tolerance. Most instrumentation software package vendors do not mention the problem, but a similar tolerance should be applied for any time intervals derived in this way. If more accurate intervals are needed, separate timing circuits should be used.

Control outputs

When dealing with instrumentation, it is often as important to control the operation of external equipment as it is to collect information. With control commands, used in programs written by the user, very flexible control can be exercized, but some of the user-oriented packages, designed for nonprogrammers, also provide quite powerful control measures.

The CHROMATOCHART package, for example, is set up to produce output signals suitable for controlling gradient pumps. Two pumps can be controlled at one time to produce desired mixtures of two fluids. Gradient profiles, created by using the package, can be used to define the mixtures, and these profiles are constructed and changed with the help of a display menu.

As for more generalized control, the Analog Connection plug-in board is able to produce digital (on–off) output signals as well as to accept inputs, and the package gives users very flexible control over the input conditions that will cause outputs to be generated. Not only can various logical combinations of digital inputs be used to control these outputs, but the outputs themselves can be used to define the logic of other outputs. Effects of analog inputs can be included (by specifying set points and dead bands) in determining these logical output controls.

Graphical output

Often, showing instrumentation information in graphical form is essential to interpretation and use. Some packages include their own graphical capabilities. Others ignore graphics or would have the user turn to auxiliary programs for graphical output.

Graphics are essential to spectrum analysis, and the SPECSYSTEM package is well provided in this area. The package assembles and stores large arrays of spectral data. These data arrays then are displayed in the form of geometrical surfaces with dimensions of time, frequency, and amplitude. Alternatively, curves representing the data for individual spectrum channels can be displayed.

The Analog Connection package has no built-in capabilities for graphics. Instead users are instructed to feed stored data to the Lotus 1·2·3 package for creating graphs, printing formatted reports, and doing spreadsheet-type analyses.

Among the packages that provide instrumentation command for programming, ASYST includes graphical commands that can be called upon to create displays as needed.

AMPRIS: INSTRUMENTATION COMMANDS FOR APPLE COMPUTERS

Each of the AMPRIS packages provides a set of instrumentation control commands for a specific plug-in interface board from the vendor.

The AMPRIS-02 package ($125) is for use with the A102 8-bit, 16-channel analog input board ($299). The AMPRIS-13 package ($225) is for the A113 12-bit, 16-channel analog input board ($550). The AMPRIS-03 package ($125) is for the AO03 analog output board with 2, 4, or 8 output channels ($195, $275, $437). The AMPRIS-09 and 09 INT packages ($195, $125) are for the D109 input-output board ($330). All packages and boards are for use with Apple II and IIe computers. The vendor is Interactive Structures, 146 Montgomery Ave., PO Box 404, Bala Cynwyd, PA 19004. (215) 667-1713.

The AMPRIS commands are all written in assembly language, and they can be called in programs written in the BASIC language.

Time intervals, used in controlling the interfaces, are derived from internal computer timing. Because of the inherently unpredictable nature of this timing technique, the vendor recommends that all times should be considered to have a 5 percent worst-case tolerance. The digital interface card includes a timer implemented in circuitry to provide more accurate inter-event delays.

Analog input commands include those for taking single and serial readings and for averaging inputs from specific channels. Multiple channels, sequentially or in arbitrary sequence, also can be scanned on command.

Analog output commands provide for sending prescribed single signals or series of signals to specific channels, and there is a command to send such signals to a set of sucessive channels.

For digital devices, those with on–off output, there are commands to take single or multiple input readings from specific channels and to generate digital outputs on specific channels. Other commands control the operations of interval timers and of shift registers used for serial–parallel conversions of digital data. There also are commands to control interrupt signals from devices that request immediate attention.

Five binary utility commands are provided with each AMPRIS package to ease the process of doing logical, Boolean manipulation of numerical quantities.

ANALOG CONNECTION: HANDLING ANALOG AND DIGITAL SIGNALS

Analog Connection logs and displays the maximum, minimum, and average values of analog signals and signal differences. Digital (on–off) inputs and outputs are handled. It provides flexible user specification of both inputs and outputs.

The package runs on IBM/PC/XT/AT computers. It works with a plug-in circuit board; the price of the board and software combination is $690. A similar combination for the Apple II, II+, and IIe sells for $490. The vendor is Strawberry Tree Computers, 949 Cascade Dr., Sunnyvale, CA 94087. (408) 746-3083.

This package operates through well-designed and easy-to-use-menus and parameter display screens. There are two main menus, one for setting up the operation of the package, the other (called the display menu) for actually using the package to monitor or manipulate instrumentation readings.

Good control over inputs

Up to 16 input signals can be handled by each plug-in circuit board. The board converts analog inputs to digital form so they can be used by the computer.

The package gives the user control over important parameters for each input, including the range that will be used, the units that will be displayed, types of thermocouples that will be used, scale and offset factors, and alarm limits. By following the menu procedures, all these parameters can be set up in an organized manner. There is virtually nothing for the user to memorize, so that neophytes and occasional users of this package should have an easy time following these procedures whenever needed.

One major stumbling block can arise. The sixth choice on the set-up menu list includes selecting the number of channels that will be actively used from each plug-in board. If the user makes any changes in this choice, the menu display warns "This will erase most of your setup. Make this selection first." However, for the unwary user who has already worked through dozens of screens to set up input parameters, this warning may come too late. It is not sufficiently emphasized in the manual.

If the plug-in board contains an analog clock, its output can be used to control the timing of readings and outputs. If not, the digital clock set up by the computer's system software can be used for this purpose. One of the set-up display screens gives the user control over this choice.

It does take some experience to most effectively use some aspects of this package. For example, the user can specify the resolution (smallest detectable change) the cicuit board will recognize in the inputs. The rate at which readings are taken also can be specified, as can the A/D conversion filtering time. These adjustments, as well as the number of active input channels, can affect the time needed to cycle through a set of readings.

For user convenience, the input channels can be assigned descriptive names, a valuable feature in readily interpreting displayed results.

Flexible digital outputs

The plug-in board is able to produce digital (on–off) output signals as well as accept inputs, and the package gives users very flexible control over the input conditions that will cause outputs to be generated. Not only can various logical combinations of digital inputs be used to control these outputs, but the outputs themselves can be used to define the logic of other outputs. Effects of analog inputs can be included (by specifying set points and dead bands) in determining these logical output controls.

The package will display and store readings, differences between readings, maximum, minimum, and average values for each input and output under software control. It also will log these values at intervals selected by the user.

A potentially helpful feature of the package is the capability to feed easily stored data to the Lotus 1·2·3 package for creating graphs, printing formatted reports, and doing spreadsheet-type analyses. However, it would have been preferable to have some of these facilities built into the Analog Connection package itself.

An awkward manual

A potentially confusing aspect of display-screen operation is the way that multiple-choice items work. To change from one choice to another, the user has to select repeatedly the menu item containing the multiple choice. At each selection the chosen option shifts. For example, to specify when data logging is to begin, one of the menu items offers six possibilities: Immediate ON, MIN, HR, 12HR, and DAY. Initially the first of these is shown highlighted, indicating it is the current choice. If the user selects that menu item, the highlighting will shift to the MIN; a second selection of that same menu item will shift the highlighting to the HR, and so forth. This method of operation is not adequately explained in the documentation for the package.

To find out exactly what these various multiple choices actually mean, it is necessary to read the user manual, and this points up futher shortcomings in the documentation. Occasional users, who cannot remember that HR means logging will begin on the next even hour, may well want to look up that meaning in the manual. It is hard to find, buried on page 33, and the index to the user manual gives little help.

ASYST AQUISITION MODULE: AN INSTRUMENTATION LANGUAGE

ASYST is a computer language in the same sense that BASIC or FORTRAN are computer languages. It was designed with engineering and scientific problem solving in mind and has a variety of capabilities, including those in the areas of graphics and mathematics. This Aquisition module provides about 77 commands for using the computer to control aquisition of data from instrumentation devices and to manipulate the data for use by the computer.

The package works together with the ASYST System, Graphics Statistics module; both are required to support the functions of the Aquisition module, and both run on IBM/PC/XT/AT computers. The combined price of the two modules is $1,290 from Macmillan Software Co., 866 Third Ave., New York, NY 10022. (212) 702-3428.

ASYST is not a package designed to conveniently control instruments. It provides the user with no menus of available functions, nor does it have predesigned formats to organize its use. Instead, like the BASIC language, it simply presents the user with an OK prompt and waits for commands to be keyboarded. As with any other language, these commands must reside primarily in the user's head, which means that the ASYST language must be learned before it can be used.

Will it be used?

As with any other computer language, regardless of its intrinsic merits, the success or failure of ASYST will rest on the extent to which it is adopted and used by programming-minded engineers and scientists. If and when these users write programs that can, in turn, be put to use by the vast majority of engineers who have little or no interest in doing programming, then ASYST will indeed become a significant engineering tool.

ASYST appears to be a unique language, capable of rapidly implementing procedures that standard languages such as BASIC may process at a crawl. However, ASYST has its own shortcomings. It does not seem oriented toward programmers, the very people it must please if it is ever to amount to anything. For example, programs written in ASYST seem to work efficiently, but they are very difficult to understand and therefore they will be hard to debug and modify. Since a large portion of most programming projects goes into correcting and changing what has been written, it seems unlikely that ASYST will gain much popularity among those who do this work.

Customized universality

ASYST strives for universality. The aquisitions module is presented as a universal way to control an IBM/PC/XT/AT connected to instrumentation devices. What this actually involves, however, is making ASYST work with specific existing interface equipment, particularly interface boards that can be plugged into those computers.

At the time this review was written, ASYST included customized routines for using the Data Translation DT2801 board, the Techmaster Lab Master, Lab Tender, and DADIO boards, and some others as well. The Keithley DAS Series 500 System also can be handled, but its use required the creation of an additional settling-time command for ASYST. There are other ifs, ands, and buts about using specific interfaces. For example, depending on how the DADIO board is set up, the operation of ASYST input-output commands may become abnormal.

CHROMATOCHART: CONTROLLING CHROMATOGRAPHS

CHROMATOCHART collects data from up to four chromatographs, massages the data, displays stripcharts, and stores the data. It also produces printed reports on the characteristics of integrated peaks and compares unknown to standard data.

CHROMATOCHART is used with the ADALAB plug-in board (see the review of the QUICK I/O package later in this report). The package runs on Apple II+ and IIe computers and sells for $800 from Interactive Microware, PO Box 139, State College, PA 16804. (814) 238-8294.

CHROMATOCHART operates from a series of screen-displayed menus. The user makes selections by keying in the number of the desired item on the menu. As users get into these menus they will discover some unusual features. For example, the symbols $, :, or # indicate that the user must key in text, integer numbers between −32767 and +32767, or a power of two. For a package that tries to cater to user needs, this is hardly a convenience.

A more useful feature is that the user is presented with a screenful of information and choices, makes them or changes them until satisfied, then can look the screen over and enter it into active use. The keyboard arrow keys are used to maneuver around the screen while editing.

In this manner, the user can set up needed file names and identification of chromatograph runs. For data aquisition, the user is asked to provide memory address locations for the runs, the number of points to be averaged and summed, the chart delay, width, scale, and offset. Similarly, the user specifies the reports that are to be produced on another screen.

Users also can set up a schedule of control events that are to occur in the course of monitoring the inputs and operating pumps. The package can produce output signals suitable for controlling gradient pumps. Two pumps can be controlled at one time to produce desired mixtures of two fluids. Gradient profiles, created by using the package, can be used to define the mixtures. These profiles are constructed and changed with the help of another display menu.

Integration routines, menu operated as are most of the package features, measure peak area, retention time, height, width at half-height, and skew.

PC-488: INTERFACE WITH FIRMWARE

PC-488 is for use by engineers who are adept at programming. PC-488 is a circuit board that plugs into IBM/PC/XT computers. The software (firmware) is packaged within the circuitry on this board; it provides an instrumentation extension to BASIC, Pascal, C, and other languages used on these computers. The PC-488 commands implement the functions provided by the circuit board, and the instrumentation interface provided by the board conforms to the IEEE-488 standard.

The PC-488 board sells for $395 from Capital Equipment Corp., 10 Evergreen Ave., Burlington, MA 01803. (617) 273-1818.

At its simplest, the PC-488 can be used to connect a printer to the IBM/PC/XT, but it is intended as a means to hook in a whole string of instruments or other bus-connected devices through a single interface to the computer.

The programs that implement PC-488 commands are written in IBM/PC assembly language and can be used by PC-DOS system software or called from the computer language the user employs.

The commands themselves are high level in the sense that they implement instrumentation functions. Their essential function is to control the communications between the computer and other devices attached to an IEEE-488 instrumentation bus.

On an IEEE-488 bus all communications are centrally controlled. With the PC-488, an IBM/PC/XT is equipped to function as such a central controller.

Sequences of commands and data are sent using the TRANSMIT command. LISTEN and TALK designate the instrumentation device that will do the sending and those that will be receiving. SEND and ENTER establish the IBM/PC as a talker or a listener. SPOLL and PPOLL commands for serial and parallel polling of devices are attached to the bus. Other commands perform such functions as direct transfer of data between the IBM/PC memory and the bus. In all there are about 15 of these PC-488 commands.

QUICK I/O: INSTRUMENTATION INTERFACE SETUP AND COMMANDS

Quick I/O is a package for use by BASIC programmers and is used with the ADALAB plug-in interface board in Apple II+ and IIe computers. The software, together with the ADALAB board, sells for $495 from Interactive Microware, PO Box 139, State College, PA 16804. (814) 238-8294.

Quick I/O includes a convenient means of testing the ADALAB board and its time and alarm functions, but the main feature of the package is the set of instrumentation control commands it provides.

These commands can be keyboarded to immediately control devices connected to the ADALAB board. Usually, however, they will be imbedded in programs written in the Applesoft BASIC language. From a programmer's point of view, the commands are simply and uniformly structured. They are able to handle inputs or outputs (each board has a D/A output and an A/D input) and to handle single devices or several connected in parallel to a single board.

In any case, one must be a programmer to use these commands for monitoring or controlling instrumentation devices. The vendor is, however, beginning to produce packages directed to the needs of users who do not wish to write programs in order to control instrumentation with a computer. An example is the CHROMATOCHART package discussed elsewhere in this report.

RTFSA: REAL TIME FREQUENCY SPECTRUM ANALYZER

With the help of a plug-in interface board, RTFSA allows a personal computer to be used as a spectrum analyzer with 10-KHz bandwidth. The package accepts filtered or unfiltered data, performs fast Fourier transform analysis, and plots the magnitude of frequency spectra.

RTFSA runs on Apple II+ computers. It makes use of the AI13 interface board from Interactive Structures (see discussion of the AMPRIS package). Available as an option ($500) is a plug-in circuit card containing an anti-aliasing filter, a frequency counter–tachometer, and circuit-based multiplication to speed up the operation of the package. RTFSA sells for $400 (there are educational discounts) from Real Time Microsystems, PO Box 3081, Troy, NY 12181. (518) 274-7149.

Most of the functions this package performs are displayed on a 15-choice menu that appears when RTFSA is started. Through this menu, the user specifies from which circuit board slot, and which one of 16 A/D channels in that slot, input will be taken. The user also sets up the analysis by specifying other items on the menu such as A/D (voltage) range, frequency range, calibration factor, and threshold level at which data sampling of input signals will begin.

The same menu is used to shape the output plot, determining whether it will consist of discrete dots or continuous traces, what label the plot will bear, and the speed at which a synchronous marker on the frequency axis will be displayed.

The analysis itself is started when the user selects the CURRENT PARAMETERS OK menu item. With this selection, the package automatically samples and transforms input data and produces an output plot.

If the incoming signal falls out of the user-selected voltage range, the package will display an OUT OF RANGE message. The plotted data are automatically scaled to provide maximum resolution.

Normally, the display will continue to reflect changing input data, but, while in the plotting mode, the user can exercise control choices, including freezing the current display, producing a printed copy of the current display, and printing or storing on disk the currently displayed spectral data.

SPECSYSTEM: AUDIO SPECTRUM ANALYZER SOFTWARE

SPECSYSTEM, and the software (firmware) that comes with the accompanying plug-in board, control the operation of an Audio Spectrum analyzer and display its output data. SPECSYSTEM can be used by nonprogrammers.

SPECSYSTEM is for use with Apple II and II+ computers together with the APX252 Audio Spectrum Analyzer plug-in board ($595). The SPECSYSTEM package sells for $199 from Eventide, One Alsan Way, Little Ferry, NJ 07643. (201) 641-1200.

SPECSYSTEM and the spectrum analyzer with which it works are useful for a variety of acoustical applications, including analysis of noise, engine and machinery diagnostics, music, sound, and voice analysis.

Firmware on the board

Included on the APX252 board are software (firmware) commands that can be called by the BASIC language or invoked directly from the computer keyboard. Some of these commands perform immediately-useful analysis and display functions when keyboarded. This type of use requires little or no knowledge of programming. In addition, for those who are programmers, these and other commands can be called from BASIC programs.

The RTA command displays a bar-chart representation of the frequency components produced by the spectrum analyzer from the input signals. Other commands exert more detailed control; thus SCAN performs the spectrum analysis, BARS produces the bar-chart display, AVERAGE causes the SCAN function to use averaged data. Other commands control the frequency range of the display, the amplification of input signals, and other interface and display functions.

Surface displays

The SPECSYSTEM package uses the repertoire of APX252 commands to perform more complex operations. Its basic function is to assemble and store large arrays of spectral data. These data arrays are displayed in the form of geometrical surfaces with dimensions of time, frequency, and amplitude. Alternatively, curves representing the data for individual spectrum channels can be displayed.

SPECSYSTEM leads the user through menus that guide creation of storage areas (buffers) for the incoming spectral data, set up the parameters of the plots, create reverberation displays, control transfers of data from screen to computer memory and to disk storage, and allow convenient use of the computer's other display functions. All this can be managed by users who have no knowledge of programming.

TEMPERATURE PLOTTER: FOR STUDENTS

TEMPERATURE PLOTTER monitors up to four temperatures, storing and plotting the results.

TEMPERATURE PLOTTER runs on the Apple II+, IIe, and IIc. The package sells for $39.95. Temperature probes for use with the package sell two for $35 or four for $50. The vendor is Vernier Software, 2920 SW. 89th St., Portland, OR 97225. (503) 297-5317.

This package was designed for use with classroom experiments but may prove useful in some engineering instrumentation environments. Operation is simple and straightforward. From a main menu, the user selects the desired function, then responds to a series of screen-displayed questions and requests to implement that function.

For example, to monitor temperatures and save the results in memory, the user selects the Monitor Temperatures menu item, then types Y (yes) to a storage-wanted question, and finally types the desired length of time between recorded readings and the total number of readings desired.

The main limitations to the package are that each reading takes a minimum of one second, only four probes can be monitored, and a total of only 512 readings can be stored. There also are limitations to the accuracy of the temperature transducers and the calibration procedures.

The Analog Devices AD590 temperature transducer, in the probes for which this package is designed, can be used over the range of −55 to +150°C. The devices are available with maximum error of 0.2°C in the range of 25–50°C. Units supplied by Vernier Software typically will have maximum errors twice as large as this, therefore, users who have higher accuracy needs will have to purchase their own transducers.

Calibration is done by fitting the endpoints of a straight line to data from the transducer. This becomes more accurate as the temperature difference from one end of the line to the other is decreased.

Results can be displayed in Celsius, Kelvin, or Fahrenheit units. The package starts with the assumption that Celsius units are wanted. The user selects the desired units, and the package automatically makes the necessary conversions. However, the package always displays the currently selected units, so data retrieved from storage may be incorrectly labeled unless the user checks the units in which those data were originally stored and makes the appropriate adjustment.

The graphs produced by the package are simple point plots, with a different character used for the points of each of the four possible temperature plots.

Chapter 19
Quality Control Packages

A personal computer may be an ideal instrument for creating and presenting quality control charts and that is the function for which the Q.C.PLOT package was designed. Q.C.PLOT handles variable and attribute control charts and distribution charts. It also performs a number of mathematical transformations on data sets and can display or print critical parameters in tabular format.

Like Q.C.PLOT, the QualityAlert package also performs transformations and produces charts and tables. In addition it includes several statistical tests of normality.

The NWA QUALITY ANALYST package aims at being a comprehensive tool for statistical quality control analysis and charting. Several different types of control charts can be produced, transformations can be performed, and an impressive variety of statistical routines are available.

Descriptive statistics include the mean, standard deviation, variance, standard error, and 12 other statistical measures. The package also can produce an analysis-of-variance table for a one-way design, with either equal or unequal numbers of subclass items. There also is an analysis-of-variance table for a design that involves cross-classification of data by blocks or subjects and treatments. Linear correlation between sets of data can be carried out to produce a correlation matrix. Several types of discrete distribution functions can be calculated, including binomial, negative binomial, hypergeometric, and Poisson distributions.

Least-squares coefficients of single-variable regressions are computed. Linear, exponential, and logarithmic function types can be handled. Means tests include the correlation coefficient and the Pearson product-moment correlation as well as ordinary, single-sample, and paired-sample t-statistics.

Altogether, NWA QUALITY ANALYST comes with 35 different routines recorded on as many as seven different floppy disks. Since particular users may find that specific portions of the package work best for them, the vendor suggests that they collect groups of such routines and keep each group on a separate disk. In operation, users create data files, then call on these files to create quality control charts or to perform statistical tests on the data.

Menus and commands

Operation of the QualityAlert package is controlled by choices from menu displays. For example, main menu choices include analyses of variable data, percent defectives, and counts defects and Pareto analysis also can be chosen. Each choice leads the user to other menus where there is a further choice of the type of data display or chart. There is a helpful uniformity in that every main menu choice that involves data analysis leads the user to the same data entry and editing procedures.

All Q.C.PLOT functions likewise operate from displayed menu choices, but the operation of the NWA QUALITY ANALYST is somewhat different. Although the various routines can be accessed through menu choices, users must know, or find in the manual, the detailed commands and formats needed to actually operate these routines.

When NWA QUALITY ANALYST is started up, experienced users can go directly to the routine they want to use first by keyboarding its correct name. Users must know or find in the manual, the detailed commands and formats needed to actually operate these routines. If the user cannot remember which commands are available, pressing the ? key will cause a command list to be displayed.

Manipulating the data

The ability of a quality control package to efficiently produce desired charts and data listings depends, in part, on the means provided to users for manipulating their data.

The Q.C.PLOT package expects that there will be comments on the data, and it therefore includes a procedure to assign descriptive parameter codes to text entries. There is a procedure for assigning such codes to footnote text entries as well.

Q.C.PLOT provides for a variety of mathematical data transformations that can be applied to every data item in a file. These include arithmetic, exponential and logarithm, ranking and ordering. The only trigonometric transformation is arcsin. In addition, arithmetic can be performed between specified fields. For example, five fields can be added to produce a summed field. In this way fields, or entire files, can be added, subtracted, multiplied, or divided, one by another, or by constants.

A search and select capability in Q.C.PLOT allows specified records to be located, extracted from a file, and placed into a new file. The search can be specified by an exact matching-value, or by a range of matching-values. Records can be selectively deleted from a file with similar matching procedures.

Similar manipulation capabilities are offered by the NWA QUALITY ANALYST. There is an Extract routine that allows users to single out defined subsets of a stored data file. Selection is by column and can specify a number range or can be based on descriptive parameters, such as shift, time, or operator.

Data transformations also are performed by this package. These can be on selected columns or between the columns. Available are arithmetic, trigonometric, exponential, logarithmic, and absolute-value transformations.

The Merge routine allows rows and columns to be inserted. deleted, or extracted. Msdat handles missing data. Msort does sorting on columns containing alphabetic, numeric, time, or date information. Multiple-key ascending or descending sorts can be run. Singcol converts multiple-column files to single-column format for statistical analysis. Xor is for logical (exclusive OR) file merging.

QualityAlert provides just five mathematical data transformations, including square root and natural logarithm. Editing procedures in this package allow users to remove individual subgroups, or sets of adjacent subgroups, from a stored file. New subgroups also can be added, and data items within a subgroup can be changed.

Charts and plots

Control charts of individual values or of moving averages and moving ranges can be created with the QualityAlert package. Percent defectives analysis can produce histograms and control charts for counts of defectives found in constant-size samples. Control charts can be produced, as well, for counts of defectives in variable-size samples. QualityAlert also does count-of-defects analysis for constant- or variable-size samples. Histograms and control charts can be produced for the constant-size samples, control charts only for variable-size samples. Cumulative-sum control charts can be produced for variable-type data, and Pareto histograms, displaying the causes of defects and the overall scrap loss due to each, also can be produced by this package.

With Q.C.PLOT variable control charts, users can elect to display mean and range, or mean and standard deviation, individual values, subgroup sums, or trends. Available attribute control charts include P-charts, %P-charts, NP-charts, C-charts and U-charts. Distribution bar charts (or polygon charts) produced by this package can have probability curves as overlays. Either frequency or proportion charts can be produced. The choices of probability curves includes hypergeometric, binomial, Poisson, normal, and exponential.

NWA QUALITY ANALYST allows users to set up variable control charts for individual measurements, for subgroup ranges, and for averages. There also are sigma charts that display standard deviations. Attribute control chart routines create charts for nonconformities or for nonconformities per unit. There also are charts that show number defective or fraction defective. This package allows simple scatterplots to be created and provides automatic scaling for these plots. Frequency histograms can be displayed in numeric or graphical form.

Using foreign files

NWA QUALITY ANALYST files are kept in a format very similar to the widely used ASCII standard, allowing files from other packages, such as word processors, to be imported and used here with only minor editing. To allow use of popular spreadsheet, data base, and integrated package formats, there are routines to do conversions to and from the DIF and SYLK file exchange formats. In a similar manner, QualityAlert files can be transferred to and from other packages that use the DIF format.

Tabular data entry

Users must enter a considerable amount of data before these packages can do their work. To ease this process, all three packages provide screen-displayed tabular formats for keyboarded data.

Q.C.PLOT handles input data one row at a time. To set up Q.C.PLOT input the user must specify its structure in some detail, giving the number of items per row of data and the number of characters per item. The package then displays a column of item numbers and one of item names. The actual items for the first row of data are then entered into a third column. The user moves from item to item by pressing the Return key, corrects the data if needed, and then goes on to a similar entry display for the next row of data.

In QualityAlert, data items are handled in subgroups, with the subgroup size specified by the user. Items are entered one to a line, and the computer emits a beep when the last item in the subgroup is completed. If mistakes are made, they can be corrected by using the arrow keys and typing correct characters over those that are incorrect.

The NWA QUALITY ANALYST assumes that all data are in row and column format. The user can key in the table of data one row at a time, separating the column items with spaces. At the end of each row the Return key is pressed, which moves the display cursor to the beginning of the next row. Mistakes are corrected by backspacing and typing over items. The first column entered is expected to contain a number label for the each row. Other columns can contain descriptive entries made up of alphabetic characters.

This package also offers an alternative, spreadsheet style of data entry. Here a full-screen row and column format is set up by the package and the user can move from one cell to another to enter the data. The entry sequence can be along the rows or the columns, and the arrow keys also can be used for full freedom of movement.

Laboratory quality control

LABsolv-QC is a more specialized package designed to help organize the activities of a testing lab and to report on laboratory test results. In addition to keeping files containing sample data and results, LABsolve-QC can be used to create and maintain files on lab customers, product and customer specifications, product information, tests performed, lab personnel, and manufacturing operations sequences.

Control over package options is exercised through menu displays. Users maneuver over these and over data entry displays by means of the keyboard arrow keys.

Each result entry is stored as a record in the package database. The package assembles results data items and sample information to produce various reports that can be called up by the user. For example, reports are available listing samples by log-in or log-out date or by test-class; there are reports on samples outstanding and results totals plus about a dozen other reports on the various files kept with the package.

LABSOLV-QC

LABsolv-QC is designed to help organize the activities of a testing lab and and to analyze laboratory test results.

LABsolv-QC runs on IBM/PC/XT computers. It sells for $2,495, including a run-time version of the Informix data base. The vendor is McGrow Pridgeon Computer Services, 1 Investment Place, Baltimore, MD 21204. (800) 445-4334, (301) 821-8300.

LABsolv-QC keeps track of a variety of information on test lab operations. In addition to files containing sample data and results, it is designed to create and maintain files on lab customers, product and customer specifications, product information, tests performed, lab personnel, and manufacturing operations sequences.

Control over package options is exercised through menu displays. Users maneuver over these and other screen-entry displays by means of the keyboard arrow keys. A Sample Logging menu offers an Add/Entry/Query option that presents a screen-displayed Sample Entry Form. About 15 general description items are filled in on this screen to open a sample file.

Unitary results items

Results data items that will be used in the sample file are entered, one at a time, through a separate result-logging procedure. On a Sample Result Entry form, users identify the entry date, sample, customer, specification, product, test technician, test involved, test limits, and assign a result ID.

Only a single result value and percent deviation can be entered on this form. For each additional value in a sequence of tests, the user can call up a copy of the previously entered form and edit it with a new result ID, result value, and percent deviation.

Each result entry is stored as a record in the package database. The package assembles results data items and sample information to produce various reports that can be called up by the user. For example, reports are available listing samples by log-in or log-out date or by test-class; there are reports on samples outstanding and results totals plus about a dozen other reports on the various files kept with the package. There seems to be no facility for users to set up their own report formats.

NWA QUALITY ANALYST

NWA QUALITY ANALYST aims at being a comprehensive tool for statistical quality control analysis and charting. It has a convenient spreadsheet-style data entry procedure and flexible file manipulation facilities. Several different types of control charts can be produced, and an impressive variety of statistical routines are available.

NWA QUALITY ANALYST runs on IBM/PC and other computers that use languages compatible with MicroSoft interpreter BASIC. The package sells for $495 from Northwest Analytical, 520 NW. Davis, Portland, OR 97209. (503) 224-7727.

NWA QUALITY ANALYST comes with 35 different routines recorded on as many as seven different floppy disks, from which the user will want to make safe copies. Since this is a tool kit, and different users will make differing uses of it, the vendor suggests that users collect groups of routines that they may use together and keep each group on a separate disk.

Users create data files, then call on these files to create quality control charts or to perform statistical tests on the data.

Ready access to the routines

When the package is started up, experienced users can go directly to the routine they want to use first by keyboarding its correct name. However, others, and almost anyone who uses the package only occasionally, will probably want to get at the package capabilities through the screen-displayed menus.

Available through these menus are nine routines for file creation and manipulation, four routines for file conversion, five mathematical utiltity routines, eight for creating control charts, and nine statistical analysis routines.

Also included is a routine that sets up each keyboard Function key to call in another of the packages routines without going through the menus. Users can select which key will call which routine, thus providing a rapid way to navigate through the package.

First menus, then commands

Although the various routines can be accessed through menu choices, users must know, or find in the manual, the detailed commands and formats needed to actually operate these routines.

The Edit routine is used to enter data into a file or to modify already-stored data. The first thing this routine does is to display a request for a file name. If the user enters a name that is not already in use, the package will not respond but will expect lines of data to be keyed in.

The package assumes that all data are in tabular, row and column format, and they must be arranged in this way before the user keys them in. Column headings are ignored. The user keys in the table of data one row at a time, separating the columns with a space. At the end of the first row the Return key is pressed, which sends the display cursor to the beginning of the next row. During this entry procedure, mistakes can be corrected only by moving backward, a space at a time, with the help of the Backspace key, and then typing over the incorrect data.

The first column entered is expected to contain a number label for each row. Other columns can contain descriptive entries made up of alphabetic characters; these can be later set aside when numerical processing of the data takes place. When the user enters a blank line, the package displays the message "END OF FILE" and waits for a command from the user to store or dicard the data.

To edit an existing data file, the user enters its name at the beginning of the Edit routine. The file row and column contents are then immediately displayed, and the Edit commands (there are about 18 of them) can be used to move about through the data and insert, delete, and change them as desired.

One limitation of the display is that, if the rows are longer than 80 characters, they will wrap around with the leftover tail of the row displayed at the beginning of the neat line. This destroys the users ability to scan down the displayed columns. It can be avoided, but only by breaking the data file into sections,which will have to be reassembled before they are used for calculation or graphical display.

If the user cannot remember what to do, giving the ? command will cause a command list for the Edit routine to be displayed. Another helpful feature of this ? display is a count of the total number of rows in the data file (for long files, users must page from one screen to another to reach the end) and the amount of computer memory currently available for storage of the file that is being worked on.

Full-screen entry alternative

Those who prefer a spreadsheet style of data entry can make use of the Sedit routine as an alternative to Edit. Here a full-screen row and column format is set up by the package and the user can move from one cell to another to enter the data. The entry sequence can be along the rows or the columns, and the arrow keys can be used for full freedom of movement.

This spreadsheet mode also can be used for data file editing, providing greater freedom in this process than that available through the Edit routine. As with Edit, use of the ? command will display the various other keyboard commands available with this routine.

File manipulation

Operation of the other package routines follows similar command-initiated procedures. The Extract routine allows users to single out defined subsets of a stored data file. Selection is by column and can specify a number range or can be based on descriptive parameters, such as shift, time, or operator. Data transformations also are performed by this module. These can be on selected columns or between the columns. Available are arithmetic, trigonometric, exponential, logarithmic, and absolute-value transformations.

The Merge routine allows rows and columns to be inserted, deleted, or extracted. Msdat handles missing data. Msort does sorting on columns containing alphabetic, numeric, time, or date information. Multiple-key ascending or descending sorts can be run. Singcol converts multiple-column files to single-column format for statistical analysis. Xor is for logical (exclusive OR) file merging.

The Report routine prints all or selected columns of data files in a format specified by the user. Report and column headings can be set up in a flexible manner.

Foreign files

Files in this package are kept in a format very similar to the widely used standard ASCII format, allowing files from other packages, such as word processors, to be imported and used here with only minor editing. The Cleanup routine converts ASCII files to NWA QUALITY ANALYST format and also does the reverse conversion. Flip transposes data rows and columns. To allow use of popular spreadsheet, database, and integrated package formats, Difx does conversions to and from the DIF format, and Sylkx does conversions to and from the SYLK format.

Control chart routines

There are four routines for setting up variable control charts. Ichart creates charts for individual measurements, Range creates control charts for subgroup ranges, Schart is for sigma charts that display standard deviations, and Xchart creates control charts for averages.

Attribute control chart routines include Cchart, which creates a control chart for nonconformities, Npchart for control charts that show number defective, Pchart for fraction-defective control charts, and Uchart, which creates control charts for nonconformities per unit.

Calculation of standard deviation and range for data files with subgroups containing only one data item can be done with the help of the Movave routine. This routine gets its name from the fact that it also calculates moving averages. The Reduce routine makes similar calculations for multi-item subgroups.

Simple scatterplots can be created with the Graph routine. If automatic scaling of a scatterplot is desired, the Qgraph routine can be used.

A variety of statistics

The package provides an impressive variety of statistical routines. ANOVA calculates an analysis-of-variance table for a one-way design and can handle either equal or unequal numbers of subclass items. ANOVAR produces an analysis-of-variance table for a design that involves cross-classification of data by blocks, or subjects, and treatments, in other words, a repeated-measures or subjects-by-treatments design.

A linear correlation between sets of data is carried out by the Porvar routine, which produces a correlation matrix. Several types of discrete distribution functions can be calculated by the Dist routine, including binomial, negative binomial, hypergeometric, and Poisson distributions.

The Hgram routine does frequency histograms. These can be displayed in numeric or graphical form. Frequency counts for data sets are calculated and displayed by the Pareto routine.

Least-squares coefficients of single-variable regressions are computed by the Regress routine. Linear, exponential, and logarithmic function types can be handled.

Descriptive statistics are produced by the Sumstat routine. Included here are the mean, standard deviation, variance, standard error, and 12 other statistical measures.

Means tests calculated by the Ttests routine include the correlation coefficient and the Pearson product-moment correlation as well as ordinary, single-sample, and paired-sample t-statistics.

Well-designed manual

The extensive features of this package call for a manual with adequate explanation of operation details, and the vendors of this package have produced just such a manual. Operation of each module is explained with clear text and reproductions of the displays users will see on their computer screens. A detailed table of contents and an index guide users to the material they need.

QUALITYALERT

QualityAlert is designed to analyze variable and attribute data for statistical quality control.

QualityAlert runs on IBM/PC/XT computers and sells for $795 from Penton Software, 420 Lexington Ave., #2846, New York, NY 10017. (212) 878-9600.

Operation of this package is controlled by choices from menu displays. Main menu choices include analyses of variable data, percent defectives, and counts defects and Pareto analysis also can be chosen. Each of these choices leads the user to other menus where there is a further choice of the type of data display or chart that is desired.

Data are stored in disk files named by the user and called into play when an analysis is desired. Files can be transferred to and from other packages that use the DIF format employed by popular spreadsheets. Every main menu choice that involves data analysis leads the user to the same data entry and editing procedures.

Data in subgroups

Data entry is set up by responding to a series of screen-displayed queries. For example, data items are stored in subgroups (samples) so the user must respond to a query on the desired subgroup size (1–20 items per subgroup are allowed). Another query asks about provision for comments with each data subgroup.

With such information, the package sets up the screen for entry of the actual data items. The user enters one item, then presses the Return key, which moves the display cursor to the next line. This entry sequence is repeated for each item in the first subgroup. The computer emits a beep when the last item in the subgroup is completed. The user then goes on to enter the remaining data subgroups (up to 250 such groups are allowed in each data file).

If mistakes are made while entering a data item, they can be corrected by using the left arrow key to move the cursor back to the error, then typing the correct characters over those that are incorrect. Screen-displayed data also can be changed, before being stored, by pressing the Esc key to enter an editing mode, moving to the appropriate item with the help of the arrow keys, and then doing further keyboard editing. To abort a data entry procedure, or other package procedures, users can press the M key in response to any prompt and be returned to the previous menu.

Flexible file editing

When all subgroups have been keyboarded and the user feels the data are correct, pressing the Ctrl and X keys simultaneously will cause them to be stored. However, this storage is only in temporary memory. The user is queried about moving the data to disk storage and can do so by supplying the file name under which it is to be kept. Material stored in this manner can be be placed in a new file or added to the beginning or the end of an existing file.

Editing procedures allow users to remove individual subgroups, or sets of adjacent subgroups, from a stored file. New subgroups also can be added and data items within a subgroup can be changed.

File retrieval procedures are slowed by compulsory queries designed to help the user to avoid inadvertantly calling for the wrong file. Editing procedures also involve some cumbersome procedures. To remove subgroups from a file, or make changes in a specific subgroup, the user has to page through the file from the beginning. Only four subgroups are displayed at a time. Changes are made by responding to queries such as "Want to change any of these subgroups?" This can be a lengthy process when a user wants to get at the middle of a file containing a large number of subgroups. It would be an improvement if users could specify the desired subgroup by number, and the package would then automatically search out and display the group of four that contains that subgroup.

Normality tests and transformations

Five data transformations, including square root and natural logarithm, are provided to allow the user to try to fit stored data to a normal distribution. Original and transformed data also can be tested for normality. The tests performed are Geary's Z, the Anderson–Darling test, and there are also tests for skewness and kurtosis.

A variety of charts

To plot control charts of individual values or of moving averages and moving ranges, variable data with just one value in each subgroup must be used. The user selects the desired type of control chart from a displayed menu, then responds to a sequence of displayed queries. Thus, for an individual-values control chart, queries on the number of values to be used per moving average (1–20 are allowed) and the desired control limits have to be answered.

The package provides for percent defectives analysis, with histograms and control charts produced for counts of defectives found in constant size samples. Control charts can be produced, as well, for counts of defectives in variable size samples.

Count-of-defects analysis can be performed for constant- or variable-size samples. Again, histograms and control charts are produced for the constant-size samples, only control charts for variable-size samples.

Pareto histograms, displaying the causes of defects and the overall scrap loss due to each, also can be produced by this package. Cumulative-sum control charts can be produced for variable-type data.

In each case, the desired charts are displayed on the computer screen, and a printed copy of what is shown there can be produced by responding appropriately to the final query in each sequence.

Q.C.PLOT

Designed to help create charts used in statistical quality control work, Q.C.PLOT tabulates critical parameters as well as handling variable and attribute control charts and distribution charts.

Q.C.PLOT runs on Apple II, IIe, II+, and IIc computers and sells for $400 from Human Systems Dynamics, 9010 Reseda Blvd., #222, Northridge, CA 91324. (818) 993-8536.

Before data can be charted they have to be entered into the computer. Data files are created by working, as with all Q.C.PLOT functions, from displayed menu choices. The Random Access Files menu allows users to set up the files, enter data, merge files, make subsets, and print or delete the files.

To set up a file, the user must specify its structure in detail, giving the number of fields per record (up to 21 fields per record) and a name and the number of characters for each field. In other words, users must have not only the data but the format of the columns and rows of data they wish to enter before they sit down at the computer keyboard. Each row of data will become a computer-stored record; each column item a field entry.

When specifying file structure, the user first enters a name for the file following the screen display FILE NAME: and then fills in a number following FIELDS PER RECORD:. Pressing the Return key moves the display cursor from one entry to another. The package then presents the user with a three-column display, with columns for field number, name, and width (number of characters). The number of fields shown here correspond to the fields-per-record entered by the user. Field names and widths are keyed in as the user moves from one entry to another by pressing the Return key.

Row-by-row entry

When the data entry menu selection is made, the package displays columns of field numbers and names for the first record (first row of data). There is a third column on this display where the user fills in the data item for each field, moving from item to item by pressing the Return key. Essentially, what the user does is enters a row of data into a column of this kind of display.

Below the display is an Edit (Y,N) query. The user looks over the entered data and keys in Y if any corrections are needed. If N is the response, the data are accepted as correct. A similar entry display is presented for each row of data, and, when all rows are completed, the package stores the data on a disk.

The package expects that there will be comments on the data, and there is a procedure to assign codes to comment text entries. There also is a procedure for assigning codes to footnote text entries.

Already-stored files can be reviewed and edited by a procedure in which displays like those for data entry are shown, and the user selects which field to edit or delete.

Data transformations and selections

The package provides for a variety of data transformations, which can be applied to every data item in a file. These include arithmetic, exponential and logarithm, and ranking and ordering. The only trigonometric transformation is arcsin. In addition, arithmetic can be performed between specified fields. For example, five fields can be added to produce a summed field. In this way fields, or entire files, can be added, subtracted, multiplied, or divided, one by another or by constants.

A search and select capability allows specified records to be located, extracted from a file, and placed into a new file. The search can be specified by an exact matching-value or by a range of matching-values. Records can be selectively deleted from a file with similar matching procedures.

Many graph options

For variable control charts, users set the central value and control limits. This can be done by direct keyboard input of the desired values, or the values can be calculated from the data using three times the standard deviation for the limits or using ASTM factors. If users wish, central value and limit lines will be displayed on the control charts. Users can elect to mean and range, or mean and standard deviation, individual values, subgroup sums, or trends.

Available attribute control charts include P-charts, %P-charts, NP-charts, C-charts, and U-charts. As with variable control charts, central values and control limits can be selected or calculated.

Distribution bar charts (or polygon charts) can have probability curves as overlays. Either frequency or proportion charts can be produced. The choices of probability curves includes hypergeometric, binomial, Poisson, normal, and exponential. Data can be taken from stored files or entered from the keyboard. Frequency distribution tables and a descriptive statistics report also can be produced.

The charts can be displayed or printed. They can be saved on disk and retrieved for future use or for modification. One data file can be substituted for another for a specified chart. A summary of chart specifications can be displayed or printed. Appropriate parameters for each type of chart can be displayed or printed.

Chapter 20
Project Management Packages

Close to one hundred project management packages for personal computers have become available during the past few years. With such riches before us, it is possible to define some desirable characteristics for any particular use and then search for the right package with some confidence that it exists.

One of the first characteristics to look at is the number of tasks or activities a package can include in a given project. Those packages that handle fewer than about twenty tasks can be safely ignored. It would be simpler to plan such little projects with paper and pencil than to use a computer.

At the other extreme, it may be best to handle very large projects with thousands of tasks on large, mainframe computers. Among the personal computer packages surveyed for this chapter, those that promise to handle the largest number of activities are Plantrac (65,000) and Primavera Project Planner (10,000). These packages and others such as PAC MICRO and PMS II can be used on a personal computer and also can communicate with project management software housed on larger computers.

As we move from packages that handle few activities to those that handle many, we might expect that prices would steadily increase. Such is not the case. The Pathfinder package from Morgan Computing, for example, handles up to 3,160 activities and costs just $80, whereas many other packages are priced at a dollar or more an activity. However, the Morgan package produces simple Gantt charts; these show the start and end of activities but no interrelationships between them; and that is about all it does. If users want more, they will have to turn to other packages.

Some users may want to see PERT charts or other network charts that illustrate the dependencies of one activity on another, or they may want the package to produce information on factors such as cost and manpower. Frequently users want packages that will do progress tracking as well as project planning.

Look for helpful data entry

In terms of actual use of a project management package, users will spend most of their time simply entering data to describe the project tasks or activities. For that reason, the data entry facilities offered by these packages are an important consideration. A package may have wonderful calculation, reporting, and display capabilities, but if the process of data entry is too arduous, that package may see little actual use.

Some very helpful data entry techniques have been developed by software vendors who work on business packages for personal computers. A few of the available project management packages have taken full advantage of these techniques.

For example, the Project package, developed by a leading personal computer software company, has users enter data onto an Activity screen that shows a Gantt chart. An entry window is used to fill in the data for each activity. Entry of items is accomplished by pressing the Tab key to move from one item to another in the entry window. As data for each task are completed, the Gantt chart is automatically updated, so the user sees the developing effect on the overall project of adding each activity. A very similar activity entry method is used by the Timeline package.

An aspect of data entry that is ignored by many of these packages is the use of internal checks to verify the validity and accuracy of what is keyboarded. Mainframe data processing software has traditionally paid close attention to such input controls, since data for large computers are often entered by clerks who sit all day at the keyboard typing material that may be meaningless to them.

Usually, personal computer project management packages will be operated by the professionals who want to use the results these packages produce. Nevertheless, internal computer checks on such things as consistent use of numerals and alphbetic letters and on duplication of activity descriptions can be very helpful when there is a lot of information to input. The Plantrac package is one that uses some traditional mainframe techniques to try to keep the input accurate. In this package there are well-designed entry screen displays with internal checks on consistency and duplication of tasks. In addition, as each activity is entered, the package automatically prints out the data and stores them on disk.

In most of the project management packages discussed in this chapter, little attention seems to have been given to the plight of the user faced with inputting masses of activity data. Thus Gantt-Pack invites users to input data for as many as five hundred activities but offers a most inconvenient format. Data entry is onto a crowded screen that becomes closely packed with entries. It is arranged in rows and columns, and only the tops of the columns are labeled, so it is often hard to see which item is on which line. The detail formats required for items to be entered are not very easy to learn, and there is no audible warning when the package rejects incorrect entries. In surprising contrast, however, the data editing facilities for this package are both convenient and rapid in operation.

Another example of poorly arranged input methods is the Demi-Plan package, which prompts for the needed data. But following the entry of each activity description, the user is returned to the main menu, where the data entry option must be chosen again if another task is to be entered. Data entry itself is an unforgiving process. Users enter a six-item task description. If there are any mistakes in the entry, they have to start all over again.

Adding and deleting should be simple

It is common wisdom that plans are made only to be changed, and project management plans are no exception. Even the most perceptive and careful planner will have to make many changes to get a complex project schedule to conform to the realities it must encounter. An important feature of these packages, therefore is, their ability to easily accommodate addition, deletion,and moving of activities.

With the Timeline package, changes are easily made. The cursor is moved to the desired chart location and a Function key is pressed, bringing up the entry window for that location. Any desired changes are made through that window, and the results are immediately reflected in the chart.

The Harvard Project Manager offers an exemplary method for removing activities. On the PERT chart for this package, milestones are shown as boxes and activities as boxes in reverse video. To delete a box, the user simply moves the cursor to it and presses a couple of keys. It disappears and the chart is redrawn. When new activities are added, the package automatically adjusts the project plan.

Not all packages provide this kind of automatic adjustment. Some require completely manual readjustment with all additions and deletions; others are semiautomatic. In Pertmaster, when activities are added, the user gives the relationships to existing activities, and these are incorporated into the project plan. Activities also can be deleted, but the user must then manually rearrange the relationships among the remaining activities. The reverse is true for the PRO-JECT 6 package. When an activity is deleted by the user, the package automatically removes any references to it that may appear in other activities. But when users add an activity, they must edit the other activities to include such references.

Calendars and other constraints

The real world places certain limits on any project plan. These constraints commonly have to do with time, people (called "resources" in the parlance of project management), and money.

The Timeline package has a particularly flexible approach to relating the time sequencing of one activity with another. Activities can be marked with a specific start or end date, or ASAP (as soon as possible), or ALAP (as late as possible), or they can be designated as Span activities. The Span activities are flexible in time, able to shrink or expand to accommodate the other activities. Users can decide which tasks to sequence or join to one another. Partial joining (starting the next task when the preceding one is only partly completed) also is allowed.

The days of a project can become difficult to describe, for work often will not be done on weekends and holidays, nor on vacations, and all of these may differ from one project to another, one year to another.

Many project management packages therefore incorporate calendars. For example, the Garland Publishing Pathfinder package uses a calendar file in which each day is represented by a code. The built-in calendar runs from 1975–1990; to handle other years the user must use a text editor to build a new calendar.

Some packages provide for multiple calendars. Thus, the Pertmaster package allows users to set up any number of calendars with different workday, holiday, and shift schedules, then select which calendar will be used with any given project. The package provides convenient calendar editing features.

Resource leveling is hard to find

Some project management packages take no account of people and other limited resources that may or may not be available to carry out project activities, leaving it to the planner to incorporate such considerations in activity durations and sequencing, but most of the packages track such resources in one way or another.

Ideally, a project management package should be able not only to indicate when limited resources are close to being exhausted but to redistribute the resources to achieve *best* results. Very few of the packages described in this chapter even attempt such redistribution, called *resource leveling*. One that does is the Primavera Project Planner, which averages out the availability of up to six vital resources for each task. When the package finds that the available resources are not sufficient for all active tasks, it will try to reassign resources automatically without compromising the critical path or the project completion date.

A simpler type of resource leveling is performed by the Project Management package from the Institute of Industrial Engineers. The package includes a resource allocation module that takes both activity time and available resources into account in calculating the start and finish time for each activity, so that a limited resource is efficiently used. Up to two hundred activities can be handled, but only one limited resource. The module does not yield optimum results, but the solution will improve with additional iterations.

Costs and cost tracking

Not all project management packages concern themselves with costs, but most make monetary considerations an integral part of package operations

In the MicroGANTT package, resources are always related to tasks that are billable or have billable components. The user can assign hourly and fixed costs, along with availability and time attributes, to each resource. An hourly billing rate can be assigned to an entire project, and this rate can be overridden for specific tasks. This information allows the program to calculate the resource requirements by time period for each task. The PAC MICRO package has an optional cost module used to track budgeted versus actual costs, resources, and durations. Very useful bar chart displays are used to compare the actual and budgeted amounts.

Project Scheduler 5000 reports include a list of resources and associated costs. Users can graph resource costs with or without labor costs. When actual costs are input to the package, users can graph planned versus actual cost data, and they also can graph cost comparisons of planned, revised-planned, and actual costs to date.

What-if analysis

By changing the relationships between project activities and by adjusting constraints of time, cost, and resources, planner scan produce scenarios of possible outcomes for their projects. Some packages make this kind of "what-if" analysis particularly easy by providing for the kind of easy addition, deletion, and changing of activities that has already been discussed.

Plantrac is specifically oriented to this what-if analysis. It has very flexible capabilities for modifying the planning parameters and observing the results. Constraints of time, personnel, and cost can be readily imposed and removed to examine their effects on the overall project.

Flexible reports

For many users, the printed reports produced by project management packages are very significant, for such reports can be a means of communication with those who will carry out the projects and with decision makers who may have ultimate control over planning decisions.

Usually the packages turn out only a few reports, and these are fixed in format. However, some, such as PRO-JECT 6, have a more flexible reporting facility. With this package, users can choose to select or sort many arrangements of data for a report. Formatted report definitions can be set up and reused as needed. Documents created by word processing packages can be incorporated into the reports.

The MicroGANTT package has a feature called Select that will scan a list of projects and extract the tasks that match codes previously assigned by the users. This allows reports to be organized by employee, machine, job number, client, or any category the user chooses.

For those who deal with government agencies, it may be essential that the reports they produce conform to the specifications of those agencies. Two of the packages discussed in this chapter have been designed with this in mind. The PMS II package produces reports that meet ER–1–1–11 and DOD 7000–2 specifications. Primavera Project Planner also produces reports that meet the contracting requirements of the Department of Defense and other government agencies.

THE CONFIDENCE FACTOR

The Confidence Factor offers a critical path module that handles up to 100 tasks with ten resource categories. It also performs decision-making functions.

The Confidence Factor runs on IBM/PC computers. It sells for $389 from Simple Software, 2 Pinewood, Irvine, CA 92714. (714) 857-9179.

Critical path calculations are only part of what this package offers. It has a decision tree module for organizing considerations that go into decisions. There is a risk simulation module for making predictions and a linear programming module as well. Other modules are Best Course of Action for optimal choices and Yes/No Decisions for ranking alternatives.

The critical path module operates with time units and the user decides whether these will represent days, weeks, months, or years. The actual units are never displayed by the package.

Up to ten personnel resources, with time-unit costs applied to each task in a project, can be handled.

The package calculates the critical path, personnel requirements, and costs, producing a Gantt chart, project data report, job cost report, start-finish report, resource allocation report, and smoothing report.

The user manual is skimpy and not too helpful.

CPM/PERT

CPM/PERT produces a Gantt chart and an activity schedule report that includes critical path length and may include a probability of on-time completion.

CPM/PERT runs on IBM/PCs and sells for $249 from Elite Software Development, PO Box 1194, Bryan, TX 77806. (409) 775-1782.

The CPM in the name refers to the Critical Path Method. Users can use the CPM mode, then upgrade their operations to PERT by adding two additional time estimates for each activity. The PERT method allows computation of a probability for completing a project on time.

Program operation is controlled from a master menu. The package prompts the user for inputs. There is a *beginner* operational mode in which added HELP information is displayed. These displays can be suppressed by operating in the *expert* mode.

The capacity of this package depends on available computer main memory. With 128K, up to 600 variables can be handled. Only day units of time can be used. Project models must have one and only one terminating node.

Holidays can be set at fixed dates or on certain days of the week within a given date range. Vacation and weekend no-work schedules are set using up to 26 calendar files, and each activity can reference a different calendar file.

Legible, detailed, and clear reports are generated, including a critical path report, holiday and calendar reports, and an activity cost report.

Gantt charts are produced on the printer; users can specify the characters to be used for the various chart features. Since previous calculation results are not stored, they are repeated when a Gantt chart is called for, and this can take minutes of computation time.

DEMI-PLAN

Demi-Plan is a low-cost package that will handle up to 100 tasks but is inconvenient to operate. The package will produce task-resource listings and Gantt charts.

Demi-Plan runs on IBM/PC/XT and TRS-80 I, III, and 4 computers and sells for $49.95 from Demi-Software, 62 Nursery Rd., Ridgefield, CT 06877. (203) 431-0864.

Demi-Plan operates from menus. When the user selects the Add option, the package prompts for the needed data. Following the entry of each task description, the user is returned to the main menu to select Add again if another task is to be entered.

Data entry itself is an unforgiving process. Users enter a six-item task description; if there are any mistakes in the entry, they have to start all over again, and that is the way the entire data entry process is handled.

Starting times for projects and tasks are expressed as the number of days from the beginning of the project, and it is up to the user to make the necessary conversions.

Saving project data on disk with this package means also ending program operation. The package simply quits, and the user has to reset the computer.

EMPACT

Up to 250 tasks can be handled by EMPACT with an unlimited number of resource categories.

EMPACT runs on IBM/PCs and sells for $149.95 from Applied MicroSystems, PO Box 832, Roswell, GA 30077. (404) 475-0832.

This package is quite simple to operate, and the HELP displays quickly clear up most misunderstandings about how it works.

Data input displays prompt the user for start and end dates, but the package does not calculate work days, so users are required to do this themselves. Since this involves taking account of various month lengths, it is an arduous procedure for mental arithmetic. However, the package does offer a stored calendar. Users can select any month and year and get an immediate calendar display.

Those working with lengthy input for a project are living dangerously since the package keeps all the data in main memory, storing it on disk only when the user moves to another project.

Reports produced by the package are variations of a format that shows descriptions, dates, and codes for each activity. An analysis report displays the earliest and last scheduled dates, along with the number and percentage of tasks completed for the entire project. Gantt charts with graphic symbols also are available.

EX-PERT/80

EX-PERT/80 handles up to 999 tasks but only one resource category.

EX-PERT/80 runs on IBM/PC, TRS-80 I and III, and Apple II and IIe computers. It sells for $115 from Decision Support Software, 1300 Vincent Pl., McLean, VA 22101. (703) 442-7900.

Data entry is clumsy. After entering a task name the package prompts for a string of six comma-separated entries. If there are any mistakes, the user must reenter all six. There is an editing mode, but one reviewer found it defective.

EX-PERT/80 has no built-in calendar function. It does take optimistic and pessimistic estimates of task duration from the user. These are used to calculate probabilities for various project completion dates.

Outputs are, with the exception of the Gantt chart, directed to the screen for display. On an IBM/PC, the only way to get a print out of the displayed material, is to use the computer's PrtSc key.

The package will not tolerate nodes on Gantt charts that are dangling, not connected in both directions, except the first and last nodes. If the diagram topology is unsatisfactory, the package will crash.

Other outputs include a graph that shows the amount of resources needed for each project day. There is also a report that shows task beginnings and endings, and a listing can be produced of tasks that must or may be active over a specified range of days.

The user manual is clearly written; however, one text serves for all computers on which the package may run. As a result, there are a few confusing instructions about use of the keyboard.

GANTT-PACK

Gantt-Pack produces Gantt and critical milestone charts.

Gantt-Pack runs on IBM/PC, CP/M, and TRS-80 computers. It sells for $395 from Gantt Systems, 495 Main St., Metuchen, NJ 08840. (201) 494-4752.

Data entry is onto a crowded screen that becomes closely packed with entries. It is arranged in rows and columns, and only the tops of the columns are labeled, often making it difficult to see which item is on which line. Formats required for items to be entered are not very easy to learn. There is no audible warning when the package rejects incorrect entries. Data editing is convenient and rapid in operation.

The package allows the user to choose among many different measures of time when the data entry format is set up. It does not take holidays automatically into consideration.

When making menu selections, the user has to enter simultaneously the first letter in the name of a menu section and a letter denoting the desired menu subsection, a procedure that is difficult to learn to use.

The package assumes that the user will memorize the names of the available files, since the name of the desired file must be typed before other package procedures can begin.

Gantt-Pack produces a Gantt chart, a critical milestone chart that shows current status of tasks, and listings of original and calculated data. The charts are not set up to display the names of events, only numbers. There are no resource management reports.

The user manual includes clear directions for the use of each screen display used in the package.

HARVARD PROJECT MANAGER

Harvard Project Manager is designed to make good use of a personal computer. It has an excellent tutorial. The package is limited to handling projects that will fit within the main memory of the computer. For an IBM/PC with 128K memory, that means that a project with about 200 tasks can be handled. Subprojects can be set up, within the same memory limitations, allowing users to develop a library of standard subtasks that can be used over and over again.

Harvard Project Manager runs on IBM/PC computers. It sells for $395 from Harvard Software, 521 Great Rd., Littleton, MA 01460. (617) 486-8431.

The Harvard Project Manager makes good use of windowing techniques, dividing the screen display into separate areas, each of which can show different aspects of the materials that are to be used.

The main display of this package shows four windows. The first contains a PERT chart; the second, a bar chart project schedule that highlights the project's critical path; the third a calendar; and the fourth displays a function-selection menu. Users can make a menu choice and have the chosen window fill the screen. If desired, two windows can share the screen so that, for example, the Gantt and PERT charts can be viewed together.

When starting up a new project, the user will get little assistance from this package. It is expected that the initial planning will be done with paper and pencil, then the results used to construct an initial PERT chart with the package. But once the initial entries are made, the Harvard Project Manager provides powerful means to monitor and change the data and to compute new costs and dates as the project proceeds. For example, once the initial project data are entered, the package automatically readjusts all costs and dates as new data are added.

Adding a task to the project is eased by further use of display windows. As tasks are added to the PERT chart, a window can be called up on which the user can key items of specific information about the task that are to be stored. Shuttling back and forth between windows is a simple process, often involving only a single keystroke.

On the PERT chart, milestones are shown as boxes and activities as boxes in reverse video. To delete a box, the user simply moves the cursor to it and presses a couple of keys. It disappears and the chart is redrawn. In general, as new information is entered, the displays will adjust automatically. There are some arbitrary restrictions on the PERT charts. For example, all tasks must branch from milestones, and milestones must always have at least one task before and after them.

The project calendar is flexible, allowing the user to factor in holidays and any other out-of-the-ordinary days. Output is also flexible, allowing the user to produce printed PERT charts, calendars, and reports. Where needed, these can be printed sideways so that material of almost any length can be accommodated.

Resource allocation can be done as the PERT chart is being constructed or afterwards. There is no resource smoothing, and resources cannot be allocated for only a part of a task.

An expense can be assigned to each activity, but the package does not provide for separating out fixed and variable expenses. A percent-complete figure allows the package to be used for some project tracking as well as planning.

Learning through windows

As with any package designed to perform complicated functions, there is some learning the user has to go through before the Harvard Project Manager can be mastered. For example, when the main package functions are being selected, it is the Backspace key and spacebar that control cursor movement. Once the user is operating within a given function, the cursor arrow keys are used to control selections. When it comes to editing text for descriptions, the Tab key is used to control cursor location.

To aid the user in learning these procedures, a disk-stored tutorial is provided, and this too makes use of display windows, showing explanations in one window while the user carries out the lessons in another. This excellent tutorial is coordinated with the user manual. Unfortunately, the manual itself is brief. It depends unduly on the tutorial and does not provide sufficient material for later reference. There is a reference card that summarizes package functions.

The charts produced by the package can be printed as well as displayed, and the printout can be either in the conventional vertical manner, or horizontal (which takes more time but is often visually much more pleasing). The package also produces detail reports showing all project information and calculation results and status reports listing information on selected activities. A calendar of a project can be printed.

INTEPERT

Many tasks and projects (1200–1900 of each) can be related to one another as subtasks by IntePert. Up to 26 resource categories can be handled.

IntePert runs on IBM/PC computers. It sells for $195 from Schuchardt Software Systems, 515 Northgate Dr., San Rafael, CA 94903. (415) 492-9330.

Each task is entered as a separate unit. The package prompts for the needed items. Calculation of time intervals is done by the package from the project and task dates the user supplies. When a task is inserted or deleted, the package automatically adjusts. As tasks are entered, they are automatically stored in disk memory, protecting the information from inadvertent loss.

Tasks can be divided into subtasks with up to 64 hierarchy levels allowed.

The press of a key produces a PERT chart with critical path tasks shown across the top. Another key produces a Gantt chart. Both charts can be printed, using ordinary print characters.

A detailed report lists tasks and all associated data. A schedule report lists all activities by date.

The manual includes a detailed and skillfully formulated tutorial. All the material is well written, and there is a reference section, error message explanations, and a detailed index.

MICROGANTT

MicroGANTT does budgeting and Gantt charts but has no PERT chart displays. Up to 100 tasks can be handled, and many projects can be interrelated by treating them as subprojects. An unlimited number of resource categories can be used.

MicroGANTT runs on IBM/PC computers. It sells for $395 from Earth Data, PO Box 13168, Richmond, VA 23225. (804) 231-0300.

MicroGANTT is command rather than menu driven, with the needed commands displayed at the bottom of the screen at all times.

Users are prompted for entry of data, first of overall project data, then of data for task records. The records may be billable or not. The package automatically saves data in disk storage almost immediately after they are entered.

A calendar serves to keep track of the length and order of the tasks. Each project can run as long as 61 units of time, and these units can be years, quarters, months, weeks, or days.

The package does not provide for such built-in delays as holidays and other nonworking days, nor for such contingencies as sick leave and vacations. These must be accounted for by the user.

If the user does not assign a task start date (or a prerequisite task that defines that start), the package will set the start date to the beginning of the project. Since the dates can be changed at will by the user, this assumption is simple to correct, should it be wrong.

Resources such as personnel can be allocated easily on a percentage basis, but allocating several workers with different billing rates to a single project or task becomes complicated.

The package has good provisions for cost tracking in that it allows users to assign an hourly billing rate to an entire project and override this rate for specific tasks. It also allows fixed costs to be assigned to each task.

Resources are always related to tasks that are billable or have billable components. The user can assign a cost, along with availability and time attributes, to each resource. This information allows the program to calculate the resource requirements for each task by time period. Fixed costs may be included.

Composite task feature

The package has a powerful composite task feature. It can be used to create subprojects, and each of these can contain up to 100 tasks. However, subproject information cannot be used in a fully integrated way with the main project information. For example, detailed main project resource reports may not incorporate subproject data.

Completion of a given task can be made fully or partially dependent on completion of other tasks or on the availability of a specific resource. Tasks can be assigned fixed durations, and selected tasks can be designated as milestones.

The limit of 100 tasks is not just arbitrary, it is quite practical. As tasks and composite tasks are added, program execution gradually slows down. Data for composite tasks are taken from disk storage so, if many of these are used, disk accessing can add greatly to the time the program takes.

Helpful displays and printouts

Results of entries can be observed quickly on a Gantt chart displayed on the screen. This provides a convenient way to test out various changes in schedule or tasks and see the results of different alternatives. The displayed chart gives a clear picture of the various tasks at hand, of the critical path of the project, and of how near it is to completion.

The initial planning scheme can be updated as actual results are received. The Gantt chart display is used as a basis to revise the remaining schedules. Unfortunately, it is somewhat difficult to update the project calendar, since MicroGANTT treats completed and in-progress tasks in the same way at the project time line.

In addition to the Gantt chart, the package provides a display that shows tasks by number, name, and type along with other important features of each task. A cost report breaks out fixed costs for each task for each period of the project schedule and also summarizes costs for each period. In addition, a report generator provides information keyed by task code or task description.

For a more detailed presentation of costs, the package produces printouts that break out resource use as well as cost. Printing of large reports is sequenced so that completed pages can easily be assembled into larger layouts.

The package produces detailed time-cost summaries and task summaries. The fixed format reports are detailed and relatively easy to read. A report generation feature called Select is available for creation of custom reports. The charts use letter-codes to key different kinds of data. One reviewer found that it took 40 minutes to print out a complex Gantt chart, an inordinately long time.

Gantt charts can be created for small work segments that may be used repeatedly during a project. These can be stored, then retrieved and inserted in a project as needed.

Those who need a refresher course in project management techniques will find the manual supplied with the package helpful. What is missing are the instructions that novice computer users will need to get the package properly installed, on a floppy or hard disk, so it can be conveniently used, and there is no hand-holding tutorial for the timid user. The package runs from menus, and there is a screen-displayed entry form for data, so the package, aside from the manual, is relatively easy to learn and use.

MICROPERT

MicroPERT creates Gantt and PERT charts, activity, and resource lists, based on user responses to screen-displayed prompts about tasks and times. However, it does not handle any financial data. Up to 220 tasks can be handled.

MicroPERT runs on IBM/PC computers. It sells for $350 from Sheppard Software, 4750 Clough Creek Rd., Reading, CA 96002. (916) 222-1553.

The package requires that users state how many activities and events will be in a project. To aid in the paper and pencil preplanning needed to get these numbers, MicroPERT provides users with printed worksheets for events, activities, and project layout.

When the activity and event numbers are entered, the package prompts the user for event and activity data. There is no display that accompanies this data entry, but MicroPERT editing facilities are good. However, when an event is deleted, the user must update all the activities that used that event. The package automatically renumbers all subsequent events, and it will handle activity deletions automatically as well.

Time periods are set up in day, week, month, quarter, or yearly units. Holidays have to be handled as dummy activities.

Output printing options include horizontal as well as vertical printing, and presentation (slow) as well as draft-quality printing. In addition to the charts, MicroPERT produces a schedule report that can be sorted by the user.

The 210-page manual is intelligently written and includes a useful tutorial and a good index.

MICROTRAK

As many as 300 tasks can be handled by MicroTrak with up to ten resource categories.

MicroTrak runs on IBM/PC computers. It sells for $595 from SofTrak Systems, PO Box 22156, Salt Lake City, UT 84112. (801) 531-8550.

MicroTrak will handle about 300 tasks on one floppy disk. Only ten resources can be used.

Before this package begins its work, the user has to divide the project manually into task sets and also must define the interactions between tasks.

As with most project management packages, most of the time the user spends is taken up with data entry. As new activities are entered, the package automatically updates its display.

Tasks can be deleted, and the package will automatically remove all interactions with other tasks and recalculate the network.

Reports generated by the package include one on schedule, on the network, on milestones, updates, activity-within-resource, and resource-within-activity.

The user manual has an introduction to critical path scheduling that prepares the user for the terminology used with this package. Unfortunately, the manual is not too helpful beyond that point. For example, the complex network charts the package produces are not explained at all.

MILESTONE

Milestone will handle hundreds of tasks, with the exact number depending on the size of available memory. It has unique scrolling features that allow convenient reading of very large Gantt charts.

Milestone runs on IBM/PC, CP/M, and USCD Pascal computers. It sells for $250 from Digital Marketing, 2363 Boulevard Circle, Walnut Creek, CA 94595. (415) 947-1000.

Data input for this package has the user tab from one displayed data item to another and keyboard the correct values for each item.

The package does rapid scrolling that operates either horizontally or vertically. Users can jump-scroll over any chosen number of jobs or work periods. With this scrolling, it is easy to make practical use of a large Gantt chart.

The number of activities that can be handled depends on the available computer main memory. It takes an IBM/PC with 448K to handle 300 tasks. The package keeps the user informed of how much room is left for further input.

Milestone only allows the use of 9 resources and 12 holidays. Time units can be hours, days, weeks, months, quarters, or federal fiscal quarters.

Output reports include project description and job description.The time schedule output is a conventional Gantt chart with critical path shown. There is also a columnar job report.

Financial information is presented in a histogram format, allowing users to see cash flow, along with the critical path and job loading.

The user manual is put together well, with explanations of features and reports keyed to a sample problem.

PAC MICRO

PAC MICRO offers project progress tracking as well as planning. It can readily exchange information with similar mainframe-based packages from the same vendor. Up to 400 tasks and up to 100 resource categories can be handled.

PAC MICRO runs on IBM/PCs and sells for $990 from AGS Management Systems, 880 First Ave., King of Prussia, PA 19406. (215) 265-1550.

PAC MICRO works from easy-to-use displayed menus. There is an entry screen for each task. Unfortunately, there is no reference on one screen to any of the previous ones, so users can easily lose track of what they are doing.

Resource data are entered on separate screens. A master list of resources with associated costs is created. In separate steps, resource assignments to project tasks are entered. This seems to involve unnecessary duplicate data entry.

Optional printing

An optional report printing feature is available. Without it, the package offers only displays, and users are on their own in getting their computers to print what is displayed.

There are some nice display features. Reports list the project tasks in order of criticality and show predecessor and sucessor tasks. A Gantt chart is included in the output.

The portion of a project taken up by each task is displayed in the form of a pie chart. However, this kind of display is useful only when relatively few tasks are involved, since the pie sections soon become too fine to be significant.

A column chart can be used to display the number of resource units required on each day of the project. This quickly reveals overallocation of a resource.

Users can exit the main program of the package and use another program to produce a network diagram of their projects.

The cost module used to track budgeted versus actual costs, resources, and durations also is an optional feature. Very useful bar chart displays are used to compare the actual and budgeted amounts.

The user's manual is thoughtfully constructed to walk users step by step through the procedures required by this package. There are plenty of figures showing what to expect in the way of screen displays. Unfortunately, the manual is not designed to function as a reference.

PATHFINDER

As many as 500 tasks can be handled by Pathfinder. Resource categories are limited to 30 per project and just one per task.

Pathfinder runs on IBM/PC and CP/M computers. It sells for $299 from Garland Publishing, 136 Madison Ave., New York, NY 10016. (212) 686-7492.

Procedures for adding, deleting, and inserting tasks work efficiently. There is an interactive dialog for entering task information.

The package uses a calendar file in which each day is represented by a code. The built-in calendar runs from 1975–1990; to handle other years the user must use a text editor to build a new calendar.

Pathfinder produces exceptions reports to spot activities that need attention, cash flow reports, and Gantt charts. There also can be up to four project schedule reports in which users can sort and arrange activities as desired. There is no output that analyzes the interdependence among various tasks.

PATHFINDER

Pathfinder handles up to 3,160 tasks, taking simple inputs and producing Gantt charts.

Pathfinder runs on IBM/PC computers. It sells for $80 from Morgan Computing, 10400 North Central Expwy, #210, Dallas, TX 75074. (214) 739-5895.

This is a simple package that takes in only four data items for each task: beginning and end event numbers, a time to perform,and a name. It is designed to quickly produce Gantt charts, and that is just what it does, no more, no less.

Its editing features work well. If subelements of a task are deleted, the entire project will be rearranged automatically.

The package does have some awkward limitations. It will not accept unterminated activities and will not run if there is more than one activity (the final one) that has no continuing connection.

The output is a Gantt chart, which can be produced on virtually any printer including daisy wheel types. Events listings also can be produced.

PERTMASTER

Depending on available computer memory, Pertmaster can handle 1,000–2,500 tasks, with 29 resource categories per task.

Pertmaster runs on IBM/PC and CP/M computers. A version that will handle 1,500 tasks sells for $695; a 2,500-task version sells for $895. An *optional plotting package, Pertplotter, sells for $495. Westminster Software, 3000 Sand Hill Rd., Bldg. 4 #242, Menlo Park, CA 94025. (415) 845-1400.*

The package uses the IBM/PC Function keys, and a list of their uses is displayed at the bottom of the screen. Users can set up their own abbreviations for tasks and resources, which will be recognized by the package.

Users can set up any number of calendars with different workday, holiday, and shift schedules, and they can select which calendar will be used with any given project. Calendar editing features are convenient.

Time periods used by the package are user selected and flexible. They can be in seconds, minutes, hours, days, shifts, weeks, months, years, or in completed work periods.

Projects can be set up based on Gantt charts or on PERT charts. Tasks can be added, the user giving the relationships to existing tasks. Tasks also can be deleted, but the user must then manually rearrange the relationships among the remaining tasks.

There is a standard report on activities sorted by activity number, float, early start, or description. The package also produces project and period bar charts. Users can define their own reports as well.

Pertmaster can send its graphical output to plotters as well as printers, using an additional program called Pertplotter.

The user manual has good tutorial and reference sections but gives little explanation of how errors are handled.

PLANTRAC

Plantrac is a very capable package that can handle up to 25 projects with a total of 65,000 tasks and 200 resource categories per project.

Plantrac runs on IMB/PC, TRS-80, and CP/M computers. It sells for $3,000 plus $1,000 per year for updates and support from Computerline Limited, 755 Southern Artery, Quincy, MA 02169. (617) 773-0001.

The menus are not the usual kind found in personal computer packages. One menu leads to another, but it is difficult for users to keep track of how to return to a menu once it has been passed.

Data entry includes internal checks on consistency and duplication of tasks. There are well-designed screen displays on which the data are entered.

As each task is entered, the package automatically prints out the data and stores them on disk. That is methodical, but the printing activity may be disturbing to those accustomed to using their computers in peaceful silence. The printer is actually used as a logging device to record everything entered on the keyboard. There is no option by which the user can turn it off.

Output from the package includes PERT and Gantt charts and, for those who have plotters attached to their systems, project network drawings. There are many report options. Information on specific items can be displayed in bar chart or histogram format. The data can be sorted as desired for reporting purposes. Holidays and other work-week adjustments can be added when reports are being specified.

There are very flexible capabilities for modifying the planning parameters and observing the results. Constraints of time, personnel, and cost can be imposed and removed to examine their effects on the overall project.

The user manual has a section on the basics of PERT chart use, but there is no tutorial material on package operation. There are explanations of the operation of the various package modules but no index.

PLAN/TRAX

Up to 300 tasks with ten resource categories can be handled by PLAN/TRAX.

PLAN/TRAX runs on IBM/PC computers. It sells for $595 from Engineering-Science, 57 Executive Park So. NE, #590, Atlanta, GA 30329. (404) 325-0124.

The user manual gives instructions that are difficult to understand and incomplete. However, once past the learning barrier, users will find that this package is quite simple to operate. Tasks can be inserted or deleted easily and quickly.

PLAN/TRAX will not tolerate project nodes that cannot be traced back to the initial node or do not flow into the final one. It does not handle fractional time periods.

The package does not produce pert charts, but it does turn out Gantt and personnel charts.

Users can track project progress by entering the funds spent and time taken to date. The package produces charts and graphs showing the schedule and budget status of the project. One Gantt-type chart shows current activity status and projected finish dates, assuming the present conditions continue.

PMS-II

PMS-II is capable of handling sizable projects, with the number of projects limited only by disk capacity. As many as 2,750 tasks per project can be handled and 96 resource categories can be used. The package does project tracking as well as planning. PMS-II is not an easy package to learn and use.

PMS-II runs on IBM/PCs; it sells for $1,295. A companion resource-handling package, RMS-II, sells for $995. The vendor is Aha Inc., 147 South River St., Santa Cruz, CA 95061. (408) 458-9119.

The main menu for this package offers users some 14 procedure choices for entering or updating project information and producing reports.

Input data for each task are entered in response to prompts from the package. Unfortunately, the entry screen does not include any project summary or other indication that would help the user keep track of what has already been entered. For that, the user has to return to the main menu.

A variety of output reports can be produced. These include reports on activity and changes from previous activity reports, edit listings, PERT charts, funding schedules, percent-complete reports, subcontractor billing reports, and Gantt charts. Two of the reports are designed to meet government contract requirements. However, all these are printed reports and are unavailable as screen displays.

The user manual is not easy for beginners to follow. There is a sample project on the disk, but no written or displayed tutorial material accompanies it. The manual has no index that would allow it to be used as a reference source.

PRIMAVERA PROJECT PLANNER: FROM MAINFRAME TO MICRO

Primavera Project Planner is a package with powerful capacities and features, designed to meet U.S. government reporting requirements. It can handle a total of 10,000 tasks, with the number of projects limited only by disk space. If that is what you need, the high price and batch-oriented operation of Primavera will not put you off.

Equipment required is an IBM/PC with a 10-million-byte hard disk and 512K main memory. Price of the package is $2,500. It is available from Primavera Systems, 29 Bala Ave., Bala Cynwyd, PA 19004. (215) 667-8600.

Originally developed for mainframe computers, this package is generous in its parameters. It will handle projects that include up to 10,000 tasks; any task can be related to as many as 128 other tasks, whose beginning or end depends on the progress of the original task. Up to six resources, selected from an overall project total of 96 resource categories, can be associated with each task, along with six categories (and any number of subcategories) of fixed and variable costs, an events log, and constraints on the dates.

However, like many mainframe-originated programs, Primavera is essentially batch oriented. That means that to revise a project, the user has to run a separate scheduling program, then print a revised project report to see the revisions in place.This process will take about two minutes. That is not very long, but is a far cry from the fast interaction most personal computer users expect.

Resource leveling

One of the important features of the package is its ability to do resource leveling, that is, to average out the availability of up to six vital resources for each task and, in this way, to calculate the duration of the task. When the package finds that the available resources are not sufficient for all active tasks, it will try to reassign resources automatically without compromising the critical path or the project completion date.

The Critical Path Method is used to schedule tasks into the overall project. Resources and costs can be tracked by task, and the package will compare expected to actual figures. The package can set up project schedules based on plans and on actual progress data and spell out the differences between the two.

With this package, the display screen is devoted mainly to entry of project data. As each task is initially entered, the user provides a task name and number and an estimate of the task's duration. A series of seven screen-displayed forms are then provided on which the detailed characteristics of the task are entered. These forms cover project dates (which may be assigned by the user as well as computed by the package), time parameters, costs, resources to be used, successor tasks (which can be related by start-to-start, or end-to-end, or start-to-end succession), freeform notes, and codes to be used in connection with the task. Completed forms can be printed out on command.

Allows detailed specification

Each task may be assigned up to 20 codes. These serve as a form of indexing that allows a variety of special reports to be readily produced. For example, a code may be used to identify the geographic location of the task.

Then, in a large project a report showing the progress of all tasks in a given geographic area can be produced.

As with tasks, the specification of resources can be quite detailed. Resource availability can be spelled out in maximum and normal units that can be used each day. Up to five different availability periods can be assigned; during each of these, the maximum and normal availability of the resource can be set at different levels. In addition, the desired duration of use of each resource and its lag time can be specified. Given these data, the package will calculate total resources needed for the task and the project. As the work proceeds, it will also compare actual data to scheduled use.

Although little is available in the way of reports that are displayed, Primavera can produce a great variety of printed reports. These are said to be adequate to meet the requirements of the Department of Defense and other government agencies. Gantt charts provide a ready view of the project's critical path, showing dates, tasks, and durations with reporting-time units set at days, weeks or months. The relationship of a task to its predecessors and sucessors can be shown in a network pathanalysis printout. Resource reports are available to list or graph resource use, estimate resources needed for project completion, and compare resource use to project progress. Other reports spell out task and cost data.

As would be expected from a mainframe-originated package, all printouts are designed for a 132-column printer. However the package does include a routine to use compressed printing.

PROJECT

Project can handle up to 128 tasks per project with 8 resource categories per task, 64 per project.

Project runs on IBM/PC computers. It sells for $250 from Microsoft, 10700 Northrup Way, PO Box 97200, Bellevue, WA 98004. (206) 828-8080.

Project works from command choices displayed at the bottom of the screen. Users start by setting up work days and declaring holidays on a Calendar screen. Data are next entered into an Activity screen, on which a Gantt chart is shown. An entry window is used to fill in the data for each activity. Entry of items is accomplished by pressing the Tab key to move from one item to another in the entry window. Up to 128 activities can be entered on one Activity screen, and other similar screens can be linked to accommodate more activities.

Each project can be up to ten years long and can involve up to $21 million in costs. Each activity can have up to eight predecessors and can use up to eight resources.

The package requires a lot of data. Each activity has a description, duration, pedecessors, start date, and names and quantities of resources. Activities and resources can be added or removed readily, but entries often have to be completely retyped.

Resource unit costs and availability are shown in tabular format on a Resource screen. In practice, users need to move frequently from one screen to another, and each time a user moves from the Activity to the Resource screen, the package recalculates, a process that can take minutes with a fully entered large project.

The package's calendar can accommodate any ten-year period between 1900 and 2068. Holidays and weekend work can be selected as desired.

Other projects can be linked to the current one as subtasks for handling very large overall projects.

A sorting feature allows activities to be rearranged by criticality, duration, dates, or in alphabetical order.

The package produces histograms to display the daily time assigned to each resource, convenient for locating overloads and slack time.

There is a Gantt chart that prints sideways. A Project Detail report presents a page of information on each activity, including a summary of the resources assigned to that activity. A Table report gives a summary for the activities, showing early and late starts and finishes and identifying critical activities. A Resource report summarizes time and costs. However, there are no PERT charts, and there is no provision for assigning fixed costs or income to tasks.

This package is very easy to learn to use. There is a half-hour tutorial and the instructions and illustrations in the manual are clear.

PRO-JECT 6

From 75 to 250 tasks can be handled by PRO-JECT 6, depending on available computer memory. Thirty resource categories can be used, with up to five used for any given task.

PRO-JECT 6 runs on IBM/PC computers. A 75-task version sells for $99, a 150-task version for $149, and a 250-task version for $199. SoftCorp, 2340 State Rd., 580 #244, Clearwater, FL 33575. (813) 799-3984.

The data entry is straightforward, except that users must give an ADD command before entering data for each task instead of simply entering one task after another. The Tab key is used to advance from one entry item to another, and the Enter key is used to end the entry for a task, a confusing arrangement for those who are accustomed to pressing the Enter key after each item in other packages.

The package handles dates only between the years 1980 and 1989, a narrow range for some uses. If data entry is started and items are entered until the 1989 limit is exceeded, there seems to be no way to adjust to more acceptable dates without starting over.

PRO-JECT 6 accepts only integer time durations, so half-days or half-hours cannot be handled. Numbers, rather than descriptions, are used to identify tasks.

When a task is deleted by the user, the package automatically removes any references to it that may appear in other tasks. However, when users add a task, they must edit other tasks to include such references.

The package has a very flexible reporting facility. Users can choose to select or sort many arrangements of data for a report. If desired, formatted report definitions can be set up and reused as needed. Documents created by word processing packages can be incorporated into the reports.

Vertical bar charts allow users to review the use of project resources.

The user manual is clear and concise. It has a command reference section and a listing of error messages but no index.

PROJECT MANAGEMENT

Up to 500 tasks can be included in the critical path information handled by this package and up to 300 tasks in the PERT information it produces.

Project Management runs on IBM/PC computers. It sells for $170 (or $140 to members) from the Institute of Industrial Engineers, 25 Technology Park, Norcross, GA 30092. (404) 449-0460.

This package includes four separate modules. One determines the critical path, calculates free and total floats, and earliest and latest start times for up to 500 activities. Users have to draw their own Gantt charts, because this module does not do any. Once a project has been defined, activities cannot be inserted or removed, but their dates can be changed.

A PERT module calculates an estimate of the mean and variance of the time to reach each node of a project network. It also will estimate the probability of meeting scheduled completion times. The user enters three time estimates for each activity to start the calculation.

A resource allocation module takes both activity time and available resources into account in calculating the start and finish time for each activity, so that a limited resource is efficiently used. Up to 200 activities can be handled, but only one limited resource. The module does not yield optimum results, but the solution will improve with additional iterations.

A project network analysis module calculates the critical path, earliest and latest starting times, and appropriate floats for a network in which the activities (up to 300) are assumed to take place at the nodes, rather than on the lines.

PROJECT SCHEDULER 5000

Project Scheduler 5000 can handle as many as 500 to 750 tasks per project, with the number of projects limited by disk storage capacity. Up to 96 resource categories can be handled.

This package runs on IBM/PC computers. It sells for $345 from Scitor Corp., 256 Gibraltar Dr., D7, Sunnyvale, CA 94089. (408) 730-0400.

This package works from menus, with somewhat annoying delays between menu displays. The menu screens are not well organized, nevertheless the package seems to function well.

As each task data item is entered, the cursor moves to the next item on the screen. When the data for a task are complete, there is a pause of few seconds while the package redraws and redisplays the Gantt chart. When the user adds, inserts, or deletes tasks, the package handles the details automatically and adjusts the rest of the project as needed.

Projects can be easily modified. Each task and milestone is assigned a number on the Gantt chart. A new task can be inserted and the entire project renumbered. Tasks also can be moved about.

The package provides for detailed resource information on each task. Up to 96 different resources can be handled for each project. Users can enter both resource labor codes and fixed costs. The package will calculate the project cost on a daily basis and display it along the bottom of the Gantt chart. As many as 30 holidays can be scheduled for a project.

Project Scheduler 5000 produces Gantt charts, a schedule of resources and costs, and several reports, but no PERT chart. The reports include a list of resources and associated costs, and a work week definition and calendar. There is a report option that allows users to put together a wide variety of tailored reports.

There are similar flexible choices for the graphical output of the package. Users can graph resource allocation by time period or resource costs with or without labor costs. When actual costs are input to the package, users can graph planned versus actual cost data, and they also can graph cost comparisons of planned, revised-planned, and actual costs to date.

The user manual includes a tutorial exercise that takes about two hours to complete. The manual itself is well organized and has an index.

THE PROJECT MANAGER

The Project Manager calculates start and end dates as well as total entered costs, and it produces Gantt charts. Projects with up to 399 tasks can be handled. It does not handle resources.

The Project Manager runs on IBM/PC computers. It sells for $495 from John Wiley and Sons, 605 Third Ave., New York, NY 10158. (212) 850-6000.

Users have to make up their own project diagram before beginning to enter data into this package, so they can supply the required node numbers to identify the individual tasks.

Scheduling time has to be in person days, but fractional days are allowed. Handling of dates makes no use of calendar files, and the package seems to have some difficulty handling weekends.

The package produces Gantt charts that display the entire project on a single sheet. This format gets quite cramped when lengthy projects are being shown. Five reports, none of which is headed by a title, can be produced.

Demonstration data are included on the package disk, but there are no tutorial instructions. The 95-page manual is readable, but it is not too well organized, and there is no index.

PROJECT MANAGER WORKBENCH

Project Manager Workbench does its calculations rapidly, supporting complex task interrelations and producing printed Gantt charts and reports. A network chart can be displayed. The number of projects and tasks that can be handled is limited by disk capacity, and nine resources can be used with each task.

Project Manager Workbench runs on IBM/PC computers. It sells for $1,150. A version that handles up to 300 tasks sells for $750 from Applied Business Technology, 76 Laight St., New York, NY 10013. (212) 219-8945.

Users of this package have the option of menu operation or of using direct commands from the keyboard. The menus respond so rapidly that there seems to be little reason not to use them.

Data entry is well designed for handling large projects. Task data are entered on one display screen. Another screen is used to define relationships between the various tasks. These can be grouped into activities, which can in turn be grouped into phases of the overall project.

As these relationships are entered by the user, the package displays a growing map of the defined relationships between tasks. A detailed Gantt chart can be displayed and scrolled through by the user.

An automatic scheduling feature adjusts for resource and dependency constraints. Activities are scheduled in increasing order of slack time. Reports are quickly produced and printed.

This is an easy-to-learn package that comes with on-disk samples and a 96-page tutorial. Unfortunately, the on-disk material has some bugs; the tutorial is, however, well written and helpful. The user's manual has a reference section that describes all aspects of package operation in detail. It also gives the user insight into project management concepts and provides practical suggestions.

QWIKNET

Qwiknet can handle up to 250 tasks, with 100 resource categories per project, 12 per task. It can schedule backward or forward. Pull-down windows and a mouse device are used.

Qwiknet runs on IBM/PC computers. It sells for $800 from Project Software and Development, 14 Story St., Cambridge, MA 02138. (617) 661-1444.

A mouse device is used to aid in moving the cursor around the display screen. By clicking the mouse button, users can move from one choice to another when the item where the cursor is located is a multiple-choice type. Across the top of the displays there is a line of items. With the cursor on one of these, a click of the mouse's left-button will open a pull-down menu window that allows for further choices. In this way the user can access separate displays for entering task data, relationships between tasks, and resources for the tasks. There also are separate holiday calendars and resource lists to keyboard. This fractionated data entry is easy to use, but the process can become laborious. There seem to be difficulties in adding and deleting tasks.

Qwiknet does project tracking, following actual, estimated, and target schedules and costs. The package is capable of scheduling backward from a given date. Several resources can be used with each task, and date constraints can be placed on tasks. Resources can be handled as distributed or lumped quantities.

Five reports are produced, including one listing task information, a task schedule, a Gantt chart, a target schedule listing, and a target schedule bar chart. A listing that includes costs for scheduled tasks and one that shows task dependencies, are displayed but not printed. There are no reports showing funding schedules or personnel loading.

Qwiknet does rapid calculations, responding almost instantaneously to user requests. Reports take only a few minutes to process. However, it cannot tolerate interruption during printing operations.

There is both a tutorial manual and a reference manual. The latter is well organized and designed to be read from beginning to end. There is an index but without sufficient detail.

SCHEDULE-IT, GANTT-IT, UNDER-CONTROL

This group of three packages includes software for project scheduling (Schedule-It), tracking (Gantt-It), and filing (Under-Control). Up to 1300 tasks can be handled on an ordinary (320K) IBM/PC disk. The packages do not appear to be fully operational.

The packages run on IBM/PC computers. All three sell for a combined price of $180 from A+ Software, 16 Academy St., Skanateles, NY 13152. (315) 658-6918.

Schedule-It has a calendar and handles hour, day, or month time units. It calculates critical paths, slack time, and project milestones. Gantt-It is the package that does Gantt chart displays and produces standard project management reports. Under-Control manages the files and also allows users to design their own reports to supplement or replace the standard ones that Gantt-It initiates.

One reviewer reported that the packages were not satisfactorily debugged.

Data input is done with the Schedule-It package. A split-screen display shows eight previous task entries on top and the current task at the bottom.

The documentation for Under-Control is good, but the manuals for the other two packages are skimpy.

TARGET TASK

TARGET TASK allows flexible schedule estimates. It generates Gantt charts and project management reports. The number of tasks that can be handled depends on the available computer main memory. With 512K memory, about 500 tasks can be handled.

TARGET TASK runs on IBM/PC computers and sells for $329 from Comshare, 5901 B Peachtree Dunwoodie Rd., #275, Atlanta, GA 30328. (404) 391-9012.

The menus from which this package operates are easy to use and are accompanied by HELP information that can be called upon to explain the options available to the user.

Data entries are checked by the package, which spots duplicate tasks. However, users cannot look back to view previously entered tasks.

Along with other input data on project tasks, users are asked to enter pessimistic, optimistic, and crash completion estimates of times needed for completion of each task. By making use of these estimates, charts and reports can be generated that reflect a variety of possible outcomes.

The package comes with a schedule of major holidays built in, and holidays can be added or deleted as needed.

TARGET TASK does not produce PERT charts, nor will it project labor resource requirements or show cost scheduling. What it concentrates on is calculating what happens when completion times change, and the package recalculates alternative scenarios rapidly. A cost estimate is given for completion of the project under each alternative schedule. A revised Gantt chart can be produced in seconds. Report generation is rather slow, and can grind on for a half hour or more for a reasonably complex project.

The user manual is clearly written, describes package operation, and also gives pointers on planning and PERT chart construction. It has an good index.

TASK MANAGER

Task Manager is a special-purpose data manager applied to project applications. It can handle up to 999 tasks, with one person per task. Task Manager has particularly flexible reporting capabilities. However, it does no critical path calculations.

Task Manager runs on IBM/PC computers. It sells for $395 from Quala, 23026 Frisca Dr., Valencia, CA 91355. (805) 255-2922.

Task Manager is operated from menu displays. The initial menu shows the available functions.

The input displays for this package are particularly well designed, making data entry an easy process. Arrow and other keyboard cursor controls are used to place data into displayed boxes, and keyboard inputs are automatically checked for length and other characteristics.

The package makes no provision for holidays or modified work schedules, so users will usually work with time durations rather than start and end dates.

Task Manager offers a choice of many different reports. In fact, users can define their own reports, then produce them one at a time or all at once in a batch. There is a list of sorting arrangements from which users can choose. When the desired sort order is chosen, users can then select a specific date range, individual, or department to be selected out and covered in the reports, or they can include all the data that have been input. Reports cover only 4-month or 12-month periods. Longer projects must be reported in segments.

The same kind of sorting and selection can be applied to Gantt charts. However, Task Manager does no critical path calculations. What it does display are dates, personnel, and percent complete for each task.

The user manual clearly explains package operation and has a good index.

TASKPLAN

Engineer C. Lamar Williams wrote TaskPlan in his spare time. It handles up to 50 tasks with one resource per task and 60 time periods, and it keeps track of project costs.

TaskPlan runs on IBM/PCs. It is a free package, but a $20 contribution is suggested. From Williams Software and Services, 1114 Pusateri Way, San Jose, CA 95121. (408) 227-4238.

The package comes with no manual, so it is very much a do-it-yourself experience to use it. Every time the package is started up, two screens of information are displayed, and users can print these themselves to get some sort of printed instructions to go by.

Data entry is done onto a display screen that has space for ten tasks. Start and end times and costs are entered. Cost multipliers can be used, separately for each task or once for all tasks. Data can be corrected by backspacing during entry or corrections can be made at the end of each screen before it is entered.

Tasks can be added as afterthoughts only by rerunning the package, with an option that preserves the previous data. A task can be deleted by setting its cost at zero and its time interval at the minimum .01 units.

When data entry is completed, the package will report on the costs for each time period and can display a chartlike display of incremental and cumulative costs.

One reviewer suggested that whatever TaskPlan can do might be better done with a spreadsheet or data base package.

TIMELINE

Users can plan backward from a deadline as well as forward with Timeline. An unlimited number of resources can be used with a project.

Timeline runs on IBM/PC computers. It sells for $495 from Breakthrough Software, 505 San Marin Dr., Novato, CA 94947. (415) 898-1919.

Timeline operates from a main menu that can be recalled by pressing the / key. Commands can be invoked by typing the first letter of the command or by moving the cursor to it and pressing the Return key.

Data entry is done by filling out screen-displayed forms. First, resources are designated and option and calendar forms are checked. A task-entry window is superimposed on a Gantt chart display. As activity data are entered, the Gantt chart is drawn by the package. Activities can be marked with a specific start or end date, or ASAP (as soon as possible), or ALAP (as late as possible), or they can be designated as Span activities. The Span activities are flexible in time, able to shrink or expand to accommodate the other activities. Users can decide which tasks to sequence or join to one another. Partial joining (starting the next task when the preceding one is only partly completed) is also allowed.

Changes are made easily. The cursor is moved to the desired chart location and a Function key is pressed. This brings up the entry window for that location. Any desired changes are made through that window, and the results are immediately reflected in the chart.

A resource-leveling facility spots overscheduled resources. Resource histograms graphically show allocation.

The package also produces a PERT chart that can be viewed as needed. On that chart, task descriptions are printed in the boxes, but the critical path is not shown. The display is not too well arranged and may be somewhat difficult to read.

A Gantt chart highlighting feature allows users, at the press of a key, to see all tasks that are joined to a specified task. A filtering function allows users to view only those tasks that match user-selected criteria.

Tasks can be sorted in various ways for reporting purposes. There is a detail report that gives full information about each task. Data from the packages can be sent to various spreadsheet packages using an added program available at no cost from the vendor. Tables are used to summarize information on tasks, resources, and costs.

Time Line comes with an audio tape tutorial and manual that takes users through a sample project planning exercise. The 126-page reference manual is easy to use, even somewhat entertaining, and it has a good index. Users also are provided with a quick reference card that shows Function keys and shortcuts. On-screen HELP text displays are available for most of the situations users will face.

VISISCHEDULE

The number of tasks VisiSchedule can handle depends on available computer memory. With 128K it handles 100 tasks; with 192K, 300 tasks. Nine resource categories can be used.

VisiSchedule runs on IBM/PC computers. It sells for $195 from VisiCorp, 1953 Lindy Ave., San Jose, CA 95131. (408) 946-9000.

VisiSchedule works from a nicely designed set of menus that provide needed information along with selection options. As data on the various tasks that make up a project are entered, the package constructs a Gantt chart that changes with each entry.

The package allows use of different time units but not of fractional units. It is limited to eight resource categories. More than one category and one person can be assigned to each task, and the package will compute appropriate salary rates.

Color is used in the displays, and users can elect to change these colors.

The package produces four reports, all well designed. A Job Description report gives data on job length, finish dates, slack time, criticality, personnel required, and costs. A Gantt chart includes added data on personnel levels and costs. A Project Description report summarizes the job schedule, resource categories, and holidays, and a Tabular Job report prints user-selected data.

VisiSchedule comes with adequate printed instruction and tutorial materials. The user manual reads easily and is convenient as a reference to the package commands and operations.

Index